Lecture Notes
in Control and Information Sciences 315

Editors: M. Thoma · M. Morari

Wolfgang Herbordt

Sound Capture for Human/Machine Interfaces

Practical Aspects of Microphone Array Signal Processing

With 73 Figures

Author

Dr. Wolfgang Herbordt
ATR – Advanced Telecommunications Research Institute International
Spoken Language Translation Research Laboratories
2-2, Hikaridai, Seiko-cho, Soraku-gun
Kyoto 619-0288
Japan

ISSN 0170-8643

ISBN 3-540-23954-5 **Springer Berlin Heidelberg New York**

Library of Congress Control Number: 2005920066

Springer is a part of Springer Science+Business Media

springeronline.com

Printed in The Netherlands

Typesetting: Data conversion by author.
Final processing by PTP-Berlin Protago-TEX-Production GmbH, Germany
Cover-Design: design & production GmbH, Heidelberg
Printed on acid-free paper 89/3141/Yu - 5 4 3 2 1 0

Preface

For convenient human/machine interaction, acoustic front-ends are required which allow seamless and hands-free audio communication. For maximum speech intelligibility and optimum speech recognition performance, interference, noise, reverberation, and acoustic echoes of loudspeakers should be suppressed. Microphone array signal processing is advantageous to single-channel speech enhancement since the spatial dimension can be exploited in addition to the temporal dimension.

In this work, joint adaptive beamforming and acoustic echo cancellation with microphone arrays is studied with a focus on the challenges of practical systems. Particularly, the following aspects are efficiently solved, leading to a real-time system, which was successfully used in the real world:

- High suppression of both strongly time-varying interferers, as, e.g., competing speakers, and slowly time-varying diffuse noise, as, e.g., car noise in passenger cabins of cars,
- Efficient cancellation of acoustic echoes of multi-channel reproduction systems even in strongly time-varying acoustic conditions with high background noise levels and with limited computational resources,
- High output signal quality with limited array apertures and limited numbers of microphones because of product design constraints,
- Robustness against reverberation with respect to the desired signal, moving desired sources, or array imperfections such as position errors or gain and phase mismatch of the microphones.

Detailed theoretical analysis and experimental studies illustrate the performance of the system. Audio examples can be found on the web page http://www.wolfgangherbordt.de/micarraybook/. Special focus is put on the reproducibility of the results by giving detailed descriptions of the proposed algorithms and of the parameter settings.

The intended audience of this book is both specialists and readers with general knowledge of statistical and adaptive signal processing. For any question or comment, please don't hesitate to contact the author!

Acknowledgements

I would especially like to thank my supervisor, Prof. Walter Kellermann of the Friedrich-Alexander University in Erlangen, Germany, for the unique opportunity to unify scientific and private interests in his research group in a very fruitful atmosphere with many productive discussions.

Since the beginning, this research was funded by several grants from Intel Corp., which made this work possible and which particularly led to the practical aspects work. I am especially thankful to David Graumann of Intel Corp., Hillsboro, OR, and Jia Ying of the Intel China Research Center, Beijing, China, for continuously supporting and promoting this work within and outside of Intel. I would also like to thank all the other people working with Intel who made my stays in China and in the United States unforgettable experiences.

I would like to thank Prof. Rainer Martin of the Ruhr-University in Bochum, Germany, Prof. Heinrich Niemann of the Friedrich-Alexander University in Erlangen, and Darren Ward of the Imperial College in London, UK, for their interest in my work, for reviewing this thesis, and for finding the time to participate in the defense of this thesis.

I am very thankful to everybody working in the Telecommunications Laboratory in Erlangen who made my stay here so enjoyable. I especially would like to thank my 'office mate' Lutz Trautmann for his friendship, his advises, and for being so considerate with his experiments with new algorithms for the simulation of musical instruments. I also would like to thank Herbert Buchner for many fruitful discussions about adaptive filter theory and for the great collaboration. Further, I would like to thank Ursula Arnold for her invaluable administrative support, and Rüdiger Nägel and Manfred Lindner for the construction of microphone array hardware.

I would like to thank all the people that I know through my numerous business trips for their helpful discussions and for the great moments together. Especially, I would like to thank Henning Puder of Siemens Audiologische Technik in Erlangen for proof-reading this thesis and Satoshi Nakamura of

ATR in Kyoto, Japan, for giving me the possibility to continue this research with a focus on automatic speech recognition in the ATR labs.

Finally, I am very thankful to my family and to my friends for their continuous encouragement, for their understanding, and for the infinite number of relaxed moments during the last years.

Erlangen, *Wolfgang Herbordt*
June 2004

Contents

1

Introduction

With a continuously increasing desire for natural and comfortable human/machine interaction, the acoustic interface of any terminal for multimedia or telecommunication services is challenged to allow seamless and hands-free audio communication in such diverse acoustic environments as passenger cabins of cars or office and home environments. Typical applications include audio-/video-conferencing, dialog systems, computer games, command-and-control interfaces, dictation systems, or high-quality audio recordings. Compared to speech and audio capture using a microphone next to the desired source, seamless audio interfaces cause the desired source signal to be impaired by reverberation due to reflective acoustic environments, local interference and noise, and acoustic echoes from loudspeakers. All these interferers are not only annoying to human listeners but, more importantly, they are detrimental for example when speech recognition is involved.

Acoustic echo cancellation is desirable whenever a reference of the loudspeaker signals is available, since acoustic echo cancellation allows maximum suppression of these interferers. Techniques for acoustic echo cancellation evolved over the last two decades [BDH+99, GB00], and led to the recent presentation of a five-channel acoustic echo canceller (AEC) for real-time operation on a personal computer [BK02, BBK03].

For suppression of local interferers and noise, beamforming microphone arrays are very effective since they suppress interference and noise by spatio-temporal filtering without distorting the desired signal [BW01] – in contrast to single-channel speech enhancement based on temporal filtering [Lim83]. Here, adaptive data-dependent beamformers seem to be the optimum choice since they account for the characteristics of the desired signal and of the interferers in order to maximize the suppression of interference and noise. However, adaptive data-dependent beamformers are particularly challenging for speech and audio signals in realistic acoustic conditions for several reasons:

1. Reverberation of the acoustic environment with respect to (w.r.t.) the desired signal [Ows80, CG80, WDGN82], moving desired sources, or array

imperfections as position errors or gain and phase mismatch of the microphones [God86, Jab86a, YU94] may lead to distortion of the desired signal by the adaptive beamformer.

2. The strong time-variance of typical interferers, as, e.g., competing talkers, makes the estimation of the characteristics of the interferers difficult, since it is generally not sufficient to estimate them during pauses of the desired signal.
3. The aperture of the array and the number of sensors are generally limited due to product design constraints. This reduces robustness against distortion of the desired signal and interference rejection at low frequencies [Fla85a, Fla85b].
4. Typical acoustic environments for hands-free acoustic human/machine interfaces are highly diversified, reaching from passenger cabins of cars with low reverberation times[1] T_{60} and with mostly slowly time-varying diffuse interference and noise, but with low signal-to-interference-plus-noise ratios (SINRs), to office and home environments with strong reverberation and mostly strongly time-varying directional interference (Table 1.1).

Table 1.1. Summary of characteristics of typical acoustic environments for hands-free acoustic human/machine interfaces

<table>
<tr><th>Environment</th><th>car</th><th>office/home</th></tr>
<tr><td>Reverberation time T_{60}</td><td>30–80 ms</td><td>150–500 ms</td></tr>
<tr><td rowspan="2">Noise/ interference</td><td>car noise, street noise</td><td>competing talkers, fan noise</td></tr>
<tr><td colspan="2">air conditioning (echoes of loudspeaker signals → acoustic echo cancellation)</td></tr>
<tr><td>Noise type</td><td>mostly non-directional</td><td>mostly directional</td></tr>
<tr><td>Time variance</td><td>mostly slowly time-varying</td><td>mostly strongly time-varying</td></tr>
</table>

An adaptive beamformer which addresses the problem of reverberation, moving sources, and array imperfections is presented in [HSH99]. However, adaptation of this version of the 'generalized sidelobe canceller' (GSC) [GJ82] is difficult, and often leads to cancellation of the desired signal or to transient interference and noise when the desired speaker and interference and noise are active simultaneously ('double-talk'). Therefore, we propose in this work

[1] The reverberation time T_{60} is defined by the time which is necessary for the sound energy to drop by 60 dB after the sound source has been switched off [Sab93].

a robust version of this GSC (RGSC) with low computational complexity, which effectively resolves the adaptation problems, which may be used down to low frequencies ($f \geq 200$ Hz for 4 sensors with an array aperture of 28 cm), and which may be applied to such diversified acoustic conditions as described in Table 1.1.

For practical multimedia terminals, a combination of a beamforming microphone array with acoustic echo cancellation is desirable in order to optimally suppress acoustic echoes, interference, and noise [Mar95, Kel01]. Here, we generally have to deal with the problem of maximally exploiting positive synergies between acoustic echo cancellation and beamforming, while minimizing the computational complexity of the combined system. Therefore, we compare in this work various combinations of beamforming and acoustic echo cancellation, each of them having advantages for different situations. Especially, a new structure is presented, which uses the RGSC as a basis, and which outperforms the other combinations for highly transient echo paths with low computational complexity.

This work is organized as follows: In Chap. 2, we introduce the description of propagating acoustic wave fields by space-time signals. For characterizing space-time signals captured by a sensor array, we use deterministic models and probabilistic models. Typical examples for the usage of each of the models illustrate basic characteristics of acoustic wave fields.

In Chap. 3, optimum linear multiple-input multiple-output (MIMO) filtering is presented. Generally, data-dependent sensor array processing can formally be described by optimum MIMO filters. We derive optimum MIMO filters in the discrete time domain using averaging of finite data blocks in order to optimally meet the requirements of time-varying wideband signals. The relation to optimum MIMO filtering in the discrete-time Fourier transform (DTFT) domain is shown.

In Chap. 4, the concept of beamforming with sensor arrays is introduced. We present a unified description of data-dependent wideband beamformers using MIMO optimum filtering in the discrete time domain. Vector-space interpretations in the discrete time domain and descriptions in the DTFT domain illustrate the properties of the beamformers. The discussion motivates the choice of the GSC after [HSH99] as a basis for the proposed acoustic human/machine front-end.

In Chap. 5, we derive the GSC after [HSH99] in the framework of data-dependent optimum wideband beamforming in the discrete time domain. We show that the GSC after [HSH99] can be realized such that robustness against distortion of moving desired sources in reverberant acoustic environments is assured while strongly time-varying as well as slowly time-varying interference and noise are efficiently suppressed. This version of the GSC is referred to as RGSC. Fundamental properties of the RGSC are illustrated by experiments.

In Chap. 6, various methods for combining beamforming with multi-channel acoustic echo cancellation are studied. After summarizing the problem of and the solution to multi-channel acoustic echo cancellation, we discuss how positive synergies between beamforming and acoustic echo cancellation can be obtained while minimizing the computational complexity for real-world systems. The discussion shows that different systems maximally suppress echo signals, local interference, and noise for different acoustic conditions. Experimental results for realistic scenarios using the RGSC and a stereophonic AEC illustrate the characteristics of the different combinations.

In Chap. 7, an efficient real-time realization of an acoustic human/machine front-end using the RGSC and a multi-channel AEC is proposed based on the results of the previous chapters. In order to minimize the computational complexity while optimally accounting for time-varying acoustic conditions, we derive a realization in the discrete Fourier transform (DFT) domain using multi-channel adaptive filtering in the DFT domain. Experimental results show that the proposed system efficiently suppresses interference and noise in such diverse acoustic environments as passenger cabins of cars with presence of car noise or in reverberant office environments with competing speech signals. The application of the acoustic front-end to an automatic speech recognizer confirms the high output signal quality by greatly improved word accuracies compared to far-field microphones or conventional fixed beamformers. Furthermore, the same parameter setup can be used for the RGSC in a wide variety of acoustic environments with highly varying SINRs and noise/coherence properties. Audio examples which illustrate the output signal quality of the real-time system can be found at www.wolfgangherbordt.de/micarraybook/.

In App. A, a new multi-channel method is presented for unbiased estimation of the SINR at the sensors for mixtures of non-stationary signals. This method uses a classifier for 'desired signal only', 'interference only', and 'double-talk', which is applied in Chap. 7 for controlling the adaptation of the RGSC in the DFT domain.

In App. B, the experimental setup and the acoustic environments are described.

2

Space-Time Signals

In single-channel techniques for hands-free acoustic human/machine interfaces, we deal with waveforms which are functions of the continuous time. The aim of multi-channel sound capture is to exploit the structure of propagating waves, i.e., spatial and temporal properties in order to better meet the requirements of speech enhancement. The received signals are thus deterministic functions of position and of time, and, therefore, are called *space-time signals* or *spatio-temporal signals*. They have properties which are governed by the law of physics, in particular the wave equation. Just as temporal filtering can be described by temporal impulse responses, the wave propagation in acoustic environments can be modeled using space-time filters which are described by spatio-temporal impulse responses. Often, the deterministic model of space-time signals cannot be applied to acoustic signals, since audio signals can hardly be described by functions where each time instance is assigned a unique numerical value. The deterministic model of room impulse responses is not appropriate if the spatial extension of the source cannot be neglected since such spatio-temporal impulse responses of acoustic environments can generally not be described analytically. In such situations, it is more convenient to use statistical random fields which are the extension of stochastic processes to multi-dimensional parameter spaces.

This chapter is organized as follows: In Sect. 2.1, we introduce propagating wave fields and their deterministic description as space-time signals as well as the notion of spatio-temporal impulse responses. In Sect. 2.2, we introduce spatio-temporal random fields and their statistical characterization. Typical examples for the usage of each of the models illustrate basic characteristics of acoustic wave fields.

Further information about propagating acoustic wave fields can be found in, e.g., [Kut91, Cro98]. Space-time signals are described in, e.g., [JD93, Tre02]. More details on statistical modeling of general-purpose signals can be found in, e.g., [BP71, WH85, PP02].

2.1 Propagating Wave Fields

A space-time signal is written as $s(\mathbf{p}, t)$, where $\mathbf{p}$ is the position of observation and where t denotes the continuous time. The position vector $\mathbf{p}$ is represented in a 3-dimensional (3-D) right-handed orthogonal coordinate system as illustrated in Fig. 2.1. Cartesian coordinates are denoted by (x, y, z). Spherical coordinates are denoted by (r, θ, ϕ), where r is the radius, $0 \leq \theta \leq \pi$ is the elevation angle, and $0 \leq \phi < 2\pi$ is the azimuth. Figure 2.1 shows the relationship between Cartesian coordinates and spherical coordinates.

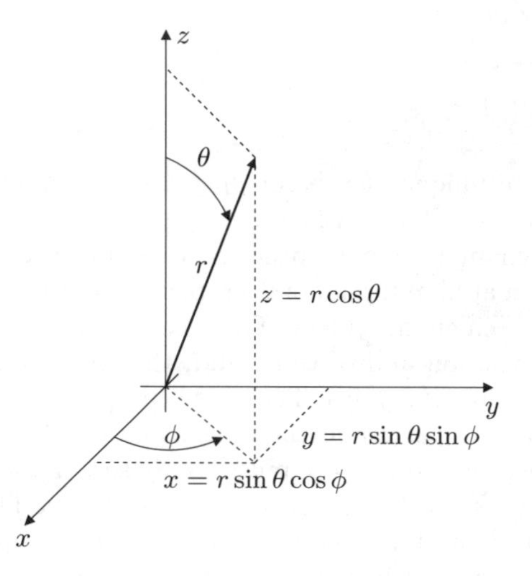

Fig. 2.1. Right-handed orthogonal coordinate system with Cartesian coordinates (x, y, z), and spherical coordinate system with coordinates (r, θ, ϕ)

In our scenario, the space-time signal $s(\mathbf{p}, t)$ describes the sound pressure of an acoustic wave field as a function of the position of observation $\mathbf{p}$ and as a function of time t. The medium of propagation of the acoustic waves should be homogeneous, dispersion-free, and lossless. Homogeneity assures a constant propagation speed throughout space and time. Dispersion occurs in non-linear media, where the interaction with the medium depends on the amplitude and on the frequency-contents of the wave. A medium is lossless if the medium does not influence the amplitude attenuation of the propagating wave. With these assumptions, the propagation of sound waves is governed by the scalar linear wave equation, which can be written as ([Kut91, Cro98, Rai00])

$$\nabla^2 s(\mathbf{p}, t) = \frac{1}{c^2} \frac{\partial^2 s(\mathbf{p}, t)}{\partial t^2}, \tag{2.1}$$

where ∇^2 is the Laplacian operator and where c is the sound velocity[1].

[1]We assume $c = 344\,\mathrm{m/s}$ for a temperature of 20°C and normal atmospheric pressure of 101 kPa. Note that the sound velocity may vary in acoustic environments,

One solution of the scalar wave equation (2.1) is the *monochromatic plane wave* which can be generally described by the complex exponential

$$s(\mathbf{p}, t) = S \exp\left(\jmath \left(\omega_0 t - \mathbf{k}_0^T \mathbf{p}\right)\right) . \tag{2.2}$$

$(\cdot)^T$ denotes transpose of a matrix or of a vector. The constant factor S is the amplitude, ω_0 is the temporal frequency of the monochromatic plane wave. The vector $\mathbf{k}_0$ is called *wavenumber vector*. It is defined as

$$\mathbf{k}_0 = \frac{\omega_0}{c}\mathbf{a}(\theta, \phi) = \frac{2\pi}{\lambda_0}\mathbf{a}(\theta, \phi) , \tag{2.3}$$

where λ_0 is the wavelength and where $\mathbf{a}(\theta, \phi)$ is a unit vector ($r = 1$) with spherical coordinates pointing into the direction of propagation of the monochromatic plane wave. With the direction of propagation being defined toward the spatial origin, $\mathbf{a}(\theta, \phi)$ is given by:

$$\mathbf{a}(\theta, \phi) = \begin{pmatrix} -\sin\theta\cos\phi \\ -\sin\theta\sin\phi \\ -\cos\theta \end{pmatrix} . \tag{2.4}$$

The direction of $\mathbf{k}_0$ expresses the direction of propagation of the monochromatic plane wave, the magnitude of $\mathbf{k}_0$ represents the number of cycles in radians per meter of length in the direction of propagation. The wavenumber vector $\mathbf{k}_0$ can thus be interpreted as the spatial frequency variable according to the temporal frequency variable ω_0. The scalar product $\mathbf{k}_0^T \mathbf{p}$ in (2.2) is the propagation delay with the origin of the coordinate system as reference. Equation (2.2) refers to a *plane wave* since, for any time instant t, the points with equal amplitude of $s(\mathbf{p}, t)$ are lying on planes which are defined by $\mathbf{k}_0^T \mathbf{p} = a$, where a is a constant. The term *monochromatic* means that $s(\mathbf{p}, t)$ consists of one single harmonic with frequency ω_0. In the context of array signal processing, monochromatic plane waves play a major role since many performance measures and many array designs are based on them. (See Chap. 4).

A second solution of the scalar linear wave equation is given by the *monochromatic spherical wave*, which describes the wave field of a point source which is located at the origin of the underlying coordinate system. That is,

$$s(r, t) = \frac{S}{r}\exp\left(\jmath \left(\omega_0 t - |\mathbf{k}_0| r\right)\right) . \tag{2.5}$$

In contrast to a plane wave, the amplitude of the monochromatic spherical wave decreases hyperbolically with the distance of observation r. The points

---

e.g., due to spatial temperature variations. The effects of these inhomogeneities are neglected in Chaps. 2–4. For the practical audio acquisition system in Chaps. 5–7, these effects are taken into account by applying methods which are robust against spatial inhomogeneities, as, e.g., simple delay&sum beamformers in combination with interference cancellation using adaptive filters. Additionally, the influence of temporal fluctuations of the temperature on the performance of audio acquisition systems has to be considered for the design of the adaptive filters [EDG03].

with constant amplitude of $s(r,t)$ are lying on spheres which are concentric to the spatial origin. Generally, the radiation of point sources is modeled by spherical waves if the positions of observation are close to the source, i.e., if the wave front of the propagating wave is perceptively curved w.r.t. the distance between the positions of observations (*near field*). The direction of propagation depends on the position of observation. For large distances, the wave field of point sources can be modeled by plane waves, since the wave front resembles a plane wave for decreasing curvature of the wave front (*far field*). The direction of propagation is thus approximately equal at all positions of observations. The transition between the near field of a point source and the far field of a point source also depends on the distance between the observation positions. Generally, for

$$r > \frac{2d^2}{\lambda_0}, \tag{2.6}$$

where d is the distance between the observation positions, the far field is assumed [Goo68].

Due to the linearity of the wave equation, the principle of superposition holds. Hence, more complicated (arbitrary periodic) wave fields, as, e.g., propagation of several sources, of wideband sources, or of spatially continuous sources, can be expressed as Fourier integrals w.r.t. the temporal frequency ω. The propagation in reflective environments, i.e., multi-path propagation, can be modeled by linear filters, which are typically not invariant in time or in position. The time-variance results from moving acoustic sources and changes in the acoustic environment. The position invariance is due to propagation paths which are not identical at all positions in the room. The linear *space-time filters* are described by *spatio-temporal impulse responses* $h(\mathbf{p}, \boldsymbol{\pi}; t, \tau)$. The spatio-temporal impulse response is a function of the position(s) of observation $\mathbf{p}$, of the position(s) of excitation $\boldsymbol{\pi}$, of the observation time(s) t, and of the excitation time(s) τ. With the excitation signal $s(\boldsymbol{\pi}, \tau)$, the observation signal $x(\mathbf{p}, t)$ is obtained by integration over all excitation positions and all excitation times, which is given by the four-dimensional (4-D) integral[2]

$$x(\mathbf{p}, t) = \int_{-\infty}^{\infty} \int_{-\infty}^{\infty} s(\boldsymbol{\pi}, \tau)\, h(\mathbf{p}, \boldsymbol{\pi}; t, \tau)\, d\boldsymbol{\pi}\, d\tau\,. \tag{2.7}$$

The observed signal is equal to $h(\mathbf{p}, \boldsymbol{\pi}; t, \tau)$ for excitation with a spatio-temporal unit impulse $\delta(\|\mathbf{p} - \boldsymbol{\pi}\|_2)\delta(t - \tau)$, where $\|\cdot\|_2$ is the L_2-norm.

In the same manner as the influence of acoustic environments can be modeled by spatio-temporal impulse responses, acoustic wave fields can be processed by space-time filters in order to obtain desired spatio-temporal characteristics. Typical space-time filters are *beamformers*. These processors combine spatial and temporal filtering in order to preserve signals from certain directions while canceling signals from other directions. Signals in the

[2] We represent the 3-D integral over the whole space w.r.t. the vector $\boldsymbol{\pi}$ by a single vector integral from $-\infty$ to ∞.

beam are passed while those out of the beam are attenuated. Spatial filtering by beamforming is thus analogous to temporal bandpass filtering. For processing signals in space and time, it is necessary to measure the propagating wave field at positions of interest $\mathbf{p}$. Generally, *arrays* of spatially distributed sensors are used, where the sensors are located at fixed positions $\mathbf{p}_m$, $m = 0, 1, \ldots, M-1$. The number of sensors of the array is denoted by M. A sensor array with equally-spaced sensors with inter-sensor spacing d is called *uniform* array. In Fig. 2.2, a 1-D sensor array is depicted for an odd number of sensors (Fig. 2.2a) and for an even number of sensors (Fig. 2.2b) with a monochromatic plane wave with wavenumber vector $\mathbf{k}_0$ arriving from direction θ.

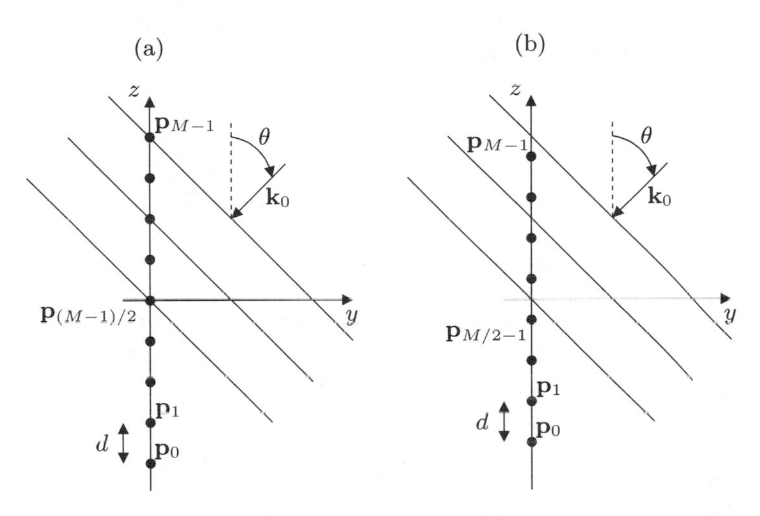

Fig. 2.2. 1-D sensor array **(a)** with an odd and **(b)** with an even number of uniformly spaced sensors M with a monochromatic plane wave with wavenumber vector $\mathbf{k}_0$ arriving from direction θ

The space-time signals at the positions $\mathbf{p}_m$ of the sensors are obtained by spatial sampling. The spatial sampling can be expressed by a multiplication of the wave field $x(\mathbf{p}, t)$ with a spatial unit impulse as follows:

$$x(\mathbf{p}_m, t) = x(\mathbf{p}, t)\, \delta(\|\mathbf{p} - \mathbf{p}_m\|_2)\,. \tag{2.8}$$

In our context, the acoustic wave field is measured by *omnidirectional* microphones, which have the same sensitivity for all frequencies of interest and for all directions. The microphones should be ideal in a sense that the recorded signals are equal to the sound pressure $x(\mathbf{p}_m, t)$ at the sensor positions $\mathbf{p}_m$. Non-ideal or directional sensor characteristics may be included into the spatio-temporal impulse responses which model the propagation in the acoustic environment. The output signals of the sensors are synchronously uniformly sampled over time at times kT_s, where k is the discrete time index and where

T_s is the sampling period. The sampling frequency is denoted by $f_s = 1/T_s$. The sampling frequency f_s is chosen such that the temporal sampling theorem [OS75] is fulfilled. We obtain the discrete-time sensor signals as

$$x_m(k) := x(\mathbf{p}_m, kT_s) \,. \tag{2.9}$$

An acoustic source is often modeled as a point source at position $\mathbf{p}_s$ with excitation signal $s(\mathbf{p}, t) = s_{\mathbf{p}_s}(t)\delta(\|\mathbf{p} - \mathbf{p}_s\|_2)$. Typical examples are human speakers, loudspeakers, and – relative to the distance between the sensors and the source – small objects which produce any kind of noise, as, e.g., computer fans. The (multi-path) propagation in the acoustic environment can then be described by a spatio-temporal impulse response $h(\mathbf{p}, \mathbf{p}_s; t, \tau)$. The observation with a sensor array after spatial and temporal sampling yields after integration of (2.7) over $\boldsymbol{\pi}$ the sensor signals

$$x_m(k) = \int_{-\infty}^{\infty} s_{\mathbf{p}_s}(\tau)\, h(\mathbf{p}_m, \mathbf{p}_s; kT_s, \tau) d\tau \,. \tag{2.10}$$

If we assume a time-invariant acoustic environment, i.e.,

$$h(\mathbf{p}, \boldsymbol{\pi}; t, \tau) := h(\mathbf{p}, \boldsymbol{\pi}; t - \tau) \,. \tag{2.11}$$

we can transform (2.10) into the discrete-time Fourier transform (DTFT) domain [OS75]. Let $S_{\mathbf{p}_s}(\omega)$ be related to the sampled sequence $s_{\mathbf{p}_s}(\kappa T_s)$ by the DTFT and by the inverse DTFT (IDTFT),

$$S_{\mathbf{p}_s}(\omega) = \sum_{\kappa=-\infty}^{\infty} s_{\mathbf{p}_s}(\kappa T_s) \exp\left(-j\omega\kappa T_s\right) \,, \tag{2.12}$$

$$s_{\mathbf{p}_s}(\kappa T_s) = \frac{T_s}{2\pi} \int_0^{2\pi/T_s} S_{\mathbf{p}_s}(\omega) \exp\left(j\omega\kappa T_s\right) d\omega \,, \tag{2.13}$$

respectively [3], and let the transfer function $H(\mathbf{p}, \mathbf{p}_s; \omega)$ be the DTFT of $h(\mathbf{p}, \mathbf{p}_s; \kappa T_s)$. By replacing in (2.10) the discrete time quantities $s_{\mathbf{p}_s}(\kappa T_s)$ and $h(\mathbf{p}, \mathbf{p}_s; \kappa T_s)$ by their IDTFTs according to (2.13), we obtain the sensor signals in the DTFT domain as follows:

$$X_m(\omega) = S_{\mathbf{p}_s}(\omega)\, H(\mathbf{p}_m, \mathbf{p}_s; \omega) \,. \tag{2.14}$$

Note that acoustic environments are generally time-varying [BDH+99, HS03]. However, we will assume in this work time-invariance of the acoustic environment, if this gives additional insights into the properties of the studied systems.

[3] A sufficient condition for the existence of the DTFT is absolute summability of the sampled sequence $s_{\mathbf{p}_s}(kT_s)$: $\sum_{k=-\infty}^{\infty} |s_{\mathbf{p}_s}(kT_s)| \overset{!}{<} \infty$. This is true for physically observable signals, i.e., for finite duration and for finite amplitude of $s_{\mathbf{p}_s}(kT_s)$.

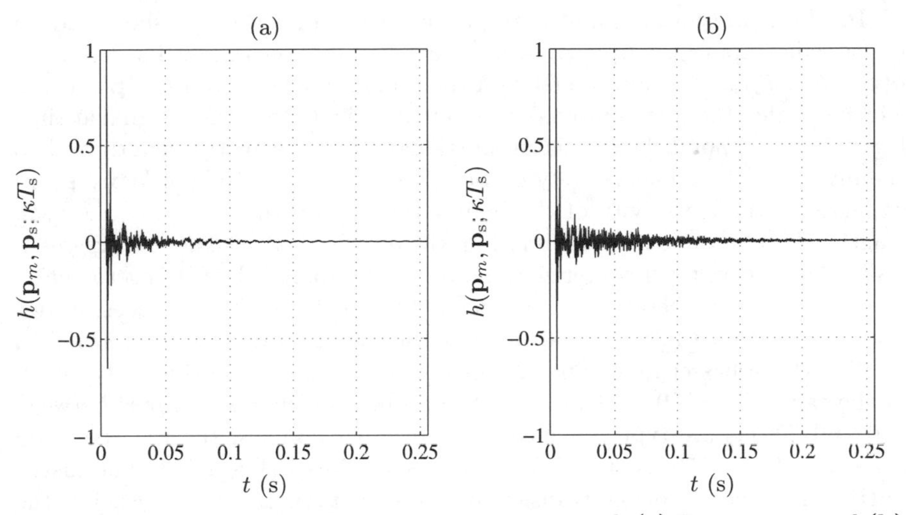

Fig. 2.3. Impulse responses of the multimedia room with **(a)** $T_{60} = 250\,\mathrm{ms}$ and **(b)** $T_{60} = 400\,\mathrm{ms}$. (See App. B.) The distance between the point source at position $\mathbf{p}_s$ and the sensor at position $\mathbf{p}_m$ is 1 m

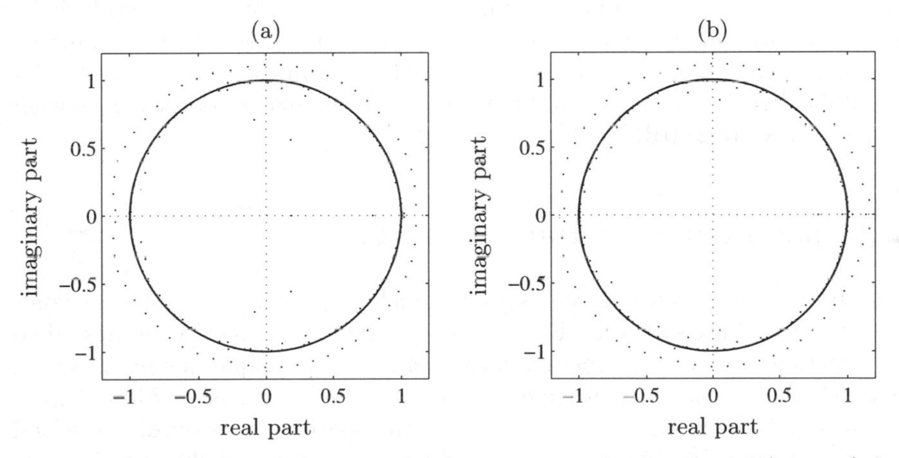

Fig. 2.4. Distribution of the zeroes of the room transfer functions $H(\mathbf{p}, \mathbf{p}_s; \omega)$ for the impulse responses of the multimedia room with **(a)** $T_{60} = 250\,\mathrm{ms}$ and **(b)** $T_{60} = 400\,\mathrm{ms}$. (See Fig. 2.3)

In Fig. 2.3, typical impulse responses $h(\mathbf{p}_m, \mathbf{p}_\mathrm{s}; \kappa T_\mathrm{s})$ are shown for a multimedia room (see App. B) with two reverberation times. The quantity $h(\mathbf{p}_m, \mathbf{p}_\mathrm{s}; \kappa T_\mathrm{s})$ is a sampled version of the room impulse response $h(\mathbf{p}_m, \mathbf{p}_\mathrm{s}; \tau)$ which specifies the contribution of the source signal to the sensor signal at time lag τ. The temporal delay τ is proportional to the geometric length and to the inverse of the sound velocity c. (See (2.5).) The amplitude of $h(\mathbf{p}_m, \mathbf{p}_\mathrm{s}; \tau)$ depends on the reflectivity of the boundaries and on the inverse of the path lengths. As a first-order approximation, the impulse response decays exponentially. This allows to characterize room impulse responses by the *reverberation time* T_{60}. The reverberation time is defined by the time which is necessary for the sound energy to drop by 60 dB after the sound source has been switched off [Sab93]. In passenger cabins of vehicles, the reverberation time is typically in the range $T_{60} = 30 \ldots 80$ ms. In office rooms, reverberation times between 150 and 500 ms are typically measured. In our examples, the reverberation times are $T_{60} = 250$ ms (Fig. 2.3a) and $T_{60} = 400$ ms (Fig. 2.3b). The distribution of the zeroes of the transfer functions $H(\mathbf{p}, \mathbf{p}_\mathrm{s}; \omega)$ are depicted in the complex plane in Fig. 2.4. It can be seen that the zeroes are almost uniformly distributed around the unit circle and most of them very close to the unit circle. This reflects the long decay rates of room impulse responses. Figure 2.4 also shows that room impulse responses have zeroes outside the unit circle, which means that the room impulse responses are generally non-minimum phase [NA79]. These characteristics of room environments are responsible for the difficulties in identifying and inverting their impulse responses or transfer functions [KHB02, HBK$^+$03b].

2.2 Spatio-temporal Random Fields

The usage of the deterministic signal model of Sect. 2.1 to acoustic signals is limited for three reasons: First, acoustic signals can hardly be described by functions where each time instance is assigned a unique numerical value. Second, the simple wave-propagation model of a point source filtered by a room impulse response cannot be used if the spatial extension of the sound source cannot be neglected relative to the distance between the sound source and the observation positions. The description of such wave fields by spatio-temporal impulse responses $h(\mathbf{p}, \boldsymbol{\pi}; t, \tau)$ is not practical, since it is hardly possible to estimate $h(\mathbf{p}, \boldsymbol{\pi}; t, \tau)$ over a continuous space $\boldsymbol{\pi}$. Third, it is not convenient to use the point-source model for modeling the wave field of a large number of sound sources, as, e.g., 'cocktail-party' noise. In all these situations, a statistical model, as described in this section, is the better choice.

In Sect. 2.2.1, the statistical model of spatio-temporal random fields is introduced. In Sect. 2.2.2, the notion of spatio-temporal and spatio-spectral correlation matrices is specified. These matrices meet the requirements of a statistical model for the statistical description of acoustic wave fields in our context. Examples for typical acoustic wave fields are given.

2.2.1 Statistical Description of Space-Time Signals

Stochastic processes statistically describe ensembles of functions that depend on a single real parameter in a given interval. Often, the parameter is the time t. For space-time signals, which depend on 4 parameters, it is more appropriate to use *random fields* with a 4-D parameter space. We will use the same (real) variable $x(\mathbf{p}, t)$ for the observation of space-time signals as for the observation of *spatio-temporal random fields.* The spatio-temporal random field is characterized by the probability density function $p_x(x(\mathbf{p}, t))$.

Throughout this work, we basically use four statistical averages to characterize a spatio-temporal random field: Corresponding to a stochastic process, the mean $\eta_x(\mathbf{p}, t)$ and the variance $\sigma_x^2(\mathbf{p}, t)$ of a spatio-temporal random field are defined as

$$\eta_x(\mathbf{p}, t) = \mathcal{E}\{x(\mathbf{p}, t)\}\,, \tag{2.15}$$

$$\sigma_x^2(\mathbf{p}, t) = \mathcal{E}\{x^2(\mathbf{p}, t)\} - \eta_x^2(\mathbf{p}, t)\,, \tag{2.16}$$

respectively, where $\mathcal{E}\{\cdot\}$ is the expectation operator. Then, the *spatio-temporal correlation function* $R_x(\mathbf{p}_{m_1}, \mathbf{p}_{m_2}; t_1, t_2)$ and the *spatio-temporal covariance function* $C_x(\mathbf{p}_{m_1}, \mathbf{p}_{m_2}; t_1, t_2)$ are used for characterizing the second-order statistical relationships of space-time random fields between two positions of observation[4] $\mathbf{p}_{m_1}$ and $\mathbf{p}_{m_2}$ and between two time instants t_1 and t_2:

$$R_x(\mathbf{p}_{m_1}, \mathbf{p}_{m_2}; t_1, t_2) = \mathcal{E}\left\{x(\mathbf{p}_{m_1}, t_1)\, x(\mathbf{p}_{m_2}, t_2)\right\}\,, \tag{2.17}$$

$$C_x(\mathbf{p}_{m_1}, \mathbf{p}_{m_2}; t_1, t_2) = R_x(\mathbf{p}_{m_1}, \mathbf{p}_{m_2}; t_1, t_2) - \eta_x(\mathbf{p}_{m_1}, t_1)\, \eta_x(\mathbf{p}_{m_2}, t_2)\,, \tag{2.18}$$

respectively. The averages $R_x(\mathbf{p}_{m_1}, \mathbf{p}_{m_2}; t_1, t_2)$ and $C_x(\mathbf{p}_{m_1}, \mathbf{p}_{m_2}; t_1, t_2)$ are related to the mean $\eta_x(\mathbf{p}, t)$ and to the variance $\sigma_x^2(\mathbf{p}, t)$ as

$$C_x(\mathbf{p}, \mathbf{p}; t, t) := \sigma_x^2(\mathbf{p}, t) = R_x(\mathbf{p}, \mathbf{p}; t, t) - \eta_x^2(\mathbf{p}, t)\,. \tag{2.19}$$

The *correlation function coefficient* $\varrho_x(\mathbf{p}_{m_1}, \mathbf{p}_{m_2}; t_1, t_2)$ is the ratio of the covariance $C_x(\mathbf{p}_{m_1}, \mathbf{p}_{m_2}; t_1, t_2)$ and the square root of the product of the variances $\sigma_x^2(\mathbf{p}_{m_1}, t_1)$ and $\sigma_x^2(\mathbf{p}_{m_2}, t_2)$:

$$\varrho_x(\mathbf{p}_{m_1}, \mathbf{p}_{m_2}; t_1, t_2) = \frac{C_x(\mathbf{p}_{m_1}, \mathbf{p}_{m_2}; t_1, t_2)}{\sqrt{\sigma_x^2(\mathbf{p}_{m_1}, t_1)\, \sigma_x^2(\mathbf{p}_{m_2}, t_2)}}\,, \tag{2.20}$$

so that $-1 \leq \varrho_x(\mathbf{p}_{m_1}, \mathbf{p}_{m_2}; t_1, t_2) \leq 1$. The statistical averages depend on the statistical properties of the sources, on the statistical properties of the propagation medium, and on the observation positions.

---

[4] We use the positions $\mathbf{p}_{m_1}$ and $\mathbf{p}_{m_2}$ in analogy to Sect. 2.1 since we observe the spatio-temporal random field, which describes the propagating waves, at the sensor positions.

According to wide-sense stationary (WSS) stochastic processes, *temporally WSS* space-time random fields are obtained with the mean $\eta_x(\mathbf{p},t)$ and the variance $\sigma_x^2(\mathbf{p},t)$ being independent of time t. Then, the correlation function $R_x(\mathbf{p}_{m_1},\mathbf{p}_{m_2};t_1,t_2)$ is invariant to a shift of the temporal origin t_1 and is only dependent on time differences $\tau = t_2 - t_1$. That is,

$$R_x(\mathbf{p}_{m_1},\mathbf{p}_{m_2};\tau) := R_x(\mathbf{p}_{m_1},\mathbf{p}_{m_2};t_1,t_1+\tau) = \mathcal{E}\left\{x(\mathbf{p}_{m_1},t_1)\,x(\mathbf{p}_{m_2},t_1+\tau)\right\} . \tag{2.21}$$

This is analogous to the time-invariance of the deterministic approach in Sect. 2.1. For *spatially homogeneous* space-time random fields, the mean $\eta_x(\mathbf{p},t)$ and the variance $\sigma_x^2(\mathbf{p},t)$ are independent of the position of observation $\mathbf{p}$. The correlation function only depends on relative positions $\boldsymbol{\pi} = \mathbf{p}_{m_2} - \mathbf{p}_{m_1}$ and not on absolute positions:

$$R_x(\boldsymbol{\pi};t_1,t_2) := R_x(\mathbf{p}_{m_1},\mathbf{p}_{m_1}+\boldsymbol{\pi};t_1,t_2) = \mathcal{E}\left\{x(\mathbf{p}_{m_1},t_1)\,x(\mathbf{p}_{m_1}+\boldsymbol{\pi},t_2)\right\} . \tag{2.22}$$

Spatially homogeneous and isotropic spatio-temporal random fields depend only on the Euclidean distance of the positions of observations, i.e., $d := \|\boldsymbol{\pi}\|_2 = \|\mathbf{p}_{m_2} - \mathbf{p}_{m_1}\|_2$. A *WSS space-time random field* is a spatially homogeneous and temporally WSS space-time random field. We write

$$R_x(\boldsymbol{\pi};\tau) := R_x(\mathbf{p}_{m_1},\mathbf{p}_{m_1}+\boldsymbol{\pi};t_1,t_1+\tau) = \mathcal{E}\left\{x(\mathbf{p}_{m_1},t_1)\,x(\mathbf{p}_{m_1}+\boldsymbol{\pi},t_1+\tau)\right\} . \tag{2.23}$$

Though acoustic wave fields are generally not temporally WSS, we will use a spectral description of spatio-temporal random fields in the DTFT domain if it gives further insights into our studies. We define the *spatio-spectral correlation function* $S_{x_{m_1}x_{m_2}}(\omega)$ between the positions $\mathbf{p}_{m_1}$ and $\mathbf{p}_{m_2}$ as the DTFT of the sampled spatio-temporal correlation function for temporally WSS random fields as[5]

$$S_{x_{m_1}x_{m_2}}(\omega) = \sum_{k=-\infty}^{\infty} R_x(\mathbf{p}_{m_1},\mathbf{p}_{m_2};kT_\mathrm{s}) \exp\left(-j\omega kT_\mathrm{s}\right) , \tag{2.24}$$

where we drop the dependency on the observation positions for spatially homogeneous and temporally WSS space-time random fields, i.e., $S_{x_{m_1}x_{m_2}}(\omega) := S_{xx}(\omega)$. Analogously to the correlation function coefficient $\varrho_x(\mathbf{p}_{m_1},\mathbf{p}_{m_2};t_1,t_2)$ in the time domain, a *spatial coherence function* $\gamma_{x_{m_1}x_{m_2}}(\omega)$ can be defined in the DTFT domain. The spatial coherence function $\gamma_{x_{m_1}x_{m_2}}(\omega)$ is equal to the spatio-spectral correlation function $S_{x_{m_1}x_{m_2}}(\omega)$ normalized by the square root of the product of the two auto-power spectral densities (PSDs) $S_{x_{m_1}x_{m_1}}(\omega)$ and $S_{x_{m_2}x_{m_2}}(\omega)$. That is,

[5] We express the dependency of the positions $\mathbf{p}_{m_1}$ and $\mathbf{p}_{m_2}$ on $S_{x_{m_1}x_{m_2}}(\omega)$ not in the argument but in the subscripts in order to have the same notation for homogeneous and non-homogeneous spatio-temporal random fields. The distinction will be clear from the given context.

$$\gamma_{x_{m_1}x_{m_2}}(\omega) = \frac{S_{x_{m_1}x_{m_2}}(\omega)}{\sqrt{S_{x_{m_1}x_{m_1}}(\omega)\, S_{x_{m_2}x_{m_2}}(\omega)}}. \tag{2.25}$$

$S_{x_{m_1}x_{m_1}}(\omega)$ and $S_{x_{m_2}x_{m_2}}(\omega)$ must be different from zero and must not contain any delta functions [BP71]. The normalization constrains the *magnitude squared coherence* $|\gamma_{x_{m_1}x_{m_2}}(\omega)|^2$ to the interval $0 \leq |\gamma_{x_{m_1}x_{m_2}}(\omega)|^2 \leq 1$. The spatial coherence describes the spatio-temporal correlation of the temporally WSS random processes $x_{m_1}(k)$ and $x_{m_2}(k)$. For $|\gamma_{x_{m_1}x_{m_2}}(\omega)|^2 = 0$, the signals $x_{m_1}(k)$ and $x_{m_2}(k)$ are uncorrelated. For $|\gamma_{x_{m_1}x_{m_2}}(\omega)|^2 = 1$, the signals $x_{m_1}(k)$ and $x_{m_2}(k)$ are completely correlated. In this case, since $\gamma_{x_{m_1}x_{m_2}}(\omega)$ is invariant to linear filtering, the relation between $x_{m_1}(k)$ and $x_{m_2}(k)$ can be described by a linear system $h(\mathbf{p}_{m_1}, \mathbf{p}_{m_2}; kT_\mathrm{s})$ with a transfer function $H(\mathbf{p}_{m_1}, \mathbf{p}_{m_2}; \omega)$ with the input signal $x_{m_1}(k)$ and with the output signal $x_{m_2}(k)$:

$$\gamma_{x_{m_1}x_{m_2}}(\omega) = \frac{H^*(\mathbf{p}_{m_1}, \mathbf{p}_{m_2}, \omega)\, S_{x_{m_1}x_{m_1}}(\omega)}{\sqrt{\left(S_{x_{m_1}x_{m_1}}(\omega)\right)^2 |H(\mathbf{p}_{m_1}, \mathbf{p}_{m_2}; \omega)|^2}} = \frac{H^*(\mathbf{p}_{m_1}, \mathbf{p}_{m_2}; \omega)}{|H(\mathbf{p}_{m_1}, \mathbf{p}_{m_2}; \omega)|}, \tag{2.26}$$

where $(\cdot)^*$ denotes conjugate complex. For obtaining this result, we assume wide-sense ergodicity and we write (2.21) as a time average $\int_{-\infty}^{\infty} x(\mathbf{p}_{m_1}, t_1) \cdot x(\mathbf{p}_{m_2}, t_1 + \tau)\, d\tau$. We replace $x(\mathbf{p}_{m_2}, t_1 + \tau)$ by a convolution integral of $x(\mathbf{p}_{m_1}, t_1 + \tau)$ and $h(\mathbf{p}_{m_1}, \mathbf{p}_{m_2}; \tau)$. After temporal sampling, integration, and transformation into the DTFT domain using (2.24), we obtain the desired result (2.26). It follows $|\gamma_{x_{m_1}x_{m_2}}(\omega)|^2 = 1$ for a linear relationship between $x_{m_1}(k)$ and $x_{m_2}(k)$. If $0 < |\gamma_{x_{m_1}x_{m_2}}(\omega)|^2 < 1$, three possibilities exist: First, additional uncorrelated noise may be present[6]. Second, the system that relates $x_{m_1}(k)$ and $x_{m_2}(k)$ is not linear. Third, $x_{m_2}(k)$ is the output signal due to an input signal $x_{m_1}(k)$ but also due to other input signals [BP71]. The estimation of coherence functions, statistical properties, and applications are discussed in, e.g., [Car87, Gar92].

2.2.2 Spatio-temporal and Spatio-spectral Correlation Matrices

In this section, spatio-temporal and spatio-spectral correlation matrices are introduced. First, the structures of spatio-temporal and spatio-spectral correlation matrices are presented and basic properties are summarized. Second, we deal with the estimation of spatio-temporal and spatio-spectral correlation matrices. Third, the narrowband assumption, which is often used to de-

[6]Note that despite a linear relation between $x_{m_1}(k)$ and $x_{m_2}(k)$, it might hold $|\gamma_{x_{m_1}x_{m_2}}(\omega)|^2 < 1$ with a dependency of $\gamma_{x_{m_1}x_{m_2}}(\omega)$ on $h(\mathbf{p}_{m_1}, \mathbf{p}_{m_2}; \tau)$. This happens if the observation time is shorter than the true length of the impulse response $h(\mathbf{p}_{m_1}, \mathbf{p}_{m_2}; \tau)$ [Mar95, BMS98]. The reduction of the coherence may be interpreted as presence of uncorrelated noise.

sign and to analyze beamformers, is introduced. Finally, typical examples for spatio-temporal and spatio-spectral correlation matrices are given.

Definition and Properties

For many array processing techniques, it is convenient to express the characteristics of the measured (sampled) spatio-temporal random fields by correlation functions computed between all pairs of sampled sensor signals $x_{m_1}(k)$, $x_{m_2}(k)$, with $m_1, m_2 = 0, 1, \ldots, M-1$, in a given temporal interval of length $(N-1)T_s$. These spatially and temporally sampled spatio-temporal correlation functions can be advantageously captured in *spatio-temporal correlation matrices*. The equivalent in the DTFT domain for temporally WSS spatio-temporal random fields is called *spatio-spectral correlation matrix*.

Spatio-temporal Correlation Matrices

Let the spatio-temporal correlation matrix of size $N \times N$ between the pair of sensors m_1 and m_2 for time lags $k_1, k_2 = 0, 1, \ldots, N-1$ be defined as

$$\begin{aligned}\left[\mathbf{R}_{x_{m_1}x_{m_2}}(k)\right]_{k_1,k_2} &= R_x(\mathbf{p}_{m_1}, \mathbf{p}_{m_2}; (k-k_1)T_s, (k-k_2)T_s) \\ &= \mathcal{E}\left\{x_{m_1}(k-k_1)\, x_{m_2}(k-k_2)\right\} . \end{aligned} \tag{2.27}$$

$[\cdot]_{k_1,k_2}$ denotes the element in the k_1-th row in the k_2-th column of a matrix or for a column vector (for $k_2 = 1$). This allows to write the spatio-temporal correlation matrix of size $MN \times MN$ for all pairs of sensor signals as

$$\mathbf{R}_{\mathbf{xx}}(k) = \begin{pmatrix} \mathbf{R}_{x_0x_0}(k) & \mathbf{R}_{x_0x_1}(k) & \cdots & \mathbf{R}_{x_0x_{M-1}}(k) \\ \mathbf{R}_{x_1x_0}(k) & \mathbf{R}_{x_1x_1}(k) & \cdots & \mathbf{R}_{x_1x_{M-1}}(k) \\ \vdots & & \ddots & \vdots \\ \mathbf{R}_{x_{M-1}x_0}(k) & \mathbf{R}_{x_{M-1}x_1}(k) & \cdots & \mathbf{R}_{x_{M-1}x_{M-1}}(k) \end{pmatrix} . \tag{2.28}$$

Capturing N subsequent samples of the sensor data $x_m(k)$ into a vector

$$\mathbf{x}_m(k) = (x_m(k),\, x_m(k-1),\, \ldots,\, x_m(k-N+1))^T , \tag{2.29}$$

and combining the vectors $\mathbf{x}_m(k)$, $m = 0, 1, \ldots, M-1$, in a stacked sensor data vector

$$\mathbf{x}(k) = \left(\mathbf{x}_0^T(k),\, \mathbf{x}_1^T(k),\, \ldots,\, \mathbf{x}_{M-1}^T(k)\right)^T , \tag{2.30}$$

we can equally write the spatio-temporal correlation matrix as

$$\mathbf{R}_{\mathbf{xx}}(k) = \mathcal{E}\left\{\mathbf{x}(k)\,\mathbf{x}^T(k)\right\} . \tag{2.31}$$

The matrix $\mathbf{R}_{\mathbf{xx}}(k)$ has the following properties: The correlation matrices $\mathbf{R}_{x_mx_m}(k)$ on the main diagonal are evaluated at single positions $\mathbf{p}_m$. The matrix $\mathbf{R}_{x_mx_m}(k)$ thus reduces to a temporal correlation matrix. The off-diagonal matrices $\mathbf{R}_{x_{m_1}x_{m_2}}(k)$, with $m_1 \neq m_2$, characterize both spatial and temporal correlation. The correlation matrix $\mathbf{R_{xx}}(k)$ is block symmetric, since $\mathbf{R}_{x_{m_1}x_{m_2}}(k) = \mathbf{R}^T_{x_{m_2}x_{m_1}}(k)$ (for real-valued space-time random fields). For WSS space-time random fields, the spatio-temporal correlation matrix is independent of time k, i.e., $\mathbf{R_{xx}} := \mathbf{R_{xx}}(k)$. The matrices $\mathbf{R}_{x_{m_1}x_{m_2}} := \mathbf{R}_{x_{m_1}x_{m_2}}(k)$ are Toeplitz. Unless the sensor array has symmetries, $\mathbf{R_{xx}}$ lacks further symmetries. Consider for example a 1-D line array with uniform sensor spacing in a WSS random field. Then, a constant difference $m_1 - m_2$ implies a constant spatio-temporal correlation function, and $\mathbf{R_{xx}}$ is Toeplitz.

If the spatio-temporal correlation function is *separable*, then, the spatio-temporal correlation function for spatially and temporally WSS random fields can be written as the product of a spatial correlation function $R_x^{(s)}(\boldsymbol{\pi})$ and a temporal correlation function $R_x^{(t)}(\tau)$, i.e., $R_x(\boldsymbol{\pi};\tau) = R_x^{(s)}(\boldsymbol{\pi})\, R_x^{(t)}(\tau)$. This implies that spatial and temporal correlation is mutually uncorrelated. The temporal correlation function does not vary spatially, and vice versa. This allows to write the spatio-temporal correlation matrix $\mathbf{R}_{x_{m_1}x_{m_2}}$ as the product of the spatial correlation coefficient $\varrho_{x_{m_1}x_{m_2}} := \varrho(\mathbf{p}_{m_1}, \mathbf{p}_{m_2}, t_1, t_2)$ and the temporal correlation matrix $\mathbf{R}_{xx} := \mathbf{R}_{x_{m_1}x_{m_1}}$, $m_1 = 0, 1, \ldots, M-1$. That is,

$$\mathbf{R}_{x_{m_1}x_{m_2}} = \mathbf{R}_{xx}\, \varrho_{x_{m_1}x_{m_2}}\,, \tag{2.32}$$

or, with the *spatial correlation matrix* $[\boldsymbol{\varrho}]_{m_1,m_2} = \varrho_{x_{m_1}x_{m_2}}$,

$$\mathbf{R_{xx}} = \boldsymbol{\varrho} \otimes \mathbf{R}_{xx}\,, \tag{2.33}$$

where $\otimes$ is the Kronecker product [Gra81].

Spatio-spectral Correlation Matrices

For temporally WSS space-time random fields, we define a *spatio-spectral correlation matrix* $\mathbf{S_{xx}}(\omega)$ of size $M \times M$ with the spatio-spectral correlation functions $S_{x_{m_1}x_{m_2}}(\omega)$ as elements as

$$[\mathbf{S_{xx}}(\omega)]_{m_1,m_2} = S_{x_{m_1}x_{m_2}}(\omega)\,. \tag{2.34}$$

For a spatially homogeneous and temporally WSS space-time random field, we can write $\mathbf{S_{xx}}(\omega)$ as

$$\mathbf{S_{xx}}(\omega) = S_{xx}(\omega)\, \boldsymbol{\Gamma}_{\mathbf{xx}}(\omega)\,, \tag{2.35}$$

where the elements of the *spatial coherence matrix* $\boldsymbol{\Gamma}_{\mathbf{xx}}(\omega)$ of size $M \times M$ are the spatial coherence functions $\gamma_{x_{m_1}x_{m_2}}(\omega)$ after (2.26):

$$[\boldsymbol{\Gamma}_{\mathbf{xx}}(\omega)]_{m_1,m_2} = \gamma_{x_{m_1}x_{m_2}}(\omega)\,. \tag{2.36}$$

A separable spatio-temporal correlation function $R_x(\boldsymbol{\pi};\tau)$ gives with (2.33)

$$\mathbf{S_{xx}}(\omega) = S_{xx}(\omega)\, \boldsymbol{\varrho}\,, \tag{2.37}$$

For separability of $R_x(\boldsymbol{\pi};\tau)$, it follows from (2.35) and from (2.37) that the spatial coherence matrix $\boldsymbol{\Gamma}_{\mathbf{xx}}(\omega)$ and the spatial correlation matrix $\boldsymbol{\varrho}$ are equivalent.

Estimation of Spatio-temporal and of Spatio-spectral Correlation Functions

If we do not use models, we generally have to estimate the spatio-temporal and spatio-spectral correlation matrices. For wide-sense ergodic spatio-temporal random fields, the correlation matrices can be estimated using temporal averaging.

Spatio-temporal Correlation Functions

We assume a wide-sense ergodic space-time random field which is described by the signals $x_{m_1}(k)$ and $x_{m_2}(k)$ in the discrete time domain. For power signals, a time-averaged estimate of the spatio-temporal correlation function can be defined as

$$\hat{R}_x(\mathbf{p}_{m_1}, \mathbf{p}_{m_2}; \kappa T_\mathrm{s}) = \frac{1}{2K+1} \sum_{k=-K}^{K} x_{m_1}(k+\kappa) x_{m_2}(k) \,, \tag{2.38}$$

where $\hat{R}_x(\mathbf{p}_{m_1}, \mathbf{p}_{m_2}; \kappa T_\mathrm{s})$ is averaged over a data block of length $2K+1$. The quantity $\hat{R}_x(\mathbf{p}_{m_1}, \mathbf{p}_{m_2}; \kappa T_\mathrm{s})$ is an unbiased and consistent estimate of $R_x(\mathbf{p}_{m_1}, \mathbf{p}_{m_2}; \kappa T_\mathrm{s})$. For Gaussianity of $x_{m_1}(k)$ and $x_{m_2}(k)$, $\hat{R}_x(\mathbf{p}_{m_1}, \mathbf{p}_{m_2}; \kappa T_\mathrm{s})$ is the maximum-likelihood estimate of $R_x(\mathbf{p}_{m_1}, \mathbf{p}_{m_2}; \kappa T_\mathrm{s})$ [Goo63].

Spatio-spectral Correlation Functions

For applying spatio-spectral correlation functions to short-time stationary wideband speech and audio signals, we estimate $S_{x_{m_1} x_{m_2}}(\omega)$ by second-order periodograms [Bri75] using data-overlapping and windowing in the time domain and exponential averaging in the DTFT domain. Let the second-order windowed periodogram at block time r be defined as

$$\begin{aligned} I_{x_{m_1} x_{m_2}}(r, \omega) = & \left(\sum_{k=0}^{K-1} w_k x_{m_1}(k + rR) \exp\left(-j\omega k T_\mathrm{s}\right) \right) \times \\ & \times \left(\sum_{k=0}^{K-1} w_k x_{m_2}(k + rR) \exp\left(-j\omega k T_\mathrm{s}\right) \right)^* , \end{aligned} \tag{2.39}$$

where the block time r is related to the discrete time k by $k = rR$. The length of the window in the discrete time domain K is related to the block overlap R by $R = \lfloor K/\alpha \rfloor$, where $\alpha \geq 1$ is the block overlap factor and where $\lfloor \cdot \rfloor$ rounds the argument toward the next smaller integer value. The windowing function w_k is normalized, i.e., $\sum_{k=0}^{K-1} |w_k|^2 = 1$. It allows to control the frequency resolution and the sidelobe leakage [Har78]. The second-order windowed periodogram $I_{x_{m_1} x_{m_2}}(r, \omega)$ is an asymptotically unbiased estimate of $S_{x_{m_1} x_{m_2}}(\omega)$ for $\omega T_\mathrm{s} \neq 0 (\mathrm{mod}\, 2\pi)$ [Bri75]. The estimate $\hat{S}_{x_{m_1} x_{m_2}}(r, \omega)$ of $S_{x_{m_1} x_{m_2}}(r, \omega)$ at block time r is obtained as

$$\hat{S}_{x_{m_1}x_{m_2}}(r,\omega) = \frac{1}{R_{\text{eff}}} \sum_{i=1}^{r} \beta^{r-i} I_{x_{m_1}x_{m_2}}(i,\omega) \,. \tag{2.40}$$

The factor $0 < \beta < 1$ is the exponential forgetting factor for averaging over the block time in the DTFT domain. The exponential averaging allows to account for the short-time stationarity of speech and audio signals. The normalization factor $R_{\text{eff}} = (\beta^r - 1)/(\beta - 1)$ can be interpreted as the effective memory for averaging the second-order windowed periodograms $I_{x_{m_1}x_{m_2}}(r,\omega)$.

Narrowband Assumption

Often, wideband beamformers are realized in the DFT domain using the *narrowband assumption* [Hod76, Com88, VB88, Tre02]. The wideband sensor signals are decomposed into narrow frequency bins, and the frequency bins are processed independently from each other.

For zero-mean wide-sense ergodic space-time random fields, we have for the variance $\sigma_x^2(\mathbf{p},t)$ of the m-th sensor signal from (2.16) and from (2.21)

$$\sigma_{x_m}^2 := \sigma_x^2(\mathbf{p}_m, kT_\text{s}) = R_x(\mathbf{p}_m, \mathbf{p}_m; 0) = \frac{T_\text{s}}{2\pi} \int_0^{2\pi/T_\text{s}} S_{x_m x_m}(\omega)\, d\omega \,. \tag{2.41}$$

We decompose the wideband sensor signal into narrowband signals and consider only one frequency bin with center frequency ω_0 and bandwidth $2\Delta\omega$. The spectrum of each narrowband signal is thus zero outside of the spectral band $[\omega_0 - \Delta\omega, \omega_0 + \Delta\omega]$ so that the variance of the m-th sensor signal in the frequency bin with center frequency ω_0 is given by

$$\sigma_{x_m}^2(\omega_0) = \frac{T_\text{s}}{2\pi} \int_{\omega_0-\Delta\omega}^{\omega_0+\Delta\omega} S_{x_m x_m}(\omega)\, d\omega \,. \tag{2.42}$$

For $\Delta\omega$ being sufficiently small so that $S_{x_m x_m}(\omega)$ is constant over the interval $[\omega_0 - \Delta\omega, \omega_0 + \Delta\omega]$, (2.42) can be approximated by

$$\sigma_{x_m}^2(\omega_0) \approx \frac{\Delta\omega T_\text{s}}{\pi} S_{x_m x_m}(\omega_0) \,, \tag{2.43}$$

where $S_{x_m x_m}(\omega_0)$ must not contain a delta function[7]. This approximation is called narrowband assumption. For calculating the variance $\sigma_{x_m}^2(\omega_0)$ of a sensor signal in one of the frequency bands, it is thus not necessary to evaluate the integral in (2.42), but $\sigma_{x_m}^2(\omega_0)$ is approximated by the PSD of the sensor signal at the center frequency of the given frequency interval. This often simplifies the design and the analysis of wideband beamformers.

The narrowband assumption requires that the frequency bins are mutually uncorrelated. In real applications, the PSD $S_{x_m x_m}(\omega)$ has to be estimated by,

[7] If $S_{x_m x_m}(\omega_0)$ contains a delta function, the integration in (2.42) has to be performed in order to calculate $\sigma_{x_m}^2(\omega_0)$.

e.g., averaging periodograms $I_{x_m x_m}(r, \omega)$ over observation windows of finite length according to (2.40). However, due to the finite length of the observation windows, the frequency bins are generally not mutually uncorrelated so that the narrowband assumption holds only asymptotically for observation windows of infinite length [Hod76, Tre02].

Examples

Next, we consider typical examples of space-time random fields which will be encountered in our context. We derive spatio-spectral correlation functions and give expressions for the spatio-spectral correlation matrices of the space-time random fields.

Monochromatic Plane Wave in an Acoustic Free-Field

We consider an ergodic monochromatic plane wave in an acoustic free-field after (2.2). The plane wave captured by a sensor at position $\mathbf{p}_m$ is then described by

$$x_m(k) = S \exp\left(j\left(\omega_0 k T_\mathrm{s} - \mathbf{k}_0^T \mathbf{p}_m\right)\right) . \tag{2.44}$$

The spatio-temporal correlation function between two sensors is obtained from (2.38) for $K \to \infty$ as

$$R_x(\mathbf{p}_{m_1}, \mathbf{p}_{m_2}; \kappa T_\mathrm{s}) = \lim_{K\to\infty} \frac{1}{2K+1} \sum_{k=-K}^{K} x_{m_1}(k+\kappa) x_{m_2}(k) . \tag{2.45}$$

Substituting $x_{m_1}(k)$ and $x_{m_2}(k)$ by (2.45) and transforming $R_x(\mathbf{p}_{m_1}, \mathbf{p}_{m_2}; \kappa T_\mathrm{s})$ into the DTFT domain yields the spatio-spectral correlation function

$$S_{x_{m_1} x_{m_2}}(\omega) = \frac{2\pi S^2}{T_\mathrm{s}} \delta(\omega T_\mathrm{s} - \omega_0 T_\mathrm{s}) \exp\left(-j\mathbf{k}_0^T\left(\mathbf{p}_{m_1} - \mathbf{p}_{m_2}\right)\right) , \tag{2.46}$$

where only the range $0 \le \omega \le 2\pi$ is considered. The term

$$S_{ss}(\omega) = \frac{2\pi S^2}{T_\mathrm{s}} \delta(\omega T_\mathrm{s} - \omega_0 T_\mathrm{s}) \tag{2.47}$$

is the PSD of the source signal. The spatial correlation is described by the exponential term.

The spatio-spectral correlation matrix $\mathbf{S}_{\mathbf{xx}}(\omega)$ of the monochromatic plane wave is given by (2.34) with elements (2.46). Let the *steering vector* be defined as

$$\mathbf{v}(\mathbf{k}, \omega) = \left(\exp\left(j\mathbf{k}^T \mathbf{p}_0\right), \exp\left(j\mathbf{k}^T \mathbf{p}_1\right), \ldots, \exp\left(j\mathbf{k}^T \mathbf{p}_{M-1}\right)\right)^H , \tag{2.48}$$

where the time delays $\tau_m = \mathbf{k}^T \mathbf{p}_m$ are the *inter-sensor propagation delays* w.r.t. the spatial origin, so that the steering vector incorporates the geometry of the array and the DOA of the source signal. This allows to write $\mathbf{S}_{\mathbf{xx}}(\omega)$ as

$$\mathbf{S}_{\mathbf{xx}}(\omega) = S_{ss}(\omega)\,\mathbf{v}(\mathbf{k}_0,\omega_0)\mathbf{v}^H(\mathbf{k}_0,\omega_0)\,. \tag{2.49}$$

We thus have a spatially homogeneous wave field with spatial coherence matrix

$$\mathbf{\Gamma}_{\mathbf{xx}}(\omega) = \mathbf{v}(\mathbf{k}_0,\omega_0)\mathbf{v}^H(\mathbf{k},\omega)\,. \tag{2.50}$$

Point Source in a Reverberant Environment

For a point source $s_{\mathbf{p}_\mathrm{s}}(t)$ at position $\mathbf{p}_\mathrm{s}$ in a reverberant environment, the sensor signals are given by (2.10). Assuming a temporally WSS source signal and time-invariant room impulse responses, the spatio-temporal correlation function $R_x(\mathbf{p}_{m_1},\mathbf{p}_{m_2};\kappa T_\mathrm{s})$ between two sensor signals can be determined by (2.45). Substituting in (2.45) $x_{m_1}(k)$ and of $x_{m_2}(k)$ by (2.10) and applying (2.24), we obtain for the spatio-spectral correlation function between the two sensor signals

$$S_{x_{m_1}x_{m_2}}(\omega) = H(\mathbf{p}_{m_1},\mathbf{p}_\mathrm{s},\omega)S_{ss}(\omega)H^*(\mathbf{p}_{m_2},\mathbf{p}_\mathrm{s},\omega)\,, \tag{2.51}$$

where $S_{ss}(\omega)$ is the PSD of the source signal $s_{\mathbf{p}_\mathrm{s}}(kT_\mathrm{s})$.

Let the vector $\mathbf{h}_\mathrm{f}(\mathbf{p}_\mathrm{s},\omega)$ of length M capture the transfer functions between the source position $\mathbf{p}_\mathrm{s}$ and all sensor positions, i.e.,

$$\mathbf{h}_\mathrm{f}(\mathbf{p}_\mathrm{s},\omega) = \left(H(\mathbf{p}_0,\mathbf{p}_\mathrm{s},\omega),\, H(\mathbf{p}_1,\mathbf{p}_\mathrm{s},\omega),\, \ldots,\, H(\mathbf{p}_{M-1},\mathbf{p}_\mathrm{s},\omega)\right)^T\,. \tag{2.52}$$

The spatio-spectral correlation matrix $\mathbf{S}_{\mathbf{xx}}(\omega)$ can then be written as

$$\mathbf{S}_{\mathbf{xx}}(\omega) = S_{ss}(\omega)\,\mathbf{h}_\mathrm{f}(\mathbf{p}_\mathrm{s},\omega)\mathbf{h}_\mathrm{f}^H(\mathbf{p}_\mathrm{s},\omega)\,. \tag{2.53}$$

Spatially and Temporally White Fields

For spatially white, non-homogeneous, and temporally non-stationary, space-time random fields, the correlation matrices $\mathbf{R}_{x_{m_1}x_{m_2}}(k)$ for $m_1 \neq m_2$ vanish. This yields a block-diagonal spatio-temporal correlation matrix

$$\mathbf{R}_{\mathbf{xx}}(k) = \mathrm{diag}\left\{\mathbf{R}_{x_0x_0}(k),\, \mathbf{R}_{x_1x_1}(k),\, \ldots,\, \mathbf{R}_{x_{M-1}x_{M-1}}(k)\right\}\,, \tag{2.54}$$

where $\mathrm{diag}\{\cdot\}$ is the diagonal matrix formed by the listed entries. For spatial and temporal wide-sense stationarity, (2.54) simplifies to

$$\mathbf{R}_{\mathbf{xx}} = \mathrm{diag}\left\{\mathbf{R}_{xx},\, \mathbf{R}_{xx},\, \ldots,\, \mathbf{R}_{xx}\right\}\,, \tag{2.55}$$

or, equivalently, with (2.34) in the DTFT domain

$$\mathbf{S}_{\mathbf{xx}}(\omega) = S_{xx}(\omega)\,\mathbf{I}_{M\times M}\,, \tag{2.56}$$

where $\mathbf{I}_{M\times M}$ is the identity matrix of size $M\times M$. For spatially and temporally white WSS space-time random fields with variance σ_x^2, the temporal auto-correlation matrices $\mathbf{R}_{x_mx_m}$ are diagonal,

$$\mathbf{R_{xx}} = \sigma_x^2 \mathbf{I}_{MN \times MN} , \tag{2.57}$$

which may be written in the DTFT domain as

$$\mathbf{S_{xx}}(\omega) = \sigma_x^2 \mathbf{I}_{M \times M} . \tag{2.58}$$

A typical spatially and temporally white field is the sensor noise, which is generated by the electronics of the transducers and the analog/digital-conversion.

Diffuse Fields

Diffuse (or *spherically isotropic*) wave fields are obtained if an infinite number of mutually uncorrelated temporally WSS point sources with identical PSD $S_{ss}(\omega)$ are uniformly and continuously distributed on a sphere[8] with infinite radius r. The correlation function between two positions of observations $\mathbf{p}_{m_1}$ and $\mathbf{p}_{m_2}$ of this spatially homogeneous, spatially isotropic, and temporally WSS space-time random field is given by

$$R_x(\|\mathbf{p}_{m_2} - \mathbf{p}_{m_1}\|_2; \tau) = \frac{1}{\pi} \int_0^\infty S_{ss}(\omega) \frac{\sin \omega \|\mathbf{p}_{m_2} - \mathbf{p}_{m_1}\|_2 / c}{\omega \|\mathbf{p}_{m_2} - \mathbf{p}_{m_1}\|_2 / c} \cos \omega\tau \, d\omega , \tag{2.59}$$

[Eck53, Jac62, Cox73b, JD93, Tre02]. The argument of the Fourier integral (2.59) is equal to the spatio-spectral correlation function between two sensors with inter-sensor spacing $d := \|\mathbf{p}_{m_2} - \mathbf{p}_{m_1}\|_2$. The spatial coherence function between the two sensors with inter-sensor spacing d is obtained by replacing $S_{x_{m_1} x_{m_2}}(\omega)$ in (2.25) by the argument of the Fourier integral (2.59) and by considering the homogeneity of the diffuse wave field, i.e., $S_{x_{m_1} x_{m_1}}(\omega) = S_{x_{m_2} x_{m_2}}(\omega) = S_{ss}(\omega)$. It yields,

$$\gamma_{\mathrm{si}, x_{m_1} x_{m_2}}(\omega) = \frac{\sin \omega \|\mathbf{p}_{m_2} - \mathbf{p}_{m_1}\|_2 / c}{\omega \|\mathbf{p}_{m_2} - \mathbf{p}_{m_1}\|_2 / c} . \tag{2.60}$$

For small sensor spacings and for low frequencies, the spatial coherence function is close to one. It reduces with increasing frequency and with increasing distance $d := \|\mathbf{p}_{m_2} - \mathbf{p}_{m_1}\|_2$. It is equal to zero for half of the wavelength, $\lambda/2 = \omega/(2c)$, being multiples of the distance d.

The spatio-spectral correlation matrix of the diffuse wave field for a uniform linear array with equal inter-sensor spacing d is thus given by

$$\mathbf{S_{xx}}(\omega) = S_{ss}(\omega) \, \mathbf{\Gamma}_{\mathrm{si}}(\omega) , \tag{2.61}$$

where the elements of the spatial coherence matrix of the diffuse wave field, $\mathbf{\Gamma}_{\mathrm{si}}(\omega)$, are defined as

$$[\mathbf{\Gamma}_{\mathrm{si}}(\omega)]_{m_1, m_2} = \frac{\sin(\omega d (m_1 - m_2)/c)}{\omega d (m_1 - m_2)/c} . \tag{2.62}$$

Figure 2.5 illustrates (2.62) for different sensor spacings d.

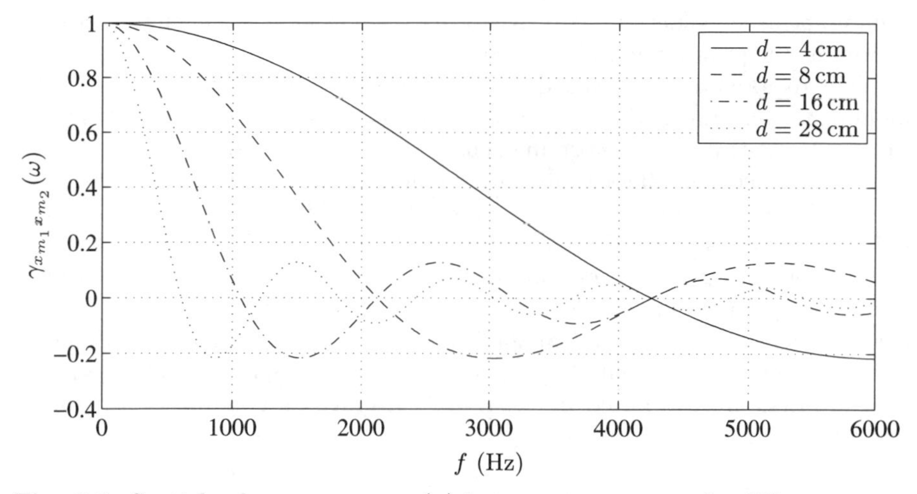

Fig. 2.5. Spatial coherence $\gamma_{x_{m1}x_{m2}}(\omega)$ between two sensors for different sensor spacings d over frequency f for a diffuse wave field

Typically, very late reflections (late reverberation) in reverberant acoustic environments or car noise inside the passenger cabin of a car are modeled as diffuse wave fields [Mar95].

2.3 Summary

In this chapter, we have introduced a deterministic and a statistical description of acoustic signals that will be used throughout this work. Starting from the scalar linear wave equation, we introduced monochromatic plane waves and monochromatic spherical waves. While plane waves can be used to describe signals in the far field of a sensor array, spherical waves emitted by point sources can be applied to model signals in the near field of a sensor array. More generally, acoustic sources and the propagation of their wave fields can be modeled by spatio-temporal signals and by spatio-temporal impulse responses, respectively.

Since acoustic space-time signals often cannot be described in a deterministic way, a statistical signal model is generally used to capture the (temporal) source signal characteristics. For point sources, as, e.g., human speakers or loudspeakers, the modeling of the propagation by spatio-temporal impulse responses is sufficient. However, for spatially extended sources, as, e.g., noise of air conditioners or the noise in the passenger cabin of a car, spatio-temporal

[8] The spatio-temporal correlation for – on a sphere – arbitrarily distributed, mutually uncorrelated sources is derived in, e.g., [Tre02].

random fields are generally used to capture the temporal signal characteristics as well as the propagation in the acoustic environment.

Due to its simplicity, the deterministic signal model, in particular the model of monochromatic plane waves, is advantageous for illustrating basic characteristics of beamforming sensor arrays analytically. For developing speech enhancement algorithms, this model can generally not be used because of the time-variance and because of the wideband nature of speech, noise, and audio signals. Here, the statistical description is more appropriate. The (second-order) statistical properties of space-time signals captured by a sensor array can be efficiently captured in spatio-temporal or spatio-spectral correlation matrices. In practical applications, spatio-temporal correlation matrices are generally estimated by temporal averaging. Spatio-spectral correlation matrices are estimated by recursively averaged second-order windowed periodograms.

3

Optimum Linear Filtering

In this chapter, we introduce the concept of optimum linear filtering for multiple-input multiple-output (MIMO) digital systems for solving multi-channel linear digital filtering problems. A linear MIMO system is depicted in Fig. 3.1.

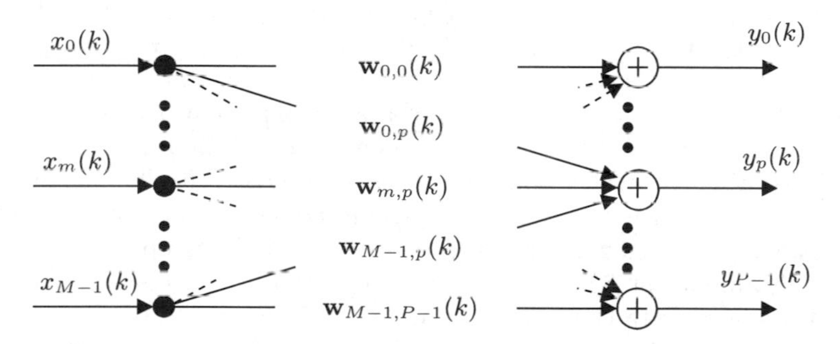

Fig. 3.1. General linear MIMO system

In our context, the real-valued multi-channel input data is obtained by spatial and temporal sampling of propagating wave fields as introduced in Chap. 2. The function of MIMO linear systems is to alter the filter input signals $x_m(k)$ in order to provide filter output signals $y_p(k)$ with desired waveforms. For maximum flexibility, each output signal is formed by a combination of all input signals. The linear filters are realized such that the output signals fit a set of prescribed desired signals in some optimum fashion. The *method of least-squares* (LS) invokes the *least-squares error* (LSE) criterion to determine the optimum linear filters. With the LSE optimization criterion, linear filters are determined such that the sum of squares of the difference between the desired waveforms and the output signals is minimized over a given time interval. It thus involves time averages, yielding a dependency on the length of the observation time interval.

The method of least-squares is a deterministic approach. It can be viewed as the deterministic analogon to the statistical *minimum mean-squared error* (MMSE) optimization criterion, yielding the *Wiener filter* for Gaussian distributed signals. Filters which are optimum in the MMSE sense are derived from ensemble averages, which can be replaced by time averages (over time intervals of infinite length) for wide-sense ergodic signals. Using time intervals of finite length, MMSE optimization reduces to LSE optimization. MMSE optimum filters are also referred to as *optimum linear filters* in the literature, while the deterministic approach is called *least-squares linear filter*. In this work, we use the term 'optimum filter' for all classes of filters which fulfill a given optimization criterion. The distinction between the classes of optimum filters is made in the given context. The term *optimum* means best with some well-defined assumptions and conditions. A change of the optimization criterion or of the statistics of the processed signals also changes the corresponding optimum filter. Therefore, an optimum filter may perform poorly if the statistics of the input signals do not correspond to the assumed statistics used for its derivation. The applications of optimum filters can be classified into four categories: system identification, equalization, signal prediction, and interference cancellation [Hay96].

For tracking time-varying signal characteristics and/or time-varying systems, optimum filters are often calculated iteratively by recursive *adaptation algorithms*, yielding the concept of *adaptive filtering*. The filter coefficients of adaptive filters adjust to the statistics of the incoming data, with the result that time-varying signal characteristics can be tracked without redesigning the filters. This capability makes adaptive filters a powerful tool for realizing systems which deal with non-stationary signals in non-stationary environments. The complexity is often reduced compared to the direct computation, which is especially important for real-time systems. Adaptive filters are generally designed to converge to the Wiener solution for WSS signals.

In the context of speech and audio signal acquisition, we generally deal with time-varying signals and time-varying acoustic environments. For realizing optimum filters for this scenario, it is appropriate to introduce the signal processing methods starting from LS optimization, which involves data blocks of finite length. The array signal processing literature often starts from narrowband signals in the DTFT domain for introducing the optimum filtering concept. In this work, we rigorously introduce the LS optimum filters first, since it better fits our signal model. After that, the relation to the better known statistical analogon for narrowband signals in the DTFT domain is established for illustrating the properties of these LS solutions.

This chapter is organized as follows: In Sect. 3.1, we introduce the concept of and the solution to optimum linear finite impulse response (FIR) filtering using LS optimization. The relation with MMSE optimization for wide-sense ergodic signals in the DTFT domain is shown. In Sect. 3.2, we illustrate typical applications of optimum linear filters as they are discussed in this work. The realization using adaptive algorithms is discussed in Chap. 7. More details about linear optimum filters can be found in, e.g., [Ste73, LH74, GvL89, Hay96, MIK00]. Compared to the existing literature, where generally single-input single-output (SISO) or multiple-input single-output (MISO) systems are discussed, we formally extend the description of optimum linear filters to the MIMO case based on [BBK03].

3.1 Generic Multiple-Input Multiple-Output (MIMO) Optimum Filtering

In this section, the optimization problem is described. The structure of a MIMO optimum filter is given in Sect. 3.1.1. This structure is common to LSE optimization (Sect. 3.1.2) and MMSE optimization (Sect. 3.1.3).

3.1.1 Structure of a MIMO Optimum Filter

The structure of a linear MIMO optimum filter is depicted in Fig. 3.2. The linear MIMO system with M input channels and P output channels is described by the $MN \times P$ matrix $\mathbf{W}(k)$, which captures MP column vectors $\mathbf{w}_{m,p}(k)$ of length N with filter coefficients $w_{n,m,p}(k)$, $n = 0, 1, \ldots, N-1$, i.e.,

$$[\mathbf{W}(k)]_{m,p} = \mathbf{w}_{m,p}(k), \tag{3.1}$$

where $m = 0, 1, \ldots, M-1$, and $p = 0, 1, \ldots, P-1$. The system $\mathbf{W}(k)$ is driven by M input signals $x_m(k)$. The input signals $x_m(k)$ are captured in a vector $\mathbf{x}(k)$ of length MN according to (2.30). The vector of P output signals

$$\mathbf{y}(k) = \mathbf{W}^T(k)\,\mathbf{x}(k) \tag{3.2}$$

of the MIMO system, where

$$\mathbf{y}(k) = \left(y_0(k),\, y_1(k),\, \ldots,\, y_{P-1}(k)\right)^T, \tag{3.3}$$

is subtracted from the column vector $\mathbf{y}_{\mathrm{ref}}(k)$ of length P,

$$\mathbf{y}_{\mathrm{ref}}(k) = \left(y_{\mathrm{ref},0}(k),\, y_{\mathrm{ref},1}(k),\, \ldots,\, y_{\mathrm{ref},P-1}(k)\right)^T, \tag{3.4}$$

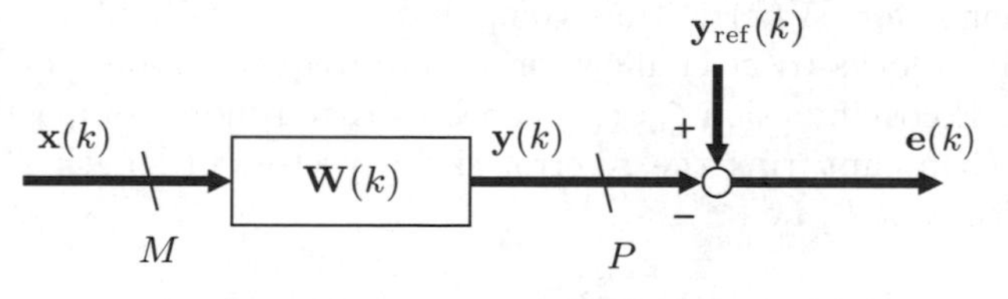

Fig. 3.2. Optimum MIMO filter

which captures the reference signals $y_{\mathrm{ref},p}(k)$ for the P output channels of the MIMO system. This yields the error vector $\mathbf{e}(k)$,

$$\mathbf{e}(k) = (e_0(k),\, e_1(k),\, \ldots,\, e_{P-1}(k))^T \,, \tag{3.5}$$

as follows:

$$\mathbf{e}(k) = \mathbf{y}_{\mathrm{ref}}(k) - \mathbf{y}(k)\,. \tag{3.6}$$

The error signals $\mathbf{e}(k)$ are used for formulating cost functions in order to determine $\mathbf{W}(k)$ according to some optimization criterion.

3.1.2 Least-Squares Error (LSE) Optimization

In this section, the concept of LSE optimization is introduced. First, the optimization problem is specified. Second, the normal equation of the LSE criterion is derived, which is required for solving the LS problem. Third, the LS problem and its solution in the vector space are given. Such vector-space illustrations will be used throughout the work, since they allow for a demonstrative interpretation of the presented methods.

Formulation of the Optimization Criterion

We introduce the LSE optimization criterion for MIMO systems for signal segments of length K, which overlap by a factor $\alpha \geq 1$. The number of 'new' samples per block is denoted by $R = \lfloor K/\alpha \rfloor$. The LSE cost function for MIMO systems for joint optimization of all output channels is defined as the sum of K squared values of the error signal vector $\mathbf{e}(k)$:

$$\xi_{\mathrm{LSE}}(r) = \sum_{k=rR}^{rR+K-1} \mathbf{e}^T(k)\,\mathbf{e}(k)\,. \tag{3.7}$$

LSE optimization consists of the minimization of the energy of the error signals over a block of K samples. The filter weights $\mathbf{W}(k)$ which are optimum in the LSE sense are obtained by

$$\mathbf{W}_{\mathrm{LSE}}(rR) = \mathrm{argmin}_{\mathbf{W}(rR)}\; \xi_{\mathrm{LSE}}(r)\,. \tag{3.8}$$

One optimum for $\mathbf{W}(k)$ is thus computed from a data block of length K. Overlapping blocks ($\alpha > 1$) allow for a more frequent update of $\mathbf{W}_{\mathrm{LSE}}(rR)$.

The LSE cost function $\xi_{\mathrm{LSE}}(r)$ can be written more compactly in matrix notation when capturing the K error signal vectors of length P in a matrix $\mathbf{E}(k)$ of size $K \times P$, i.e.,

$$\mathbf{E}(k) = (\mathbf{e}(k),\, \mathbf{e}(k+1),\, \ldots,\, \mathbf{e}(k+K-1))^T \,. \tag{3.9}$$

That is,

$$\xi_{\mathrm{LSE}}(r) = \mathrm{tr}\{\mathbf{E}^T(k)\mathbf{E}(k)\} := \|\mathbf{E}(rR)\|_F^2 \,, \tag{3.10}$$

where $\mathrm{tr}\{\cdot\}$ sums up the elements on the main diagonal of the matrix in the argument and where $\|\cdot\|_F$ is the Frobenius norm of a matrix. Let a data matrix $\mathbf{X}(k)$ of size $MN \times K$ be defined as

$$\mathbf{X}(k) = (\mathbf{x}(k), \mathbf{x}(k+1), \ldots, \mathbf{x}(k+K-1)) \,, \tag{3.11}$$

and introduce the $K \times P$-matrix $\mathbf{Y}_{\mathrm{ref}}(k)$ of reference vectors $\mathbf{y}_{\mathrm{ref}}(k)$ after (3.4) as

$$\mathbf{Y}_{\mathrm{ref}}(k) = (\mathbf{y}_{\mathrm{ref}}(k), \mathbf{y}_{\mathrm{ref}}(k+1), \ldots, \mathbf{y}_{\mathrm{ref}}(k+K-1))^T \,. \tag{3.12}$$

It follows for the error matrix $\mathbf{E}(k)$ the expression

$$\mathbf{E}(k) = \mathbf{Y}_{\mathrm{ref}}(k) - \mathbf{X}^T(k)\mathbf{W}(k) \tag{3.13}$$

$$= \mathbf{Y}_{\mathrm{ref}}(k) - \mathbf{Y}(k) \,, \tag{3.14}$$

and the LSE cost function can be written as

$$\xi_{\mathrm{LSE}}(r) = \|\mathbf{Y}_{\mathrm{ref}}(rR) - \mathbf{X}^T(rR)\mathbf{W}(rR)\|_F^2 \,. \tag{3.15}$$

Derivation of the Normal Equation

The cost function $\xi_{\mathrm{LSE}}(r)$ is minimized by setting the gradient of $\xi_{\mathrm{LSE}}(r)$ after (3.15) w.r.t. $\mathbf{W}(k)$ equal to zero. We obtain

$$\begin{aligned}
\nabla_{\mathbf{W}(rR)} &= \frac{\partial\, \xi_{\mathrm{LSE}}(r)}{\partial\, \mathbf{W}(rR)} \\
&= \frac{\partial}{\partial\, \mathbf{W}(rR)} \left\|\mathbf{Y}_{\mathrm{ref}}(rR) - \mathbf{X}^T(rR)\mathbf{W}(rR)\right\|_F^2 \\
&= \frac{\partial}{\partial\, \mathbf{W}(rR)} \left[\mathrm{tr}\left\{\mathbf{Y}_{\mathrm{ref}}^T(rR)\mathbf{Y}_{\mathrm{ref}}(rR) - 2\mathbf{W}^T(rR)\mathbf{X}(rR)\mathbf{Y}_{\mathrm{ref}}(rR) + \right.\right. \\
&\qquad \left.\left. + \mathbf{W}^T(rR)\mathbf{X}(rR)\mathbf{X}^T(rR)\mathbf{W}(rR)\right\}\right] \\
&= 2\left(-\mathbf{X}(rR)\mathbf{Y}_{\mathrm{ref}}(rR) + \mathbf{X}(rR)\mathbf{X}^T(rR)\mathbf{W}(rR)\right) \stackrel{!}{=} \mathbf{0}_{MN\times P} \,.
\end{aligned} \tag{3.16}$$

The normal equation is obtained by writing (3.16) as

$$\mathbf{X}(rR)\mathbf{X}^T(rR)\mathbf{W}(rR) \stackrel{!}{=} \mathbf{X}(rR)\mathbf{Y}_{\mathrm{ref}}(rR) \,. \tag{3.17}$$

Note that the matrix

$$\mathbf{\Phi}_{\mathbf{xx}}(k) = \mathbf{X}(k)\mathbf{X}^T(k) \tag{3.18}$$

of size $MN \times MN$ is the (non-normalized) time-averaged estimate of the correlation matrix $\mathbf{R}_{\mathbf{xx}}(k)$ and a non-normalized maximum-likelihood estimate of the correlation matrix $\mathbf{R}_{\mathbf{xx}}$ for Gaussian WSS signals $\mathbf{x}(k)$ at time k over K sampling intervals [Goo63]. The matrix $\mathbf{\Phi}_{\mathbf{xx}}(k)$ is referred to as *sample spatio-temporal correlation matrix* [Hay96]. The reliability of the estimate

$1/K \times \mathbf{\Phi}_{\mathbf{xx}}(k)$ of $\mathbf{R}_{\mathbf{xx}}(k)$ increases along with the ratio formed by the number of samples K and the product MN [RMB74][1].

Accordingly, the matrix product on the right side of (3.17) can be viewed as a (non-normalized) time-averaged estimate of the correlation matrix between the filter inputs $\mathbf{x}(k)$ and the reference signals $\mathbf{y}_{\mathrm{ref}}(k)$, defined by

$$\mathbf{R}_{\mathbf{xy}}(k) = \mathcal{E}\{\mathbf{x}(k)\mathbf{y}_{\mathrm{ref}}^T(k)\}\,. \tag{3.19}$$

For Gaussian WSS signals, $\mathbf{X}(rR)\mathbf{Y}_{\mathrm{ref}}(rR)$ is the maximum-likelihood estimate of $\mathbf{R}_{\mathbf{xy}}(k)$.

Solving the normal equation (3.17) for $\mathbf{W}(rR)$ yields the optimum LSE processor $\mathbf{W}_{\mathrm{LSE}}(k)$. If the correlation matrix $\mathbf{\Phi}_{\mathbf{xx}}(k)$ is non-singular, the LS solution can be found by inversion of $\mathbf{\Phi}_{\mathbf{xx}}(k)$.

Each of the P MISO systems (columns of $\mathbf{W}_{\mathrm{LSE}}(k)$) is optimized independently of the others since the output channels only depend on the input signals and not on the output signals of the other channels. Therefore, joint optimization of all output channels yields the same MIMO processor as independent optimization. As a result, the advantageous statistical properties of LSE estimators of MISO systems [Mil74, GP77] directly apply to MIMO systems: We assume that the data matrix $\mathbf{X}(k)$ is known without uncertainty. $\mathbf{W}_{\mathrm{LSE}}(k)$ fulfills the normal equation (3.17). Substituting in (3.13) $\mathbf{W}(k)$ by $\mathbf{W}_{\mathrm{LSE}}(k)$, we obtain the error matrix $\mathbf{E}_{\mathrm{o}}(k)$, which should be modeled as a zero-mean random process that is independent of the data matrix $\mathbf{X}(k)$:

$$\mathbf{Y}_{\mathrm{ref}}(k) = \mathbf{X}^T(k)\mathbf{W}_{\mathrm{LSE}}(k) + \mathbf{E}_{\mathrm{o}}(k)\,. \tag{3.20}$$

With these assumptions, the LS estimate $\mathbf{W}_{\mathrm{LSE}}(k)$ is unbiased [MIK00]. If we further assume that $\mathbf{E}_{\mathrm{o}}(k)$ is white with variance $\sigma^2_{\mathbf{E}_{\mathrm{o}}}(k)$, then, (a) the covariance matrix of $\mathbf{W}_{\mathrm{LSE}}(k)$ is equal to $\sigma^2_{\mathbf{E}_{\mathrm{o}}}(k)\mathbf{\Phi}_{\mathbf{xx}}(k)$ and (b) $\mathbf{W}_{\mathrm{LSE}}(k)$ is the best linear unbiased estimator (BLUE). Finally, the LSE estimator is the maximum-likelihood estimator if $\mathbf{E}_{\mathrm{o}}(k)$ further has a normal distribution. For $\mathbf{X}(k)$ being probabilistic, the statistical properties of the LSE estimator are similar [MIK00].

Solution of the Least-Squares (LS) Problem Using Singular Value Decomposition (SVD)

In the following, we, first, introduce the singular value decomposition (SVD) as a unified framework for solving the LS problem [Ste73, GvL89]. This approach provides a general and numerically stable solution for LS problems, whether the normal equations are underdetermined or overdetermined, and whether the data matrix $\mathbf{X}(k)$ is full-rank or rank-deficient. SVD also allows interpretation of LS problems in the vector space of the sensor signals. However, for practical application, other techniques for solving LS problems may

[1] In [RMB74], this last property is derived for beamformers which maximize the signal-to-noise (power) ratio (SNR).

be computationally more efficient. Second, we give closed-form solutions of the LS problem.

Singular Value Decomposition (SVD)

The SVD diagonalizes a matrix of arbitrary rank and of arbitrary dimension by pre-multiplying and post-multiplying the matrix by two different unitary matrices $\mathbf{U}_{\mathbf{X}}(k)$ and $\mathbf{V}_{\mathbf{X}}(k)$:

$$\mathbf{X}(k) = \mathbf{U}_{\mathbf{X}}(k) \begin{pmatrix} \mathbf{\Sigma}_{\mathbf{X}}(k) & \mathbf{0}_{V \times (K-V)} \\ \mathbf{0}_{(MN-V) \times V} & \mathbf{0}_{(MN-V) \times (K-V)} \end{pmatrix} \mathbf{V}_{\mathbf{X}}^T(k) , \tag{3.21}$$

where $\mathbf{X}(k)$ is defined in (3.11). The orthonormal columns of the $MN \times MN$ matrix $\mathbf{U}_{\mathbf{X}}(k)$ and of the $K \times K$ matrix $\mathbf{V}_{\mathbf{X}}(k)$ are the *left* and *right singular vectors*, respectively. The matrix $\mathbf{\Sigma}_{\mathbf{X}}(k)$ is a diagonal matrix with the V corresponding *singular values* $\sigma_{\mathbf{X},v}(k)$,

$$\mathbf{\Sigma}_{\mathbf{X}}(k) = \operatorname{diag}\{(\sigma_{\mathbf{X},0}(k), \sigma_{\mathbf{X},1}(k), \ldots, \sigma_{\mathbf{X},V-1}(k))\} , \tag{3.22}$$

where the columns of $\mathbf{U}_{\mathbf{X}}(k)$ and $\mathbf{V}_{\mathbf{X}}(k)$ are arranged such that $\sigma_{\mathbf{X},0}(k) \geq \sigma_{\mathbf{X},1}(k) \geq \cdots \geq \sigma_{\mathbf{X},V-1}(k) > 0$. The relation to the eigenvalue decomposition of $\mathbf{\Phi}_{\mathbf{xx}}(k) = \mathbf{X}(k)\mathbf{X}^T(k)$ is obtained when using $\mathbf{V}_{\mathbf{X}}^T(k)\mathbf{V}_{\mathbf{X}}(k) = \mathbf{I}_{K \times K}$. That is,

$$\mathbf{\Phi}_{\mathbf{xx}}(k) = \mathbf{U}_{\mathbf{X}}(k) \begin{pmatrix} \mathbf{\Sigma}_{\mathbf{X}}^2(k) & \mathbf{0}_{V \times (MN-V)} \\ \mathbf{0}_{(MN-V) \times V} & \mathbf{0}_{(MN-V) \times (MN-V)} \end{pmatrix} \mathbf{U}_{\mathbf{X}}^T(k) . \tag{3.23}$$

The squares of the singular values $\sigma_{\mathbf{X},v}(k)$ are thus the non-zero eigenvalues of $\mathbf{\Phi}_{\mathbf{xx}}(k)$. Equivalently, we can form the product $\mathbf{X}^T(k)\mathbf{X}(k)$. This matrix of size $K \times K$ may be interpreted as a (non-normalized) temporal correlation matrix of the input data, which is obtained by averaging over M input channels and over N samples. Such correlation matrices play an important role in the context of data-dependent optimum beamforming based on LSE criteria, as we will see later in Chap. 4. Using $\mathbf{U}_{\mathbf{X}}^T(k)\mathbf{U}_{\mathbf{X}}(k) = \mathbf{I}_{MN \times MN}$, we can write

$$\mathbf{X}^T(k)\mathbf{X}(k) = \mathbf{V}_{\mathbf{X}}(k) \begin{pmatrix} \mathbf{\Sigma}_{\mathbf{X}}^2(k) & \mathbf{0}_{V \times (K-V)} \\ \mathbf{0}_{(K-V) \times V} & \mathbf{0}_{(K-V) \times (K-V)} \end{pmatrix} \mathbf{V}_{\mathbf{X}}^T(k) . \tag{3.24}$$

$\sigma_{\mathbf{X},v}^2(k)$ and the columns of $\mathbf{V}_{\mathbf{X}}(k)$ are the non-zero eigenvalues and corresponding eigenvectors of $\mathbf{X}^T(k)\mathbf{X}(k)$, respectively.

The first V columns of $\mathbf{U}_{\mathbf{X}}(k)$ thus form an orthonormal basis of the space which is spanned by the columns of $\mathbf{X}(k)$, written as $\operatorname{span}\{\mathbf{X}(k)\}$. The last $MN - V$ columns of $\mathbf{U}_{\mathbf{X}}(k)$ are an orthonormal basis for the orthogonal complement (*nullspace*) of $\operatorname{span}\{\mathbf{X}(k)\}$, written as $(\mathbf{X}(k))^{\perp}$. Accordingly, an orthonormal basis of $\operatorname{span}\{\mathbf{X}^T(k)\}$ is obtained with the first V columns of $\mathbf{V}(k)$. The nullspace of $\mathbf{X}^T(k)$ is spanned by the last $K-V$ columns of $\mathbf{V}_{\mathbf{X}}(k)$.

Solution of the LS Problem

Let the *pseudoinverse* of the data matrix $\mathbf{X}(k)$ be defined as

$$\mathbf{X}^{+}(k) = \mathbf{U}_{\mathbf{X}}(k) \begin{pmatrix} \boldsymbol{\Sigma}_{\mathbf{X}}^{-1}(k) & \mathbf{0}_{V \times (K-V)} \\ \mathbf{0}_{(MN-V) \times V} & \mathbf{0}_{(MN-V) \times (K-V)} \end{pmatrix} \mathbf{V}_{\mathbf{X}}^{T}(k) . \tag{3.25}$$

Then, the solution of the normal equation (3.17) can be computed as

$$\mathbf{W}_{\mathrm{LSE}}(k) = \left(\mathbf{X}^{T}(k)\right)^{+} \mathbf{Y}_{\mathrm{ref}}(k) . \tag{3.26}$$

Three special cases can be identified depending on the number of singular values V, or, equivalently, depending on the rank of the data matrix $\mathbf{X}(k)$, written as $\mathrm{rk}\{\mathbf{X}(k)\} := V$:

1. $V = MN = K$: Equation (3.17) has a unique solution. The matrix $\mathbf{X}^{T}(k)$ has full rank. It follows from (3.25) that the pseudoinverse $\mathbf{X}^{+}(k)$ is equal to $\mathbf{X}^{-1}(k)$. Replacing in (3.17) the filter matrix $\mathbf{W}_{\mathrm{LSE}}(k)$ by (3.26) with $\mathbf{X}^{+}(k) := \mathbf{X}^{-1}(k)$, we see that the normal equation is solved.
2. $V = K < MN$: Equation (3.17) is an underdetermined system of linear equations with no unique solution. Since $\mathbf{X}(k)$ is of full column rank, the matrix $\mathbf{X}^{T}(k)\mathbf{X}(k)$ is non-singular. The pseudoinverse reduces to
$$\left(\mathbf{X}^{T}(k)\right)^{+} = \mathbf{X}(k) \left(\mathbf{X}^{T}(k)\mathbf{X}(k)\right)^{-1} . \tag{3.27}$$
The substitution of $\mathbf{W}(k)$ in (3.17) by $\mathbf{W}_{\mathrm{LSE}}(k)$ after (3.26), with $\mathbf{X}^{+}(k)$ defined by (3.27), shows that the normal equation is solved. Among the infinite number of solutions for underdetermined sets of linear equations (3.16), the LSE estimator $\mathbf{W}_{\mathrm{LSE}}(k)$ is unique in the sense that it provides a solution with a minimum Euclidean norm.
3. $V = MN < K$: Equation (3.17) is an overdetermined system of linear equations with a unique solution. The matrix $\mathbf{X}(k)$ is of full row rank, which means that the matrix $\mathbf{X}(k)\mathbf{X}^{T}(k)$ is non-singular. The pseudoinverse $\left(\mathbf{X}^{T}(k)\right)^{+}$ can be equivalently written as
$$\left(\mathbf{X}^{T}(k)\right)^{+} = \left(\mathbf{X}(k)\mathbf{X}^{T}(k)\right)^{-1} \mathbf{X}(k) . \tag{3.28}$$
Note that (3.28) is the direct solution of the normal equation (3.17) for invertibility of $\mathbf{X}(k)\mathbf{X}^{T}(k)$.

The LS error $\xi_{\mathrm{LSE,o}}(r)$ is defined as

$$\xi_{\mathrm{LSE,o}}(r) = \|\mathbf{Y}_{\mathrm{ref}}(rR) - \mathbf{X}^{T}(rR)\mathbf{W}_{\mathrm{LSE}}(rR)\|_F^2 . \tag{3.29}$$

$\xi_{\mathrm{LSE,o}}(r) = 0$, if the condition $\mathrm{rk}\{\mathbf{X}^{T}(k), \mathbf{Y}_{\mathrm{ref}}(k)\} = \mathrm{rk}\{\mathbf{X}^{T}(k)\}$ is met. The matrix $(\mathbf{X}^{T}(k), \mathbf{Y}_{\mathrm{ref}}(k))$ is obtained by appending the matrix $\mathbf{Y}_{\mathrm{ref}}(k)$ to the right side of the matrix $\mathbf{X}^{T}(k)$. This means that the columns of $\mathbf{Y}_{\mathrm{ref}}(k)$ must be contained in the space which is spanned by the columns of $\mathbf{X}^{T}(k)$. This will be illustrated in the following section.

Vector-Space Interpretation

The columns of $\mathbf{X}^T(rR)$ after (3.11) and of $\mathbf{Y}_{\mathrm{ref}}(rR)$ after (3.12) can be thought of as vectors in a K-dimensional vector space, called *data space*. The MN column vectors $\mathbf{X}^T(rR)$ form an MN-dimensional subspace of the data space, which is called *estimation space*. The estimate $\mathbf{Y}(rR) = \mathbf{X}^T(rR)\mathbf{W}(rR)$ of $\mathbf{Y}_{\mathrm{ref}}(rR)$ is a linear combination of the columns of $\mathbf{X}^T(rR)$. The beamformer output signal $\mathbf{Y}(rR)$ thus lies in the MN-dimensional subspace which is spanned by the columns of $\mathbf{X}^T(rR)$, written as $\mathrm{span}\{\mathbf{X}^T(rR)\}$. In general, the reference signal $\mathbf{Y}_{\mathrm{ref}}(rR)$ is outside the data space. LSE optimization finds the estimate $\mathbf{Y}(rR)$ which optimally fits the reference $\mathbf{Y}_{\mathrm{ref}}(rR)$ in the LS sense. Substitution of (3.13) into (3.17) yields the condition

$$\mathbf{X}(rR)\mathbf{E}(rR) \stackrel{!}{=} \mathbf{0}_{MN\times P} \tag{3.30}$$

for optimality of the system $\mathbf{W}(rR)$. The matrix $\mathbf{0}_{MN\times P}$ is a matrix with zeroes of size $MN \times P$. Equation (3.30) is the mathematical description of the temporal *principle of orthogonality*, which shows that the estimation space, $\mathrm{span}\{\mathbf{X}^T(rR)\}$, is orthogonal to the error signals $\mathbf{E}(rR)$, if the MIMO system $\mathbf{W}(rR)$ meets the least-squares criterion. Signal components of $\mathbf{X}^T(rR)$ which are not orthogonal to the reference signals $\mathbf{Y}_{\mathrm{ref}}(rR)$ are canceled in $\mathbf{E}(rR)$ while signal components of $\mathbf{X}^T(rR)$ which are orthogonal to $\mathbf{Y}_{\mathrm{ref}}(rR)$ are let through. This vector-space interpretation of LS optimum filtering is illustrated in Fig. 3.3. The principle of orthogonality provides a simple test for verifying whether the estimator $\mathbf{W}(rR)$ is meeting the LS criterion.

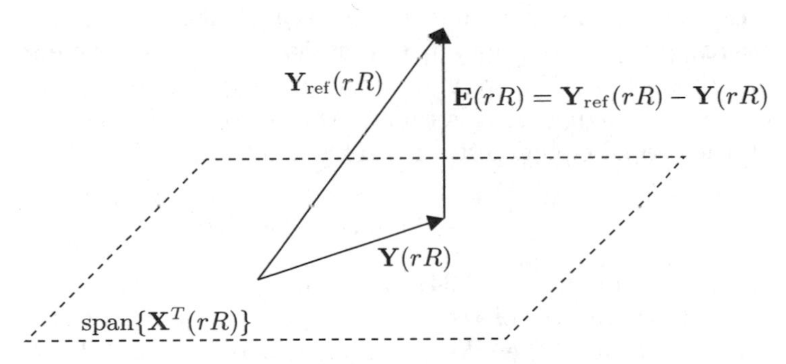

Fig. 3.3. Vector-space interpretation of LS optimum filtering

3.1.3 Minimum Mean-Squared Error (MMSE) Optimization in the DTFT Domain

In this section, the statistical analogon to the LSE estimator in the DTFT domain is derived. This formulation of optimum filters will be used in this work for illustrating fundamental properties of multi-channel audio capture techniques.

The MMSE optimization criterion is obtained by replacing in the LSE optimization criterion (3.8) the time average by the expectation operator:

$$\mathbf{W}_{\mathrm{MMSE}}(k) = \mathrm{argmin}_{\mathbf{W}(k)} E\left\{\mathbf{e}^T(k)\,\mathbf{e}(k)\right\} . \tag{3.31}$$

The normal equation is obtained by setting the gradient of the MMSE cost function

$$\xi_{\mathrm{MMSE}}(k) = E\left\{\mathbf{e}^T(k)\,\mathbf{e}(k)\right\} = E\left\{\mathrm{tr}\left\{\mathbf{e}(k)\,\mathbf{e}^T(k)\right\}\right\} \tag{3.32}$$

w.r.t. $\mathbf{W}(k)$ equal to zero. That is,

$$\begin{aligned}
&\frac{\partial \xi_{\mathrm{MMSE}}(k)}{\partial \mathbf{W}(k)} = \\
&= \frac{\partial}{\partial \mathbf{W}(k)} E\left\{\mathrm{tr}\left\{\left(\mathbf{y}_{\mathrm{ref}}(k) - \mathbf{W}^T(k)\mathbf{x}(k)\right)\left(\mathbf{y}_{\mathrm{ref}}(k) - \mathbf{W}^T(k)\mathbf{x}(k)\right)^T\right\}\right\} \\
&= \frac{\partial}{\partial \mathbf{W}(k)} E\left\{\mathrm{tr}\left\{\mathbf{y}_{\mathrm{ref}}(k)\mathbf{y}_{\mathrm{ref}}^T(k) - \mathbf{y}_{\mathrm{ref}}(k)\mathbf{x}^T(k)\mathbf{W}(k) - \right.\right. \\
&\qquad\qquad \left.\left. - \mathbf{W}^T(k)\mathbf{x}(k)\mathbf{y}_{\mathrm{ref}}^T(k) + \mathbf{W}^T(k)\mathbf{x}(k)\mathbf{x}^T(k)\mathbf{W}(k)\right\}\right\} \\
&= 2E\left\{\mathbf{x}(k)\mathbf{x}^T(k)\right\}\mathbf{W}(k) - 2E\left\{\mathbf{x}(k)\mathbf{y}_{\mathrm{ref}}^T(k)\right\} \\
&= \mathbf{R}_{\mathbf{xx}}(k)\mathbf{W}(k) - \mathbf{R}_{\mathbf{xy}_{\mathrm{ref}}}(k) \stackrel{!}{=} \mathbf{0}_{MN\times P} ,
\end{aligned} \tag{3.33}$$

where $\mathbf{R}_{\mathbf{xx}}(k)$ is the spatio-temporal correlation matrix as defined in (2.31) and where $\mathbf{R}_{\mathbf{xy}_{\mathrm{ref}}}(k)$ is the cross-correlation matrix of size $MN \times P$ between the sensor signals and the reference signals as defined in (3.19).

Assuming wide-sense ergodic sensor signals $\mathbf{x}(k)$ and wide-sense ergodic reference signals $\mathbf{y}_{\mathrm{ref}}(k)$, we can replace the expectation operator in (3.33) by temporal averaging. Transforming then (3.33) into the DTFT domain, we obtain the normal equation in the DTFT domain as follows:

$$\mathbf{S}_{\mathbf{xx}}(\omega)\,\mathbf{W}_{\mathrm{f}}(\omega) \stackrel{!}{=} \mathbf{S}_{\mathbf{xy}_{\mathrm{ref}}}(\omega) , \tag{3.34}$$

where $\mathbf{S}_{\mathbf{xx}}(\omega)$ is defined in (2.34), where $\mathbf{S}_{\mathbf{xy}_{\mathrm{ref}}}(\omega)$ is the cross-power spectral density (CPSD) matrix of size $M \times P$ between the sensor signals and the reference signals, and where $\mathbf{W}_{\mathrm{f}}(\omega)$ is a matrix of size $M \times P$ with the MIMO system with M input channels and P output channels in the DTFT domain. For existence of the optimum MMSE estimator, the spatio-spectral correlation matrix $\mathbf{S}_{\mathbf{xx}}(\omega)$ has to be positive semidefinite [MIK00]. For positive definite $\mathbf{S}_{\mathbf{xx}}(\omega)$, the solution is unique and is found by inversion of $\mathbf{S}_{\mathbf{xx}}(\omega)$:

$$\mathbf{W}_{\mathrm{f,MMSE}}(\omega) = \mathbf{S}_{\mathbf{xx}}^{-1}(\omega)\mathbf{S}_{\mathbf{xy}_{\mathrm{ref}}}(\omega) . \tag{3.35}$$

For physical applications, $\mathbf{S}_{\mathbf{xx}}(\omega)$ will generally be non-singular due to presence of spatially and temporally white noise with a diagonal spatio-spectral correlation matrix. This is in contrast to deterministic LSE estimation, where the sample spatio-temporal correlation matrix $\mathbf{\Phi}_{\mathbf{xx}}(k)$ is estimated by temporal averaging using (3.18). Though, for choosing the time interval K for temporal averaging sufficiently large, the temporally averaged estimate $\mathbf{\Phi}_{\mathbf{xx}}(k)$ will be invertible, this is not assured for short time intervals K. For $\mathbf{\Phi}_{\mathbf{xx}}(k) = \mathbf{X}(k)\mathbf{X}^T(k)$ being non-singular, $\mathbf{X}(k)$ must have full row-rank, which requires at least $K \geq MN$.

3.2 Applications of MIMO Optimum Filters

In this section, we give three examples of applications of MIMO optimum filters as they will be further discussed in this work: system identification (Sect. 3.2.1), inverse modeling (Sect. 3.2.2), and interference cancellation (Sect. 3.2.3). The fourth class of application (after [Hay96]), linear prediction, is not discussed here, since it does not play any role in our context.

3.2.1 System Identification

The principle of MIMO system identification is depicted in Fig. 3.4. The function of the optimum filter is to provide a linear model that matches the unknown system in some optimum way.

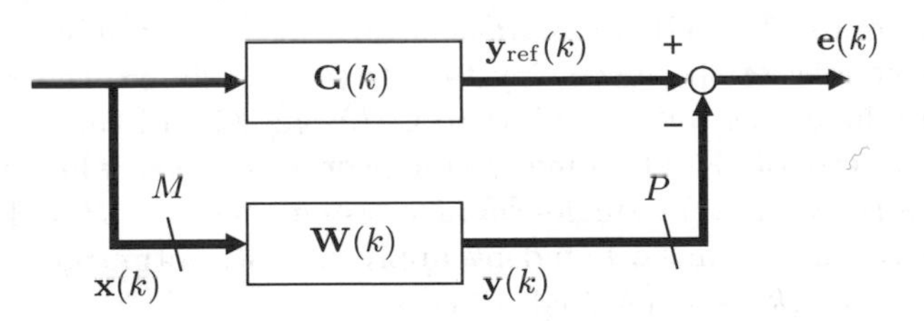

Fig. 3.4. Generic structure for MIMO system identification

We assume linearity of the system $\mathbf{G}(k)$ which should be identified. The matrix $\mathbf{G}(k)$ of size $MN_{\mathbf{g}} \times P$ is a time-varying MIMO system with M input channels and P output channels. It consists of MP filters $\mathbf{g}_{m,p}(k)$ of length $N_{\mathbf{g}}$ with elements $g_{n,m,p}(k)$, i.e.,

$$[\mathbf{G}(k)]_{mN_{\mathbf{g}},p} = \mathbf{g}_{m,p}(k)\,, \qquad (3.36)$$

where $n = 0, 1, \ldots, N_{\mathbf{g}} - 1$, $m = 0, 1, \ldots, M - 1$, and $p = 0, 1, \ldots, P - 1$. The linear MIMO modeling system $\mathbf{W}(k)$ is placed in parallel to $\mathbf{G}(k)$. Both MIMO systems are driven by the same M input signals $x_m(k)$. The output signals $\mathbf{y}_{\mathrm{ref}}(k)$ of the unknown system are subtracted from the output $\mathbf{y}(k)$ of the modeling system, which yields the error signals $\mathbf{e}(k)$. Minimization of the error signals $\mathbf{e}(k)$ according to some cost function, as, e.g., $\xi_{\mathrm{LSE}}(k)$, allows to identify the unknown system $\mathbf{G}(k)$ by the modeling system $\mathbf{W}(k)$.

The performance of system identification can be measured by the *relative mismatch*, *system error*, or *system misalignment* $\Delta W_{\mathrm{rel}}(k)$ (in dB), which is defined as the Euclidean distance between the modeling system and the unknown system normalized by the Euclidean norm of the unknown system on a logarithmic scale. That is,

$$\Delta W_{\mathrm{rel}}(k) = \\ = 10\log_{10}\frac{\sum_{n=0}^{\max\{N-1,N_{\mathbf{g}}-1\}}\sum_{m=0}^{M-1}\sum_{p=0}^{P-1}\left(g_{n,m,p}(k)-w_{n,m,p}(k)\right)^2}{\sum_{n=0}^{\max\{N-1,N_{\mathbf{g}}-1\}}\sum_{m=0}^{M-1}\sum_{p=0}^{P-1}g_{n,m,p}^2(k)} \tag{3.37}$$

The calculation of the relative mismatch requires knowledge of the system $\mathbf{G}(k)$. The impulse responses of the shorter filters of the unknown system or of the modeling system are appended with zeroes so that the total length of the considered impulse responses is $\max\{N, N_{\mathbf{g}}\}$.

3.2.2 Inverse Modeling

The principle of MIMO inverse modeling is depicted in Fig. 3.5. The optimum filter is used for providing a model of the combined system of the inverse of $\mathbf{G}(k)$ and the desired system $\mathbf{G}'(k)$. In the simplest case, $\mathbf{G}'(k)$ realizes unit impulses between each of the P' input channels and $P := P'$ output channels. If the unknown system $\mathbf{G}(k)$ has transfer functions with minimum phase, the transfer functions of the inverse system $\mathbf{W}(k)$ are simply the reciprocal of the transfer functions corresponding to $\mathbf{G}(k)$. For $\mathbf{G}(k)$ being non-minimum phase, the inverse of $\mathbf{G}(k)$ becomes unstable or non-causal, which makes other approaches necessary. For single-channel systems, i.e., $M, P = 1$, stable *approximate* solutions can be found by applying LSE optimization criteria to the error signal $e(k)$ (e.g. [WW84, PWC84]).

For a general system $\mathbf{G}'(k)$ and for $M > P$, it is shown in [MK88, Fur01] that *exact* inverse filtering can be realized by considering the M input channels

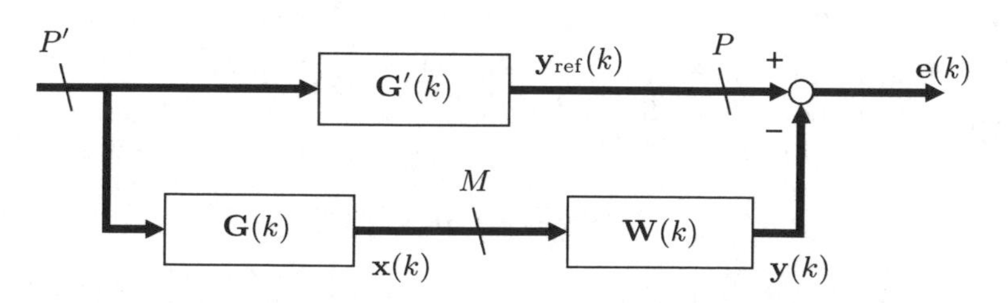

Fig. 3.5. Generic structure for MIMO inverse modeling

together using the so-called *multiple-input/output inverse theorem* (MINT). The MINT states that the exact inverse system $\mathbf{W}(k)$ can be determined by solving a system of linear equations if some conditions are fulfilled. This will be described in the following. We assume that the transfer functions which correspond to $\mathbf{G}'(k)$ and $\mathbf{G}(k)$ do not have common zeroes. Common zeroes do not need to be modeled by $\mathbf{W}(k)$, and, thus, can be eliminated before applying the described method. Especially, for $\mathbf{G}'(k) = \mathbf{G}(k)$, the modeling filters simplify to unit impulses.

The desired response of the system $\mathbf{G}'(k)$ between input p' and output p is denoted by the column vector $\mathbf{g}'_{p',p}(k)$ of length $N_{\mathbf{g}'}$,

$$[\mathbf{G}'(k)]_{p',p} = \mathbf{g}'_{p',p}(k) \,. \tag{3.38}$$

where $\mathbf{G}'(k)$ is of size $N_{\mathbf{g}'}P' \times P$. With the impulse responses $\mathbf{g}_{p',m}(k)$, block-diagonal Toeplitz matrices of size $N_{\mathbf{g}} + N - 1 \times N$ are formed as

$$\mathbf{G}_{p',m}(k) = \begin{pmatrix} \mathbf{g}_{p',m}(k) & 0 & \cdots & 0 \\ 0 & \mathbf{g}_{p',m}(k) & & 0 \\ \vdots & & \ddots & \vdots \\ 0 & & \cdots & \mathbf{g}_{p',m}(k) \end{pmatrix} , \tag{3.39}$$

and captured in an $(N_{\mathbf{g}} + N - 1)P' \times MN$ matrix $\tilde{\mathbf{G}}(k)$ as

$$[\tilde{\mathbf{G}}(k)]_{p',m} = \mathbf{G}_{p',m}(k) \,. \tag{3.40}$$

We can, thus, form a system of linear equations as

$$\mathbf{G}'(k) \stackrel{!}{=} \tilde{\mathbf{G}}(k)\mathbf{W}(k) \tag{3.41}$$

for finding the MIMO system $\mathbf{W}(k)$, which inversely models $\tilde{\mathbf{G}}(k)$ and simultaneously realizes $\mathbf{G}'(k)$. For assuring correct matrix dimensions in (3.41), the length of the filters $\mathbf{g}'_{p',p}(k)$ must be equal to

$$N_{\mathbf{g}'} \stackrel{!}{=} N_{\mathbf{g}} + N - 1 \,. \tag{3.42}$$

For $\tilde{\mathbf{G}}(k)$ to be square, the condition

$$N \stackrel{!}{=} \frac{P'(N_{\mathbf{g}} - 1)}{M - P'} \tag{3.43}$$

must be fulfilled assuming $P' < M$ [MK88, Fur01]. Interestingly, for fixed P', the necessary number of filter taps N reduces hyperbolically with the number of channels M. The inverse modeling filters are approximately by a factor $P'/(M - P')$ smaller than the filters of the unknown system $\mathbf{G}(k)$. For $P' \geq M$, (3.41) cannot be solved. If, additionally, the transfer functions which correspond to $\mathbf{g}_{p',m}(k)$ do not have common zeroes, then, the square matrix $\tilde{\mathbf{G}}(k)$ is non-singular, and the solution of (3.41) is simply obtained by inverting $\tilde{\mathbf{G}}(k)$ and by solving (3.41) for $\mathbf{W}(k)$. These properties of MIMO inverse modeling will play an important role in understanding the dependency of the behavior of optimum beamformers on the filter lengths (Chap. 5).

3.2.3 Interference Cancellation

The application of optimum MIMO filtering to interference cancellation structurally corresponds to the general MIMO system in Fig. 3.2. Interference cancellation is used in order to separate desired signals from interference and noise. Typically, *primary* signals $\mathbf{y}_{\mathrm{ref}}(k)$ consist of a mixture of desired signals, interference, and noise. Reference (or *secondary*) signals $\mathbf{x}(k)$ contain estimates of interference and noise without or with weak presence of the desired signals. The interference canceller minimizes the correlation between the reference signals and the primary signals. Thus, the error signals $\mathbf{e}(k)$ are estimates of the desired signals. If the reference signals contain components of the desired signals which are correlated with the primary channels, then, these desired signal components are removed. This leads to annoying effects, termed *cancellation of the desired signal.* Methods, which are not sensitive to the effect of cancellation of the desired signal while assuring efficient suppression of interference are called *robust.*

Equivalently, if knowledge about the statistical properties of the desired signals is available, it can be used as primary input $\mathbf{y}_{\mathrm{ref}}(k)$. The secondary inputs are then a mixture of desired signal, interference, and noise. The error signals of the MIMO system which operates in its optimum condition then contain interference and noise. The output $\mathbf{y}(k)$ of the MIMO system $\mathbf{y}(k)$ then contains the estimates of the desired signal.

Later, we will see that interference cancellation can also be seen from an inverse modeling perspective or from a system identification point of view. This gives additional insights into the behavior of optimum data-dependent beamformers, as they are discussed in Chap. 4 and in Chap. 5.

3.3 Discussion

Generally, the systems for multi-channel audio capture can be identified as MIMO systems. The input channels are given by spatially distributed sensors, the output signals correspond to the signals which are transmitted to the desired application. In these acoustic front-ends, often linear filters are used which depend on the sensor data and which are determined using some given optimization criterion.

The aim of this chapter is to summarize these optimum MIMO filters and their fundamental classes of application. The chapter should provide an understanding of and a link between the algorithms which are developed in this work.

We first described the generic structure of a MIMO optimum filter. Second, we introduced LSE optimization, which is based on temporal averaging of second-order statistics over a block of finite length of input data. Considering temporally successive data blocks, and recalculating the MIMO filter for each of these input data blocks, we obtain a system that tracks changes of the second-order characteristics of the input data. This makes LSE optimization a powerful method for time-varying acoustic conditions. For wide-sense ergodic signals and infinite observation intervals, LSE optimization is equivalent to MMSE optimization. Although stationary conditions are not met in realistic environments, MMSE optimization is an important means for analyzing optimum linear filters.

For deriving the signal processing algorithms in the following chapters, we will use LSE optimization. For illustrating the properties of these systems, we will often fall back on the statistical analogon in the DTFT domain.

The determination of optimum linear filters invokes the computation of matrix inverses. Depending on the dimensions of the data matrix $\mathbf{X}(k)$, optimum filters may be computationally very complex. For keeping the complexity in real-world applications moderate, optimum filters are generally realized by using recursive adaptation algorithms. This yields the concept of adaptive filtering.

Adaptive filters are characterized by the *speed of convergence* and the *tracking capability*. The speed of convergence specifies how fast the adaptation algorithm converges to the Wiener optimum solution for stationary operating conditions. Tracking capability means that temporal variations of the statistics of the sensor data and time-variance of the optimum filter can be followed. Tracking capability is especially important in highly non-stationary acoustic environments. Generally, the speed of convergence decreases with an increasing number of filter coefficients of the modeling filters [Hay96]. For most adaptation algorithms, the convergence speed decreases with increasing condition number of the sample spatio-temporal correlation matrix of the input signals [Hay96].

Adaptive filters are often based on gradient techniques due to their simplicity, flexibility, and robustness [WS85, Hay96, Bel01], as, e.g., the least mean-squares (LMS) algorithm. The exponentially weighted recursive least-squares (RLS) algorithm [Hay96] is computationally more intensive, but provides much faster convergence than the LMS algorithm since its mean-squared error convergence is independent of the condition number of the sample spatio-temporal correlation matrix of the input signals [Hay96]. The RLS algorithm solves the least-squares minimization problem at each iteration and is thus optimum in a deterministic sense. It can be shown that this deterministically obtained weight vector converges to the statistically optimum solution in a least-mean-squares sense for stationary conditions. Computationally efficient versions exist in the time domain [Bel01] and in the discrete Fourier transform (DFT) domain [BM00, BBK03]. These DFT-domain adaptive filters with RLS-like properties will be used in Chap. 7 for realizing an acoustic human/machine front-end for real-time applications. In our context, DFT-domain adaptive filters do not only combine fast convergence with computational efficiency, but they also provide the possibility of frequency-selective processing, which is important for the realization of data-dependent optimum beamformers (Chap. 5).

4

Optimum Beamforming for Wideband Non-stationary Signals

Array processing techniques strive for extraction of maximum information from a propagating wave field using groups of sensors, which are located at distinct spatial locations. The sensors transduce propagating waves into signals describing both a finite spatial and a temporal aperture. In accordance with temporal sampling which leads to the discrete time domain, spatial sampling by sensor arrays forms the discrete space domain. Thus, with sensor arrays, signal processing operates in a multi-dimensional space-time domain. The processor which combines temporal and spatial filtering using sensor arrays is called a beamformer. Many properties and techniques which are known from temporal FIR filtering directly translate to beamforming based on finite spatial apertures.[1]

Usually, FIR filters are placed in each of the sensor channels in order to obtain a beamformer with desired properties. Design methods for these filters can be classified according to two categories: (a) The FIR filters are designed independently of the statistics of the sensor data (*data-independent* beamformer). (b) The FIR filters are designed depending on known or estimated statistics of the sensor data to optimize the array response for the given wavefield characteristics (*data-dependent* beamformer) [VB88].

Generally, array signal processing is applied to detection and estimation problems when a desired signal is captured in the presence of interference and noise. Arrays play an important role in areas like (a) detection of the presence of signal sources, (b) estimation of temporal waveforms or spectral contents of signals, (c) estimation of directions of arrival (DOAs) or positions of multiple sources, (d) focussing on specific spatial locations for transmission. Traditionally, they are used in such diverse fields as radar, sonar, transmission systems, seismology, or medical diagnosis and treatment.

In our context, we consider the new and emerging field of space-time acoustic signal processing. We are using microphone arrays in order to focus on

[1] See [VB88] for a tutorial about beamforming relating properties of temporal FIR filtering and space-time processing.

speech signals in the presence of interference and noise. Obviously, spatio-temporal filtering is preferable over temporal filtering alone, because desired signal and interference often overlap in time and/or frequency but originate from different spatial coordinates. Spatial filtering by beamforming, ideally, allows separation without distortion of the desired signal.

In practical situations, beamforming microphone arrays have to cope with highly non-stationary environments, multi-path propagation due to early reflections and reverberation, wideband signals, and, finally, with restrictions in geometry, size, and number of sensors due to product design constraints. Therefore, adaptive realizations of data-dependent beamformers based on optimum MIMO filtering are desirable. They optimally extract non-stationary desired signals from non-stationary interference and noise for the observed sensor signals and for a given array geometry. Dereverberation is an unresolved problem if the propagation paths are not known.

This chapter focusses on optimum adaptive beamforming for non-stationary wideband signals as they are captured by microphone arrays in typical audio applications. In contrast to many traditional presentations, we explicitly drop the narrowband and stationarity assumption and formulate optimum beamformers as MIMO optimum filters using time-domain LSE criteria. We thus obtain a more general representation of various beamforming aspects that are relevant to our application. The relation of beamforming in the discrete time domain for non-stationary signals based on LSE criteria to beamforming in the DTFT domain for WSS signals based on MMSE criteria is illustrated.

The analysis of optimum adaptive beamforming will show that the realization of data-dependent beamformers using the structure of the so-called generalized sidelobe canceller is advantageous for real-time implementations due to its computational efficiency. However, despite of all advantages of traditional adaptive beamforming approaches, the problem of reverberated desired signals cannot be resolved. This makes a generalization of traditional data-dependent beamformers, as discussed in Chap. 5, necessary.

This chapter is organized as follows: In Sect. 4.1, we introduce the spatio-temporal signal model that is used in the remainder of this work. In Sect. 4.2, we present the concept of beamforming viewed from a space-time filtering perspective. Beamformer performance measures are defined and fundamental dependencies between array geometry and beamformer performance are discussed. In Sect. 4.3, data-independent beamformer designs are summarized by showing the relation between temporal FIR filter design and spatio-temporal beamformer design. Sect. 4.4 introduces the concepts of data-dependent beamformers. Relations between the different classes of data-dependent beamformers are studied, vector-space interpretations are given, and beamforming algorithms known from audio signal processing are classified. Sect. 4.5 summarizes the fundamental results of this chapter. More detailed discussions about the concepts of sensor array processing can be found, e.g., in [MM80, Hud81, Com88, VB88, JD93, Tre02].

4.1 Space-Time Signal Model

In this section, the space-time signal model that will be used in this work is specified. An overview of the scenario is depicted in Fig. 4.1.

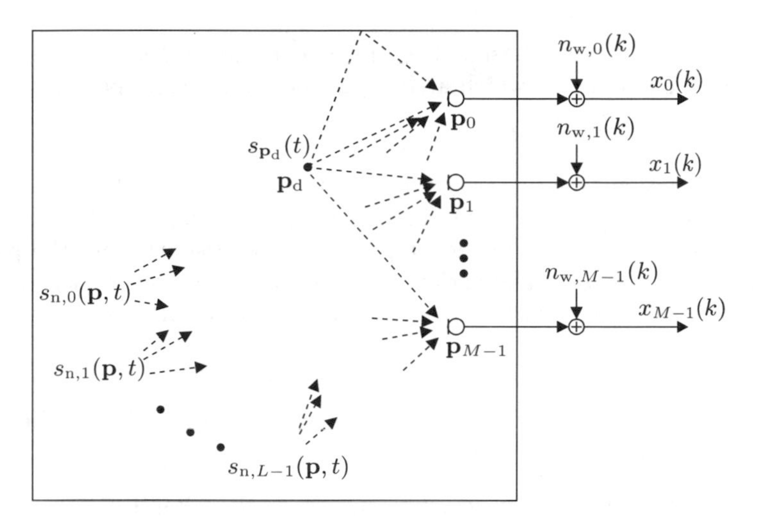

Fig. 4.1. Signal model for multi-channel speech and audio signal acquisition in noisy acoustic environments with multi-path propagation

The desired source $s_{\mathbf{p}_\mathrm{d}}(t)$ is modeled as a point source at position $\mathbf{p}_\mathrm{d}$, interference sources are described by general space-time signals $s_{\mathrm{n},l}(\mathbf{p},t)$ in order to include all kinds of interferers, as, e.g., interfering point sources or spatially diffuse interference. M microphones measure the sound pressure of the wave field of the desired source and of interference at positions $\mathbf{p}_m$. Transduction of the sound pressure and temporal sampling of the continuous-time sensor signals add spatially and temporally white noise $n_{\mathrm{w},m}(k)$ ('sensor noise'), yielding the discrete-time sensor signals $x_m(k)$.

This section is organized as follows: Section 4.1.1 describes the properties of the desired signal, Sect. 4.1.2 covers interference, Sect. 4.1.3 characterizes the sensor noise, and Sect. 4.1.4 defines the sensor signals as a superposition of the desired signal, interference, and sensor noise.

4.1.1 Desired Signal

We assume presence of a zero-mean wideband desired (point) source $s_{\mathbf{p}_\mathrm{d}}(t)$ at a spatial location $\mathbf{p}_\mathrm{d}$ in an acoustic multi-path (reverberant) environment. The desired point source should be a human speaker. The position $\mathbf{p}_\mathrm{d}$ of the desired source is known with a given tolerance. The continuous set of positions where the desired source is allowed to be located is termed *tracking region.*

This assumption about the position of the desired source is valid for, e.g., audio-visual human/machine interfaces, where the microphone array is fixed on the top of a screen or integrated into a screen. For interacting with the system, the user is located within the tracking region in front of the screen. The exact position of the user is generally not known because of speaker movements.

We define $d_m(k)$ as the signal component of the wideband desired source at the m-th microphone, which arrives via the direct signal path as

$$d_m(k) = s_{\mathbf{p}_\mathrm{d}}(kT_\mathrm{s} - \tau_{\mathrm{d},m}), \tag{4.1}$$

where attenuation is not considered. The signal $d_m(k)$ should be the *desired signal* at the m-th microphone. The delay $\tau_{\mathrm{d},m} = \|\mathbf{p}_\mathrm{d} - \mathbf{p}_m\|_2/c$ is due to propagation between the position $\mathbf{p}_\mathrm{d}$ of the desired source and the position $\mathbf{p}_m$ of the m-th sensor. The variance $\sigma^2_{d_m}(k)$ of the desired signal at the m-th sensor is defined according to (2.16) for a zero-mean space-time random process as

$$\sigma^2_{d_m}(k) = \mathcal{E}\left\{d^2_m(k)\right\}. \tag{4.2}$$

Let stacked data vectors and a data matrix w.r.t. the desired signal be defined according to (2.29), (2.30), and (3.11) as

$$\mathbf{d}_m(k) = \left(d_m(k),\, d_m(k-1),\, \ldots,\, d_m(k-N+1)\right)^T, \tag{4.3}$$

$$\mathbf{d}(k) = \left(\mathbf{d}_0^T(k),\, \mathbf{d}_1^T(k),\, \ldots,\, \mathbf{d}_{M-1}^T(k)\right)^T, \tag{4.4}$$

$$\mathbf{D}(k) = \left(\mathbf{d}(k),\, \mathbf{d}(k+1),\, \ldots,\, \mathbf{d}(k+K-1)\right), \tag{4.5}$$

respectively. We can then introduce with (4.5) the sample spatio-temporal correlation matrix $\mathbf{\Phi}_{\mathbf{dd}}(k)$,

$$\mathbf{\Phi}_{\mathbf{dd}}(k) = \mathbf{D}(k)\mathbf{D}^T(k), \tag{4.6}$$

according to (3.18), and the spatio-temporal correlation matrix with (4.4) analogously to (2.31) as

$$\mathbf{R}_{\mathbf{dd}}(k) = \mathcal{E}\left\{\mathbf{d}(k)\mathbf{d}^T(k)\right\}. \tag{4.7}$$

Although the desired (speech) signal generally has time-varying second-order statistics, we sometimes assume temporal wide-sense stationarity of the desired signal for better illustration of the characteristics of data-dependent beamformers. With temporal wide-sense stationarity of the desired signal, the spatio-temporal correlation matrix $\mathbf{R}_{\mathbf{dd}}(k)$ can be equivalently written in the DTFT domain as the spatio-spectral correlation matrix $\mathbf{S}_{\mathbf{dd}}(\omega)$ analogously to (2.34). For a spatially homogeneous and temporally WSS wave field, we introduce the PSD $S_{dd}(\omega)$ of the desired signal at the sensors, so that $\mathbf{S}_{\mathbf{dd}}(\omega)$ can be written as the product of the PSD $S_{dd}(\omega)$ and the spatial coherence matrix of the desired signal at the sensors $\mathbf{\Gamma}_{\mathbf{dd}}(\omega)$. In analogy to (2.35), we obtain:

$$\mathbf{S}_{\mathbf{dd}}(\omega) = S_{dd}(\omega)\mathbf{\Gamma}_{\mathbf{dd}}(\omega). \tag{4.8}$$

4.1.2 Interference

Next, we assume presence of zero-mean wideband *interference* $s_{\mathrm{n}}(\mathbf{p},t)$. The spatio-temporal signal $s_{\mathrm{n}}(\mathbf{p},t)$ describes the source signal(s) of a general acoustic wave field as a function of space $\mathbf{p}$ and time t. Interference includes reverberation w.r.t. the desired source, *local interferers*, and *acoustic echoes* of loudspeaker signals. According to (4.1), the desired signal consists only of the direct propagation path of the signal of the desired source. Secondary signal paths are modeled as interference. Local interferers are all kinds of undesired sound sources which contribute to the local acoustic wave field, as, e.g., competing human speakers, fan noise, street and car noise, or cocktail-party noise. Depending on the characteristics of the noise field, local interferers are modeled as point sources or as diffuse wave fields. (See Chap. 2.) Acoustic echoes are loudspeaker signals which are fed back to the microphones.[2] The loudspeakers are part of the acoustic human/machine front-end such that the loudspeaker signals can be assumed to be known. The interference $s_{\mathrm{n}}(\mathbf{p},t)$ is thus the superposition of a given number L of interferer signals $s_{\mathrm{n},l}(\mathbf{p},t)$, $l=0,\,1,\,\dots,\,L-1$, i.e.,

$$s_{\mathrm{n}}(\mathbf{p},t)=\sum_{l=0}^{L-1}s_{\mathrm{n},l}(\mathbf{p},t)\,. \tag{4.9}$$

The signal of the l-th interferer at the position of the m-th sensor is defined as $n_{\mathrm{c},l}(\mathbf{p}_m,t)$. After transduction and temporal sampling, we obtain $n_{\mathrm{c},l,m}(k):=n_{\mathrm{c},l}(\mathbf{p}_m,kT_{\mathrm{s}})$, which yields for the contribution of all interferers the superposition

$$n_{\mathrm{c},m}(k)=\sum_{l=0}^{L-1}n_{\mathrm{c},l,m}(k)\,. \tag{4.10}$$

According to (4.3)–(4.5), stacked data vectors and a data matrix w.r.t. interference are defined as

$$\mathbf{n}_{\mathrm{c},m}(k)=\left(n_{\mathrm{c},m}(k),\,n_{\mathrm{c},m}(k-1),\,\dots,\,n_{\mathrm{c},m}(k-N+1)\right)^T\,, \tag{4.11}$$

$$\mathbf{n}_{\mathrm{c}}(k)=\left(\mathbf{n}_{\mathrm{c},0}^T(k),\,\mathbf{n}_{\mathrm{c},1}^T(k),\,\dots,\,\mathbf{n}_{\mathrm{c},M-1}^T(k)\right)^T\,, \tag{4.12}$$

$$\mathbf{N}_{\mathrm{c}}(k)=\left(\mathbf{n}_{\mathrm{c}}(k),\,\mathbf{n}_{\mathrm{c}}(k+1),\,\dots,\,\mathbf{n}_{\mathrm{c}}(k+K-1)\right)\,, \tag{4.13}$$

respectively. Analogously to (4.6), a sample spatio-temporal correlation matrix

$$\mathbf{\Phi}_{\mathbf{n}_{\mathrm{c}}\mathbf{n}_{\mathrm{c}}}(k)=\mathbf{N}_{\mathrm{c}}(k)\mathbf{N}_{\mathrm{c}}^T(k) \tag{4.14}$$

and, analogously to (4.7), a spatio-temporal correlation matrix

[2] Note that acoustic echoes are modeled as interference so far. In Chap. 6, we will introduce acoustic echo cancellation as an efficient method for suppressing acoustic echoes, which makes it necessary to distinguish between acoustic echoes and interference.

$$\mathbf{R}_{\mathbf{n}_c\mathbf{n}_c}(k) = \mathcal{E}\left\{\mathbf{n}_c(k)\mathbf{n}_c^T(k)\right\} \tag{4.15}$$

w.r.t. the interference at the sensors is defined. Assuming temporal wide-sense stationarity of the interference at the sensors, the spatio-temporal correlation matrix $\mathbf{R}_{\mathbf{n}_c\mathbf{n}_c}(k)$ can be written in the DTFT domain as the spatio-spectral correlation matrix $\mathbf{S}_{\mathbf{n}_c\mathbf{n}_c}(\omega)$ analogously to (2.34). For spatially homogeneous and temporally WSS interference, we define the PSD $S_{n_cn_c}(\omega)$ of the interference at the sensors according to (2.24), and we write $\mathbf{S}_{\mathbf{n}_c\mathbf{n}_c}(\omega)$ with the spatial coherence matrix $\mathbf{\Gamma}_{\mathbf{n}_c\mathbf{n}_c}(\omega)$ of the interference at the sensors as

$$\mathbf{S}_{\mathbf{n}_c\mathbf{n}_c}(\omega) = S_{n_cn_c}(\omega)\mathbf{\Gamma}_{\mathbf{n}_c\mathbf{n}_c}(\omega), \tag{4.16}$$

in analogy to (2.35).

4.1.3 Sensor Noise

After transduction of the sound pressure by the sensors and after temporal sampling of the continuous-time sensor signals, zero-mean sensor noise $n_{\mathrm{w},m}(k)$ is present. Let data vectors be defined analogously to (4.11) and (4.12) as

$$\mathbf{n}_{\mathrm{w},m}(k) = \left(n_{\mathrm{w},m}(k),\, n_{\mathrm{w},m}(k-1),\, \ldots,\, n_{\mathrm{w},m}(k-N+1)\right)^T, \tag{4.17}$$

$$\mathbf{n}_{\mathrm{w}}(k) = \left(\mathbf{n}_{\mathrm{w},0}^T(k),\, \mathbf{n}_{\mathrm{w},1}^T(k),\, \ldots,\, \mathbf{n}_{\mathrm{w},M-1}^T(k)\right)^T, \tag{4.18}$$

respectively, and let a data matrix be defined analogously to (4.13) as

$$\mathbf{N}_{\mathrm{w}}(k) = \left(\mathbf{n}_{\mathrm{w}}(k),\, \mathbf{n}_{\mathrm{w}}(k+1),\ldots,\, \mathbf{n}_{\mathrm{w}}(k+K-1)\right). \tag{4.19}$$

Using (4.19), a sample spatio-temporal correlation matrix w.r.t. the sensor noise can be introduced analogously to (4.14) as

$$\mathbf{\Phi}_{\mathbf{n}_{\mathrm{w}}\mathbf{n}_{\mathrm{w}}}(k) = \mathbf{N}_{\mathrm{w}}(k)\mathbf{N}_{\mathrm{w}}^T(k). \tag{4.20}$$

Sensor noise can generally be assumed to be spatially and temporally white and spatially and temporally WSS, so that – according to (2.57) – the spatio-temporal correlation matrix of the sensor noise at the sensors is given by

$$\mathbf{R}_{\mathbf{n}_{\mathrm{w}}\mathbf{n}_{\mathrm{w}}} = \sigma_{n_{\mathrm{w}}}^2\mathbf{I}_{MN\times MN}, \tag{4.21}$$

with the variance

$$\sigma_{n_{\mathrm{w}}}^2 = \mathcal{E}\{n_{\mathrm{w},m}^2(k)\} \quad \forall k,\, m. \tag{4.22}$$

In the DTFT domain, the spatio-spectral correlation matrix w.r.t. sensor noise is defined according to (2.58) as

$$\mathbf{S}_{\mathbf{n}_{\mathrm{w}}\mathbf{n}_{\mathrm{w}}}(\omega) = \sigma_{n_{\mathrm{w}}}^2\mathbf{I}_{M\times M}. \tag{4.23}$$

4.1.4 Sensor Signals

The sensor signals $x_m(k)$ are the mixture of the desired signal, the interference, and the sensor noise. Interference $n_{c,m}(k)$ and sensor noise $n_{w,m}(k)$ at the m-th sensor is captured in

$$n_m(k) = n_{c,m}(k) + n_{w,m}(k) \tag{4.24}$$

and, using (4.12), (4.13), (4.18), and (4.19) in a data vector and in a data matrix

$$\mathbf{n}(k) = \mathbf{n}_c(k) + \mathbf{n}_w(k) , \tag{4.25}$$

$$\mathbf{N}(k) = \mathbf{N}_c(k) + \mathbf{N}_w(k) , \tag{4.26}$$

respectively. The sensor signals are the superposition of the desired signal and interference-plus-noise, which gives with (4.1), (4.4), (4.5), and (4.24)–(4.26)

$$x_m(k) = d_m(k) + n_m(k) , \tag{4.27}$$

$$\mathbf{x}(k) = \mathbf{d}(k) + \mathbf{n}(k) , \tag{4.28}$$

$$\mathbf{X}(k) = \mathbf{D}(k) + \mathbf{N}(k) . \tag{4.29}$$

The desired signal, the interference, and the sensor noise are assumed to be mutually temporally orthogonal, i.e.,

$$\mathbf{D}(k)\mathbf{N}_c^T(k) = \mathbf{D}(k)\mathbf{N}_w^T(k) = \mathbf{N}_c(k)\mathbf{N}_w^T(k) = \mathbf{0}_{MN \times MN} . \tag{4.30}$$

Note that this assumption is generally only asymptotically ($K \to \infty$) valid for observation intervals of infinite length. For observation intervals of finite length, temporal orthogonality will not be strictly fulfilled. Temporal orthogonality (4.30) allows to write the sample spatio-temporal correlation matrix w.r.t. interference-plus-noise, $\mathbf{\Phi}_{\mathbf{nn}}(k)$, and w.r.t. to the sensor signals, $\mathbf{\Phi}_{\mathbf{xx}}(k)$, with (4.6), (4.14), and (4.20) as

$$\mathbf{\Phi}_{\mathbf{nn}}(k) = \mathbf{N}(k)\mathbf{N}^T(k) = \mathbf{\Phi}_{\mathbf{n}_c\mathbf{n}_c}(k) + \mathbf{\Phi}_{\mathbf{n}_w\mathbf{n}_w}(k) , \tag{4.31}$$

$$\mathbf{\Phi}_{\mathbf{xx}}(k) = \mathbf{X}(k)\mathbf{X}^T(k) = \mathbf{\Phi}_{\mathbf{dd}}(k) + \mathbf{\Phi}_{\mathbf{nn}}(k) , \tag{4.32}$$

respectively. We assume in this work that the matrix $\mathbf{\Phi}_{\mathbf{xx}}(k)$ is invertible, which means that at least spatially and temporally orthogonal sensor noise is present ($\mathbf{n}_w(k) \neq 0$). Invertibility of $\mathbf{\Phi}_{\mathbf{xx}}(k)$ implies that the matrix $\mathbf{X}(k)$ has maximum row-rank. This requires $K \geq MN$, and defines a lower limit for the number of samples in one data block of length K.

The statistical analogons to (4.31) and to (4.32) are obtained when assuming that the desired signal, the interference, and the sensor noise are mutually uncorrelated[3], i.e.,

[3] Obviously, reverberation of the desired signal – which is modeled as interference here – does not meet this assumption. In Sect. 4.4.4, we will discuss measures for decorrelating the desired signal and the interference, such that this assumption is better fulfilled. Since these methods may not be appropriate for audio signal acquisition with relatively few sensors, we will model reverberation in Chaps. 5–7 as desired signal.

$$\mathcal{E}\{d_{m_1}(k_1)n_{\mathrm{c},m_2}(k_2)\} = \mathcal{E}\{d_{m_1}(k_1)n_{\mathrm{w},m_2}(k_2)\} = \mathcal{E}\{n_{\mathrm{c},m_1}(k_1)n_{\mathrm{w},m_2}(k_2)\} = 0\,. \tag{4.33}$$

Note that it is not assumed that interferer signals are mutually uncorrelated. Statistical decorrelation (4.33) gives for mutually uncorrelated interference and noise for $\mathbf{R}_{\mathbf{nn}}(k)$ and for $\mathbf{S}_{\mathbf{nn}}(\omega)$ with (4.15), (4.21), and (4.23) the expressions

$$\mathbf{R}_{\mathbf{nn}}(k) = \mathbf{R}_{\mathbf{n}_c\mathbf{n}_c}(k) + \mathbf{R}_{\mathbf{n}_w\mathbf{n}_w}\,, \tag{4.34}$$

$$\mathbf{R}_{\mathbf{xx}}(k) = \mathbf{R}_{\mathbf{dd}}(k) + \mathbf{R}_{\mathbf{nn}}(k)\,, \tag{4.35}$$

$$\mathbf{S}_{\mathbf{nn}}(\omega) = \mathbf{S}_{\mathbf{n}_c\mathbf{n}_c}(\omega) + \mathbf{S}_{\mathbf{n}_w\mathbf{n}_w}(\omega)\,, \tag{4.36}$$

respectively. For spatially homogeneous and temporally WSS interference, we further introduce

$$\mathbf{S}_{\mathbf{nn}}(\omega) = S_{nn}(\omega)\mathbf{\Gamma}_{\mathbf{nn}}(\omega)\,, \tag{4.37}$$

according to (4.8), where $S_{nn}(\omega)$ is the PSD of interference-plus-noise at the sensors. Finally, the spatio-spectral correlation matrix of the sensor signals is given with (4.7) by

$$\mathbf{S}_{\mathbf{xx}}(\omega) = \mathbf{S}_{\mathbf{dd}}(\omega) + \mathbf{S}_{\mathbf{nn}}(\omega)\,. \tag{4.38}$$

In summary, the sensor signals are a mixture of the desired signal, interference, and sensor noise. Interference represents reverberation of the desired signal, acoustic echoes, and local interference. The sensor signals are characterized by their second-order statistics, which are captured in spatio-temporal and spatio-spectral correlation matrices. Presence of spatially and temporally white sensor noise is assumed in order to assure invertibility of the spatio-temporal and the spatio-spectral correlation matrices.

4.2 Space-Time Filtering with Sensor Arrays

In this section, we first introduce beamforming using the concept of space-time filtering with sensor arrays (Sect. 4.2.1). Beamformer response and interference-independent beamformer performance measures are discussed in Sect. 4.2.2. In Sect. 4.2.3, performance measures are introduced which depend on the second-order statistics of the interference and of the desired signal. Since these measures are more appropriate for characterizing the beamformer performance for presence of non-stationary wideband signals, we will mainly rely on these performance measures in the following chapters. Finally, the influence of the beamformer properties on the array geometry is discussed in Sect. 4.2.4.

4.2.1 Concept of Beamforming

The sensor signals $x_m(k)$ are obtained by spatial and temporal sampling of the propagating wave fields by a sensor array. Space-time filtering or beamforming focusses the array on the desired source in order to separate signals from different directions, which generally have overlapping frequency content. In our scenario, the desired signal should be extracted while interference-plus-noise should be suppressed. A space-time filter is realized by placing FIR filters of length N,

$$\mathbf{w}_m(k) = \left(w_{0,m}(k),\, w_{1,m}(k),\, \ldots,\, w_{N-1,m}(k)\right)^T , \tag{4.39}$$

with tap weights $w_{n,m}(k)$, $m = 0, 1, \ldots, M-1$, into the sensor channels (Fig. 4.2). Convolution with the sensor signals $x_m(k)$ and summation over the M sensor channels yields the beamformer output signal $y(k)$:

$$y(k) = \sum_{m=0}^{M-1} \mathbf{w}_m^T(k)\mathbf{x}_m(k) . \tag{4.40}$$

Let the *tap-stacked weight vector* $\mathbf{w}(k)$ of size $MN \times 1$ be defined as

$$\mathbf{w}(k) = \left(\mathbf{w}_0^T(k),\, \mathbf{w}_1^T(k),\, \ldots,\, \mathbf{w}_{M-1}^T(k)\right)^T , \tag{4.41}$$

we can, equivalently, write for (4.40) with (4.28)

$$y(k) = \mathbf{w}^T(k)\mathbf{x}(k) . \tag{4.42}$$

The beamformer is thus a MISO system with M input channels. For $N = 1$, the FIR filters in the sensor channels reduce to simple *sensor weights*, or *shading coefficients*. For time-invariant weight vectors $\mathbf{w}_m(k) := \mathbf{w}_m$, we obtain a *fixed beamformer* $\mathbf{w}(k) := \mathbf{w}$.

For illustration of the beamforming operation, a vector-space interpretation of $y(k) = \mathbf{w}^T(k)\mathbf{x}(k)$ can be given [Cox73a]: If the vector $\mathbf{w}(k)$ is orthogonal to the vector $\mathbf{x}(k) = \mathbf{d}(k) + \mathbf{n}(k)$, then, $y(k)$ is equal to zero, and the signals are suppressed. If $\mathbf{x}(k)$ is in the vector space spanned by $\mathbf{w}(k)$,

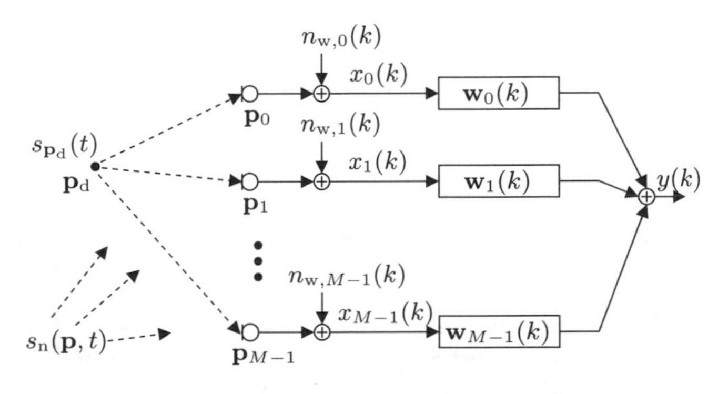

Fig. 4.2. General beamformer setup

the signals are passed. Optimum output signal quality of the beamformer is obtained if the desired signal $\mathbf{d}(k)$ is passed while interference-plus-noise $\mathbf{n}(k)$ is suppressed. Thus, beamformers are generally designed such that (a) a function of $\mathbf{d}(k)$ is in the space which is spanned by $\mathbf{w}(k)$ ('beamsteering') and (b) $\mathbf{n}(k)$ is orthogonal to $\mathbf{w}(k)$ ('interference suppression').

This vector-space interpretation suggests the classification of beamformer designs into two classes: data-independent beamformer designs and data-dependent beamformer designs. For data-independent beamformers, no information about the characteristics of the interference is necessary. Only the position of the desired source is used as a-priori information for designing the beamformer in order to obtain some desired characteristics with $\mathbf{d}(k)$ in $\text{span}\{\mathbf{w}(k)\}$. (See Sect. 4.3.) For data-dependent beamformers, information about the characteristics of the interference is additionally exploited in order to obtain $\mathbf{w}(k)$ being orthogonal to $\mathbf{n}(k)$. (See Sect. 4.4.)

The beamformer is said to be *steered* to the position $\mathbf{p}_\mathrm{d}$ of the desired source, if the filter weights $\mathbf{w}(k)$ equalize the propagation delays $\tau_{\mathrm{d},m}$ after (4.1) between the position of the desired source and the positions of the sensors such that the desired signal is time-aligned in all sensor channels. The delay of the desired signal after the filtering with $\mathbf{w}(k)$ is, for causality, equal to a constant delay τ_d' in all sensor channels, which leads to coherent superposition of the desired signal at the beamformer output (Fig. 4.3). Generally, the equalization delays $\tau_\mathrm{d}' - \tau_{\mathrm{d},m}$ are non-integer multiples of the sampling period T_s. This means that the sensor signals w.r.t. the desired source need to be aligned by fractional delay interpolation filtering [LVKL96] incorporated in $\mathbf{w}(k)$. For linear arrays along the z-axis, *broadside* steering and *endfire* steering refer to steering toward $\theta = \pi/2$ and $\theta \in \{0, \pi\}$, respectively.

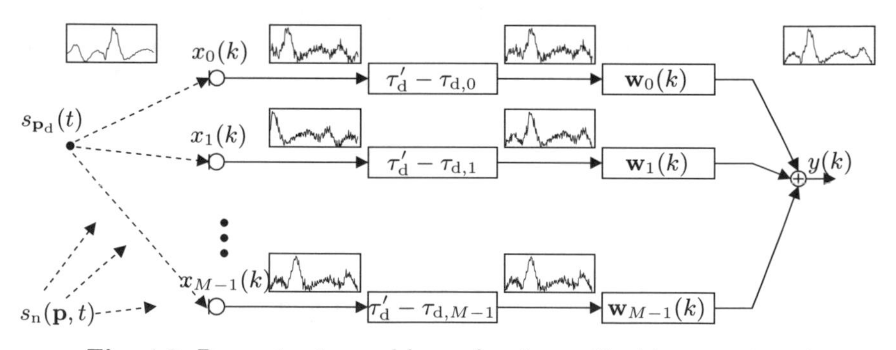

Fig. 4.3. Beam-steering and beam-forming realized in separate units

For simplifying the beamformer design, it is often convenient to separate the beam-steering filters and the beam-forming filters (Fig. 4.3). Many beamformer designs are only for sources in the far field of the sensor array. When realizing the beamsteering as a separate unit, far-field designs can also be applied to near-field beamforming, where the desired range of validity has to be taken into account. The error which is made by this simple transformation is often tolerable. Exact beamformer designs for near-field sources are obtained by using near-field/far-field reciprocity as proposed in [KAW98, KWA99]. If the position of the desired source is not a-priori known or time-varying, the location can be estimated using source localization and source tracking techniques as described in, e.g., [BW01, HB03]. If not otherwise mentioned, we assume in this chapter that the position of the desired source is known and that the beamformer is steered to the position of the desired source without mismatch. Mismatch will be considered in Sect. 4.4.2 and in Chaps. 5, 7.

4.2.2 Beamformer Response and Interference-Independent Performance Measures

In this section, we introduce interference-independent beamformer performance measures, which are generally used to characterize beamformers for monochromatic signals. First, the beamformer response is presented. For illustrating basic characteristics of sensor arrays, we second introduce the uniformly weighted beamformer for a uniformly spaced sensor array. Third, the beampattern and the power pattern of a beamformer are defined. Finally, the peak-to-zero distance, the relative sidelobe level, the directivity, the directivity index, and the array sensitivity are introduced.

Beamformer Response

Consider a monochromatic source (desired signal or interference) to be located in the far field of the array, so that its signal $s(\mathbf{p}, t)$ propagates as a plane wave with frequency ω_0 along the wavenumber vector $\mathbf{k}_0$ according to (2.2). The m-th sensor signal can be written as

$$x_m(k) = S \exp\left(j\left(\omega_0 k T_\mathrm{s} - \mathbf{k}_0^T \mathbf{p}_m\right)\right) , \tag{4.43}$$

or, equivalently, in the DTFT domain as

$$\begin{aligned} X_m(\omega) &= \sum_{k=-\infty}^{\infty} x_m(k) \exp(-j\omega k T_\mathrm{s}) \\ &= \exp(-j\mathbf{k}_0^T \mathbf{p}_m) \frac{2\pi S}{T_\mathrm{s}} \delta\left((\omega - \omega_0) T_\mathrm{s}\right) . \end{aligned} \tag{4.44}$$

Next, we assume a fixed beamformer with $\mathbf{w}_m(k) := \mathbf{w}_m$ after (4.39), and we transform $\mathbf{w}_m$ into the DTFT domain,

$$W_m(\omega) = \sum_{n=0}^{N-1} w_{n,m} \exp(-j\omega n T_\mathrm{s}) . \tag{4.45}$$

Finally, we capture the beamformer weights $W_m(\omega)$ after (4.45) into a column vector of length M as

$$\mathbf{w}_\mathrm{f}(\omega) = (W_0(\omega), W_1(\omega), \ldots, W_{M-1}(\omega))^H , \tag{4.46}$$

and we use the steering vector with the inter-sensor propagation delays after (2.48) for obtaining the DTFT of the beamformer output signal $y(k)$ as

$$Y(\omega) = \mathbf{w}_\mathrm{f}^H(\omega)\mathbf{v}(\mathbf{k}_0, \omega_0)\frac{2\pi S}{T_\mathrm{s}}\delta\left((\omega - \omega_0)T_\mathrm{s}\right) . \tag{4.47}$$

Transformation of (4.47) back into the discrete time domain yields the output signal $y(k)$ of the beamformer as

$$y(k) = \mathbf{w}_\mathrm{f}^H(\omega_0)\mathbf{v}(\mathbf{k}_0, \omega_0) S \exp(j\omega_0 k T_\mathrm{s}) . \tag{4.48}$$

The quantity

$$G(\mathbf{k}, \omega) = \mathbf{w}_\mathrm{f}^H(\omega)\mathbf{v}(\mathbf{k}, \omega) \tag{4.49}$$

is the *wavenumber-frequency response* of the beamformer or *beamformer response.* It describes the complex gain for an input plane wave[4] with wavenumber vector $\mathbf{k}$ and temporal frequency ω.

Uniformly Weighted Beamformer

As a special beamformer, we consider a *uniformly weighted beamformer* which is steered to the position of the desired source. Defining the steering vector of the desired source according to (2.48) as

$$\mathbf{v}(\mathbf{k}_\mathrm{d}, \omega) = (\exp(j\omega\tau_{\mathrm{d},0}), \exp(j\omega\tau_{\mathrm{d},1}), \ldots, \exp(j\omega\tau_{\mathrm{d},M-1}))^H , \tag{4.50}$$

where we replaced in (2.48) the delays $\tau_m = \mathbf{k}^T\mathbf{p}_m$ by the propagation delays $\tau_{\mathrm{d},m}$ w.r.t. the position of the desired source after (4.1). For a uniformly weighted beamformer with arbitrary steering direction, the sensor weights $\mathbf{w}_\mathrm{f}(\omega)$ after (4.46) are chosen as the complex conjugate steering vector $\mathbf{v}^*(\mathbf{k}_\mathrm{d}, \omega)$ after (4.50) and with the amplitude normalized by the number of sensors M. That is,

$$\mathbf{w}_\mathrm{f}(\omega) = \frac{1}{M}\mathbf{v}^*(\mathbf{k}_\mathrm{d}, \omega) . \tag{4.51}$$

For the uniformly weighted beamformer, the absolute value of all sensor weights is equal to $1/M$ ('uniform weighting'), and the phase is equalized for signals with the steering vector $\mathbf{v}(\mathbf{k}_\mathrm{d}, \omega)$ ('beamsteering'). For the DOA of the desired signal, the beamformer response is equal to one, since the phase-aligned sensor signals are summed up constructively:

$$G(\mathbf{k}, \omega) = 1/M \cdot \mathbf{v}^H(\mathbf{k}_\mathrm{d}, \omega)\mathbf{v}(\mathbf{k}_\mathrm{d}, \omega) = 1 . \tag{4.52}$$

Signals from other directions are summed up destructively so that these signals are attenuated relative to the desired signal (Fig. 4.3).

---

[4] This definition can be extended to spherical waves as described in, e.g., [JD93, Chap. 4].

Beampattern and Power Pattern

The *beampattern* or *array pattern* is defined with the beamformer response $G(\mathbf{k},\omega)$ after (4.47) as

$$B(\omega;\theta,\phi) = G(\mathbf{k},\omega)\,, \tag{4.53}$$

with $\theta \in [0;\pi]$, $\phi \in [0;2\pi[$. The beampattern is thus the beamformer response for plane waves evaluated on a sphere with spherical coordinates (θ,ϕ) and with radius c/ω [Tre02, Chap. 2]. The *power pattern* $P(\omega;\theta,\phi)$ is the squared magnitude of the beampattern:

$$P(\omega;\theta,\phi) = |B(\omega;\theta,\phi)|^2\,. \tag{4.54}$$

Consider, for example, the uniformly weighted beamformer (4.51) and a uniform sensor array with an odd number of sensors after Fig. 2.2a. From the rotational symmetry of the linear sensor array relative to the z-axis, it follows that the inter-sensor propagation delays are independent of the ϕ-coordinate. As a consequence, the beamformer response of this linear array is independent of ϕ, which means that the beamformer is not spatially selective over ϕ. The inter-sensor propagation delays for a signal arriving from direction θ are given by

$$\tau_m = \left(\frac{M-1}{2} - m\right)\frac{d\cos\theta}{c}\,. \tag{4.55}$$

The DOA of the signal of the desired source is denoted as θ_d. The inter-sensor propagation delays $\tau_{\mathrm{d},m}$ w.r.t. the desired source are obtained by replacing θ by θ_d in (4.55). The beampattern $B(\omega;\theta,\phi)$ of the uniformly weighted beamformer with uniform sensor spacing d is obtained as

$$\begin{aligned} B(\omega;\theta,\phi) &= \frac{1}{M}\mathbf{v}^H(\mathbf{k}_\mathrm{d},\omega)\mathbf{v}(\mathbf{k},\omega) \\ &= \frac{1}{M}\sum_{m=0}^{M-1}\exp\left(j\omega\left(\frac{M-1}{2} - m\right)\frac{d}{c}\left(\cos\theta_\mathrm{d} - \cos\theta\right)\right)\,. \end{aligned} \tag{4.56}$$

This truncated geometric series may be simplified to a closed form as

$$B(\omega;\theta,\phi) = \frac{1}{M}\frac{\sin\left(\omega M\tau_\mathrm{b}/2\right)}{\sin\left(\omega\tau_\mathrm{b}/2\right)}\,, \tag{4.57}$$

$$\tau_\mathrm{b} = \frac{d}{c}\left(\cos\theta_\mathrm{d} - \cos\theta\right)\,. \tag{4.58}$$

Figure 4.4 illustrates the power pattern $P(\omega;\theta,\phi) = |B(\omega;\theta,\phi)|^2$ for $M = 9$, $d = \lambda/2$, and for broadside steering ($\theta_\mathrm{d} = \pi/2$), i.e., $\mathbf{w}_\mathrm{f}(\omega) = 1/M \cdot \mathbf{1}_{M\times 1}$, where $\mathbf{1}_{M\times 1}$ is a vector of size $M \times 1$ with ones.

Interference-Independent Beamformer Performance Measures

From the power pattern, four performance measures, which are often used as design criteria for data-independent beamformers, may be derived:

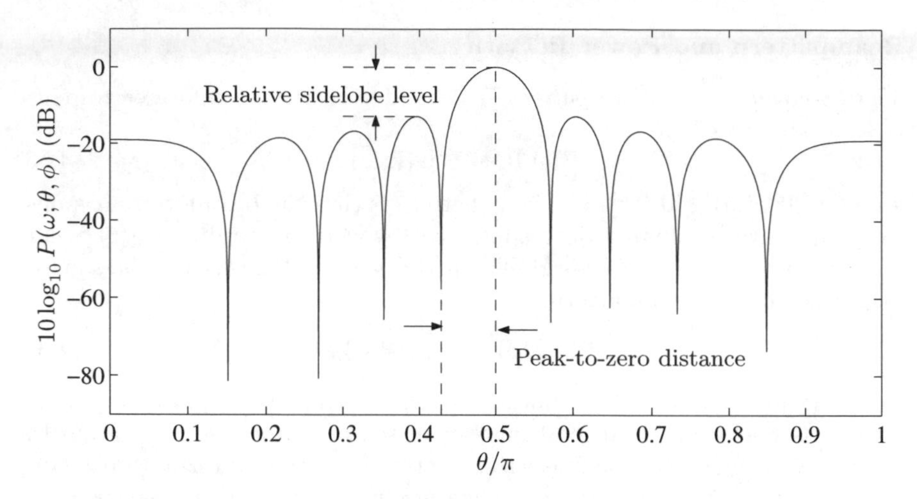

Fig. 4.4. Power pattern $10\log_{10} P(\omega;\theta,\phi)$ (in dB) of a beamformer with uniform sensor weighting, $\mathbf{w}_{\mathrm{f}}(\omega) = 1/M \cdot \mathbf{1}_{M\times 1}$, using the uniformly spaced sensor array given in Fig. 2.2 for $M = 9$, $d = \lambda/2$

Peak-to-Zero Distance

The mainlobe width may be measured for symmetric beampatterns by, e.g., the *peak-to-zero distance*[5], which will be used in this work: It is defined by the angle between the maximum of the mainlobe and the first null of the power pattern $P(\omega;\theta,\phi)$ at frequency ω. As obvious from (4.57), the first spatial null appears for the uniformly weighted beamformer for $\omega M\tau_{\mathrm{b}}/2 = \pi$, which may be written with (4.58) as

$$\cos\theta = \cos\theta_{\mathrm{d}} - \frac{2\pi c}{\omega M d}\,. \tag{4.59}$$

One may notice that the peak-to-zero distance increases with decreasing frequency ω, with decreasing array aperture Md, and with the steering direction θ_{d} moving toward endfire, i.e., $\theta_{\mathrm{d}} \to \{0,\,\pi\}$. The lowest peak-to-zero distance is obtained with broadside arrays. As a limit, no spatial null exists for the right side of (4.59) being smaller than -1. In Fig. 4.5, power patterns on a dB-scale, $10\log_{10} P(\omega;\theta,\phi)$, are illustrated for various ratios d/λ for $M = 8$ for broadside steering (upper row) and for endfire steering (lower row) for illustrating the dependency of the peak-to-zero distance on frequency, on the array aperture, and on the steering direction.

Relative Sidelobe Level

The *relative sidelobe level* is the ratio of the mainlobe level and the level of the highest sidelobe. (See Fig. 4.4.)

[5]See, e.g., [Tre02, Chap. 2] for other measures for the mainlobe width of symmetric and asymmetric beampatterns.

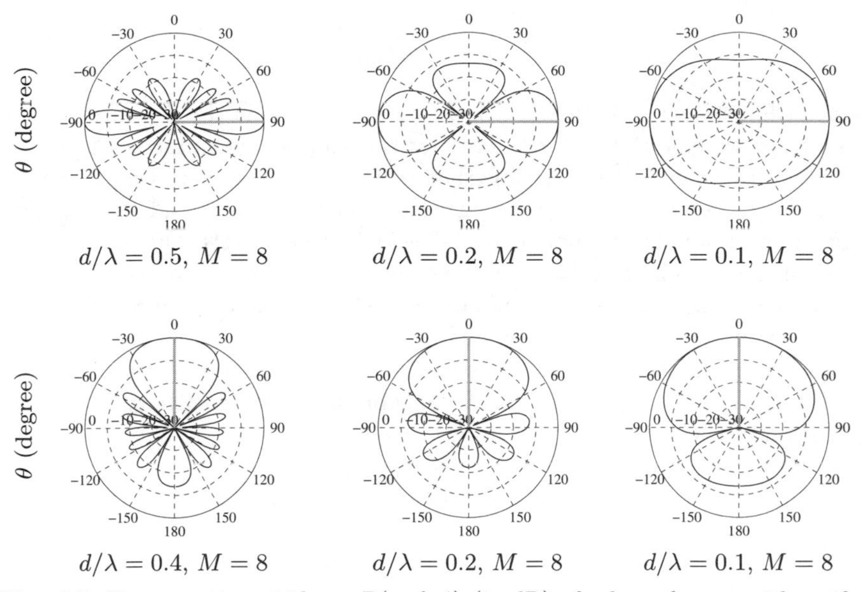

Fig. 4.5. Power pattern $10 \log_{10} P(\omega; \theta, \phi)$ (in dB) of a beamformer with uniform sensor weighting using the uniformly spaced sensor array of Fig. 2.2b for various d/λ for broadside and endfire steering. The steering direction is indicated by the straight line

Directivity and Directivity Index

The *geometric gain* or *directivity* $D(\omega)$ is defined as the ratio of the power pattern in steering direction and the power pattern averaged over all directions for a given frequency ω:

$$D(\omega) = \frac{P(\omega; \theta_{\mathrm{d}}, \phi_{\mathrm{d}})}{\frac{1}{4\pi} \int_0^{\pi} \int_0^{2\pi} P(\omega; \theta, \phi) \sin\theta \, d\theta d\phi} . \tag{4.60}$$

The *directivity index* $DI(\omega)$ is the directivity on a logarithmic scale (in dB),

$$DI(\omega) = 10 \log_{10} D(\omega) . \tag{4.61}$$

Array Sensitivity

The physical realization of beamformers is always subject to imprecision and model mismatch, as, e.g., array geometry, sensor characteristics (directivity, frequency-dependent gain and phase), position of the desired source, or sensor noise. The susceptibility of the power pattern $P(\omega; \theta, \phi)$ to these random errors can be measured by the *array sensitivity* $T_{\mathrm{f}}(\omega)$ against random errors. If we assume that the random errors are uncorrelated, zero-mean, and identically distributed random variables, the array sensitivity can be written as

$$T_\mathrm{f}(\omega) = \frac{\mathbf{w}_\mathrm{f}^H(\omega)\mathbf{w}_\mathrm{f}(\omega)}{|\mathbf{w}_\mathrm{f}^H(\omega)\mathbf{v}(\mathbf{k}_\mathrm{d},\omega)|^2} \tag{4.62}$$

[GM55, US56, McD71, CZK87]. The array sensitivity is the L_2-norm of the beamformer weight vector normalized by the squared beamformer response for the desired signal. Considering that $\mathbf{w}_\mathrm{f}^H(\omega)\mathbf{v}(\mathbf{k}_\mathrm{d},\omega) \leq 1$, where the inequality holds for mismatch of the array steering direction and the position of the desired source, we notice that $T_\mathrm{f}(\omega)$ increases with the L_2-norm of the beamformer weight vector. The susceptibility of the beamformer against uncorrelated random errors thus increases with increasing L_2-norm of $\mathbf{w}_\mathrm{f}(\omega)$.

The array sensitivity $T_\mathrm{f}(\omega)$ may be transformed into the discrete time domain, if we assume that the desired source is located in the array steering direction, i.e., $\mathbf{w}_\mathrm{f}^H(\omega)\mathbf{v}(\mathbf{k}_\mathrm{d},\omega) = 1$. We obtain the array sensitivity $T(k)$ with application of the Parseval theorem as

$$T(k) = \frac{2\pi}{T_\mathrm{s}}\mathbf{w}^T(k)\mathbf{w}(k)\,. \tag{4.63}$$

4.2.3 Interference-Dependent Performance Measures

In Sect. 4.2.2, we introduced beamformer performance measures which describe the shape of the beampattern depending on the array geometry, the steering direction, and the beamformer filter weights. In this section, performance measures are defined which take the wave field of the interference into account in order to specify the beamformer performance for a given wave field. First, the array gain and the white noise gain are defined. Second, we introduce the cancellation of the desired signal and the interference rejection of a beamformer.

Array Gain and White Noise Gain

Array Gain

The *array gain* $A(k)$ measures the ratio of beamformer output *signal-to-interference-plus-noise (power) ratio* (SINR) and the SINR at the sensors. The SINR at the sensors is defined as the ratio of the average variance of the desired signals at the sensors for a data block of length N and the average variance of interference and noise at the sensors for a data block of the same length. That is,

$$\mathit{SINR}_\mathrm{in}(k) = \frac{\mathrm{tr}\,\{\mathbf{R}_\mathbf{dd}(k)\}}{\mathrm{tr}\,\{\mathbf{R}_\mathbf{nn}(k)\}} = \frac{\mathrm{tr}\,\{\mathbf{R}_\mathbf{dd}(k)\}}{\mathrm{tr}\,\{\mathbf{R}_{\mathbf{n}_\mathrm{c}\mathbf{n}_\mathrm{c}}(k)\} + \mathrm{tr}\,\{\mathbf{R}_{\mathbf{n}_\mathrm{w}\mathbf{n}_\mathrm{w}}\}}\,, \tag{4.64}$$

where $\mathbf{R}_\mathbf{dd}(k)$, $\mathbf{R}_{\mathbf{n}_\mathrm{c}\mathbf{n}_\mathrm{c}}(k)$, $\mathbf{R}_{\mathbf{n}_\mathrm{w}\mathbf{n}_\mathrm{w}}$, and $\mathbf{R}_\mathbf{nn}(k)$, are given by (4.7), (4.15), (4.21), and (4.34) respectively. The SINR at the beamformer output $y(k) =$

$\mathbf{w}^T(k)\mathbf{x}(k)$ is defined as the ratio between the variance of the desired signal at the beamformer output,

$$\mathcal{E}\left\{\left(\mathbf{w}^T(k)\mathbf{d}(k)\right)^2\right\} = \mathbf{w}^T(k)\mathcal{E}\{\mathbf{d}(k)\mathbf{d}^T(k)\}\mathbf{w}(k) = \mathbf{w}^T(k)\mathbf{R}_{\mathbf{dd}}(k)\mathbf{w}(k) \tag{4.65}$$

and the variance of interference-plus-noise at the beamformer output,

$$\mathcal{E}\left\{\left(\mathbf{w}^T(k)\mathbf{n}(k)\right)^2\right\} = \mathbf{w}^T(k)\mathcal{E}\{\mathbf{n}(k)\mathbf{n}^T(k)\}\mathbf{w}(k) = \mathbf{w}^T(k)\mathbf{R}_{\mathbf{nn}}(k)\mathbf{w}(k) \,. \tag{4.66}$$

That is,

$$\mathit{SINR}_{\mathrm{out}}(k) = \frac{\mathbf{w}^T(k)\mathbf{R}_{\mathbf{dd}}(k)\mathbf{w}(k)}{\mathbf{w}^T(k)\mathbf{R}_{\mathbf{nn}}(k)\mathbf{w}(k)} \,. \tag{4.67}$$

The array gain is thus given by the expression

$$A(k) = \frac{\mathit{SINR}_{\mathrm{out}}(k)}{\mathit{SINR}_{\mathrm{in}}(k)} = \frac{\mathrm{tr}\left\{\mathbf{R}_{\mathbf{nn}}(k)\right\}}{\mathrm{tr}\left\{\mathbf{R}_{\mathbf{dd}}(k)\right\}} \frac{\mathbf{w}^T(k)\mathbf{R}_{\mathbf{dd}}(k)\mathbf{w}(k)}{\mathbf{w}^T(k)\mathbf{R}_{\mathbf{nn}}(k)\mathbf{w}(k)} \,. \tag{4.68}$$

For narrowband signals with center frequency ω, the array gain (4.68) may be written in the DTFT domain as a function of ω using the narrowband assumption (Sect. 2.2.2). Let ratios between the PSD of the desired signal and the PSD of interference-plus-noise at the sensors and at the output of the beamformer be defined as

$$\mathit{SINR}_{\mathrm{f,in}}(\omega) = \frac{\mathrm{tr}\left\{\mathbf{S}_{\mathbf{dd}}(\omega)\right\}}{\mathrm{tr}\left\{\mathbf{S}_{\mathbf{nn}}(\omega)\right\}} \,, \tag{4.69}$$

$$\mathit{SINR}_{\mathrm{f,out}}(\omega) = \frac{\mathbf{w}_{\mathrm{f}}^H(\omega)\mathbf{S}_{\mathbf{dd}}(\omega)\mathbf{w}_{\mathrm{f}}(\omega)}{\mathbf{w}_{\mathrm{f}}^H(\omega)\mathbf{S}_{\mathbf{nn}}(\omega)\mathbf{w}_{\mathrm{f}}(\omega)} \,, \tag{4.70}$$

respectively. Note that, for narrowband signals, $\mathit{SINR}_{\mathrm{f,in}}(\omega)$ and $\mathit{SINR}_{\mathrm{f,out}}(\omega)$ are SINRs while, for wideband signals, $\mathit{SINR}_{\mathrm{f,in}}(\omega)$ and $\mathit{SINR}_{\mathrm{f,out}}(\omega)$ are ratios of PSDs. We thus obtain for the array gain for narrowband signals the expression

$$A_{\mathrm{f}}(\omega) = \frac{\mathit{SINR}_{\mathrm{f,out}}(\omega)}{\mathit{SINR}_{\mathrm{f,in}}(\omega)} = \frac{\mathrm{tr}\left\{\mathbf{S}_{\mathbf{nn}}(\omega)\right\}}{\mathrm{tr}\left\{\mathbf{S}_{\mathbf{dd}}(\omega)\right\}} \frac{\mathbf{w}_{\mathrm{f}}^H(\omega)\mathbf{S}_{\mathbf{dd}}(\omega)\mathbf{w}_{\mathrm{f}}(\omega)}{\mathbf{w}_{\mathrm{f}}^H(\omega)\mathbf{S}_{\mathbf{nn}}(\omega)\mathbf{w}_{\mathrm{f}}(\omega)} \,. \tag{4.71}$$

For a diffuse noise field with the desired source being located in the array steering direction, the array gain $A_{\mathrm{f}}(\omega)$ is equivalent to the directivity $D(\omega)$ [Eck53, Jac62, Cox73b, Mar95].

White Noise Gain

For presence of only spatially and temporally white noise, i.e., $\mathbf{R}_{\mathbf{nn}}(k) := \mathbf{R}_{\mathbf{n}_{\mathrm{w}}\mathbf{n}_{\mathrm{w}}}$, the array gain is called *white noise gain* $A_{\mathrm{w}}(k)$:

$$A_{\mathrm{w}}(k) = \frac{\operatorname{tr}\{\mathbf{R}_{\mathbf{n}_{\mathrm{w}}\mathbf{n}_{\mathrm{w}}}\}}{\operatorname{tr}\{\mathbf{R}_{\mathbf{dd}}(k)\}} \frac{\mathbf{w}^T(k)\mathbf{R}_{\mathbf{dd}}(k)\mathbf{w}(k)}{\mathbf{w}^T(k)\mathbf{R}_{\mathbf{n}_{\mathrm{w}}\mathbf{n}_{\mathrm{w}}}\mathbf{w}(k)} \,. \tag{4.72}$$

The white noise gain in the DTFT domain for narrowband signals is equivalent to the inverse of the array sensitivity $T_{\mathrm{f}}(\omega)$. This can be seen by substituting in (4.71) $\mathbf{S}_{\mathbf{nn}}(\omega)$ by $\mathbf{S}_{\mathbf{n}_{\mathrm{w}}\mathbf{n}_{\mathrm{w}}}(\omega)$ after (4.23) and $\mathbf{S}_{\mathbf{dd}}(\omega)$ by (4.8) with $\mathbf{\Gamma}_{\mathbf{dd}}(\omega) = \mathbf{v}(\mathbf{k}_{\mathrm{d}}, \omega)\mathbf{v}^H(\mathbf{k}_{\mathrm{d}}, \omega)$ after (2.50).

As a special case for the white noise gain $A_{\mathrm{w}}(k)$, consider a uniformly weighted beamformer with $\mathbf{w} = 1/M \cdot \mathbf{1}_{M\times 1}$ and the desired signal arriving from broadside direction so that the desired signal impinges at all sensors simultaneously, i.e., $\tau_{\mathrm{d},m_1} = \tau_{\mathrm{d},m_2}\ \forall\, m_1, m_2$. Using (4.1) and (4.2), we have $d_{m_1}(k) = d_{m_2}(k)\ \forall\, m_1, m_2$ and $\sigma_d^2(k) := \sigma_{d_m}^2(k)\ \forall\, m$, respectively. The spatio-temporal correlation matrix w.r.t. the desired signal (4.7) simplifies for $N = 1$ to

$$\mathbf{R}_{\mathbf{dd}}(k) = \sigma_d^2(k)\mathbf{1}_{M\times 1}\mathbf{1}_{M\times 1}^T = \sigma_d^2(k)\mathbf{1}_{M\times M} \,. \tag{4.73}$$

The spatio-temporal correlation matrix w.r.t. the sensor noise (4.21) is obtained for $N = 1$ as $\mathbf{R}_{\mathbf{n}_{\mathrm{w}}\mathbf{n}_{\mathrm{w}}} = \sigma_{n_{\mathrm{w}}}^2 \mathbf{I}_{M\times M}$. The white noise gain $A_{\mathrm{w}}(k)$ thus reads as

$$A_{\mathrm{w}}(k) = \frac{M\sigma_{n_{\mathrm{w}}}^2}{M\sigma_d^2(k)} \frac{\sigma_d^2(k)\mathbf{w}^T\mathbf{1}_{M\times M}\mathbf{w}}{\sigma_{n_{\mathrm{w}}}^2 \mathbf{w}^T\mathbf{w}} = M \,. \tag{4.74}$$

The white noise gain gain $A_{\mathrm{w}}(k)$ for the uniformly weighted beamformer with the desired signal arriving from broadside direction is thus equal to the number of sensors M. We notice that the maximum white noise gain $A_{\mathrm{w}}(k) = M$ is only obtained with the uniform sensor weights due to the Schwartz Inequality [Ste73]. Applying other sensor weights reduces the array gain for spatially and temporally white noise fields. If the desired source does not arrive as a plane wave from broadside direction, it was stated in Sect. 4.2 that, generally, interpolation filters need to be incorporated in $\mathbf{w}$ for steering the array to the desired direction. The resulting beamformer does no longer have uniform weighting, and the white noise gain reduces relative to (4.74).

Cancellation of the Desired Signal and Interference Rejection

The array gain $A(k)$ gives no detailed information about the beamformer performance: Consider, e.g., a small value of $A(k)$. On the one hand, this may be the result of suppressing or distorting the desired signal. On the other hand, it may be due to a low suppression of interference-plus-noise. For a more detailed analysis, we introduce separate measures w.r.t. the desired signal and w.r.t. interference-plus-noise.

Cancellation of the Desired Signal

The *cancellation of the desired signal* $DC(k)$ (in dB) is defined as the ratio between the variance of the desired signal at the sensors and the variance of the desired signal at the beamformer output on a logarithmic scale. That is,

$$DC(k) = 10 \log_{10} \frac{\operatorname{tr}\{\mathbf{R}_{\mathbf{dd}}(k)\}}{\mathbf{w}^T(k)\mathbf{R}_{\mathbf{dd}}(k)\mathbf{w}(k)} \,. \tag{4.75}$$

For temporally WSS signals and for fixed beamformers, we drop the dependency on k, which yields $DC(k) =: DC$. For narrowband signals with center frequency ω, we transform DC into the DTFT domain which yields $DC_{\mathrm{f}}(\omega)$ as follows:

$$DC_{\mathrm{f}}(\omega) = 10 \log_{10} \frac{\operatorname{tr}\{\mathbf{S}_{\mathbf{dd}}(\omega)\}}{\mathbf{w}_{\mathrm{f}}^T(\omega)\mathbf{S}_{\mathbf{dd}}(\omega)\mathbf{w}_{\mathrm{f}}(\omega)} \,. \tag{4.76}$$

The cancellation of the desired signal $DC(k)$ in the experiments in Chaps. 5–7 is estimated as follows: The averaging over M sensor signals over a data block of length N by $\operatorname{tr}\{\mathbf{R}_{\mathbf{dd}}(k)\}$ in (4.75) is replaced by exponentially averaged instantaneous estimates of the variance of the sensor signals, i.e.,

$$\operatorname{tr}\{\mathbf{R}_{\mathbf{dd}}(k)\} \approx (1-\beta) \sum_{i=0}^{k} \beta^{k-i} \sum_{m=0}^{M-1} d_m^2(i) \,. \tag{4.77}$$

$0 < \beta < 1$ is an exponential forgetting factor. The normalization factor $1-\beta$ may be interpreted as the effective memory of the exponential averaging (for $k \to \infty$) [Hay96]. The forgetting factor β is chosen such that $1-\beta = 1/N$. Accordingly, the variance of the output signal of the beamformer in (4.75) is estimated by[6]

$$\mathbf{w}^T(k)\mathbf{R}_{\mathbf{dd}}(k)\mathbf{w}(k) \approx (1-\beta) \sum_{i=0}^{k} \beta^{k-i} |\mathbf{w}^T(i)\mathbf{d}(i)|^2 \,. \tag{4.78}$$

For obtaining long-term averages of $DC(k) =: DC$, we replace in (4.75) the numerator and the denominator by time averages according to

$$\operatorname{tr}\{\mathbf{R}_{\mathbf{dd}}(k)\} \approx \sum_{i=0}^{k} \sum_{m=0}^{M-1} d_m^2(i) \,, \tag{4.79}$$

$$\mathbf{w}^T(k)\mathbf{R}_{\mathbf{dd}}(k)\mathbf{w}(k) \approx \sum_{i=0}^{k} |\mathbf{w}^T(i)\mathbf{d}(i)|^2 \,, \tag{4.80}$$

respectively, so that the averaging is performed over the whole (finite) length of the input sequences of the experiment.

Estimates of $DC_{\mathrm{f}}(\omega)$ after (4.76) are calculated by using estimates of the spatio-spectral correlation matrix $\mathbf{S}_{\mathbf{dd}}(\omega)$ using second-order windowed periodograms according to (2.40).

[6]Generally, a synchronization delay between the sensor signals and the output signal of the beamformer is required. This delay will be neglected in the equations. However, the delay is present for the experiments.

Interference Rejection

The *interference rejection* $IR(k)$ (in dB) is defined as the ratio between the variance of interference-plus-noise at the sensors and the variance of interference-plus-noise at the beamformer output on a logarithmic scale. That is,

$$IR(k) = 10 \log_{10} \frac{\mathrm{tr}\{\mathbf{R}_{\mathbf{nn}}(k)\}}{\mathbf{w}^T(k)\mathbf{R}_{\mathbf{nn}}(k)\mathbf{w}(k)} . \tag{4.81}$$

For temporally WSS signals and for fixed beamformers, we have $IR(k) := IR$. For narrowband signals with center frequency ω, we transform IR into the DTFT domain which yields

$$IR_{\mathrm{f}}(\omega) = 10 \log_{10} \frac{\mathrm{tr}\{\mathbf{S}_{\mathbf{nn}}(\omega)\}}{\mathbf{w}_{\mathrm{f}}^T(\omega)\mathbf{S}_{\mathbf{nn}}(\omega)\mathbf{w}_{\mathrm{f}}(\omega)} . \tag{4.82}$$

We notice that the interference rejection $IR_{\mathrm{f}}(\omega)$ is equivalent to the array gain $A_{\mathrm{f}}(\omega)$ for $DC_{\mathrm{f}}(\omega) = 1$. For diffuse interference wave fields and $DC_{\mathrm{f}}(\omega) = 1$, we further have $IR_{\mathrm{f}}(\omega) = DI(\omega)$. For homogeneous interference-plus-noise, $\mathbf{S}_{\mathbf{nn}}(\omega)$ can be separated as $\mathbf{S}_{\mathbf{nn}}(\omega) = S_{nn}(\omega)\mathbf{\Gamma}_{\mathbf{nn}}(\omega)$ according to (4.37), and $IR_{\mathrm{f}}(\omega)$ simplifies as

$$IR_{\mathrm{f}}(\omega) = 10 \log_{10} \frac{1}{\mathbf{w}_{\mathrm{f}}^T(\omega)\mathbf{\Gamma}_{\mathbf{nn}}(\omega)\mathbf{w}_{\mathrm{f}}(\omega)} . \tag{4.83}$$

For the experiments in Chaps. 5–7, we estimate $IR(k)$, IR, and $IR_{\mathrm{f}}(\omega)$ analogously to $DC(k)$, DC, and $DC_{\mathrm{f}}(\omega)$, respectively.

4.2.4 Spatial Aliasing and Sensor Placement

The choice of the beamformer weight vector, of the steering direction, and of the number of sensors in order to obtain a desired beamformer response is one issue. Another degree of freedom is the sensor placement. The spatial sampling locations must be chosen appropriately depending on the wavelength: For collecting space-time signals with long wavelengths, large spatial areas must be sampled by the sensors.[7] The spatial extension is called *aperture* and corresponds to the temporal observation window for time-domain signals. Space-time signals with short wavelengths require small sensor spacings.

These effects can be illustrated considering the beampattern (4.57). If the wavelength is much longer than the sensor spacing d, i.e., $c/\omega \gg d$, spatial discrimination in the θ-coordinate is impossible, since $B(\omega;\theta,\phi) = 1$. For $B(\omega;\theta,\phi) = 1$, interference cannot be separated from the desired signal. Furthermore, $B(\omega;\theta,\phi)$ is periodic in ω with period $2\pi/\tau_{\mathrm{b}}$. The period τ_{b} depends on the array steering direction θ_{d}, the sensor spacing d, the sound velocity

[7] For long wavelengths, spatial selectivity can also be achieved by exploiting 'super-directivity', which allows sensor spacings $d \ll \lambda$ [GM55].

c, and the direction for which $B(\omega; \theta, \phi)$ is evaluated. As a result, spatial discrimination becomes impossible periodically in ω with period $2\pi/\tau_{\mathrm{b}}$. This effect, which is similar to temporal aliasing, is known as *spatial aliasing* due to the dependency on the θ-coordinate. The corresponding sidelobes, which have the same height as the mainlobe in the array steering direction due to this periodicity in ω are called *grating lobes.* The relationship between temporal and spatial aliasing becomes closest, if a sensor array is considered, which is steered to one endfire direction (e.g. $\theta_{\mathrm{d}} = 0$), and appearance of grating lobes is observed in the opposite endfire direction (e.g. $\theta = \pi$). Then, τ_{b} simplifies to $\tau_{\mathrm{b}} = 2d/c$, and spatial aliasing is prevented if the frequency range is limited by $\omega_{\mathrm{u}} = 2\pi c/(2d)$, or

$$d \geq 2\pi c/(2\omega_{\mathrm{u}}) := \lambda_{\mathrm{u}}/2 \,, \tag{4.84}$$

which corresponds to half-wavelength spacing. Spatial aliasing is illustrated in Fig. 4.6 for broadside steering (upper figures) and for endfire steering (lower figures) for a uniformly spaced sensor array with $M = 8$ sensors.

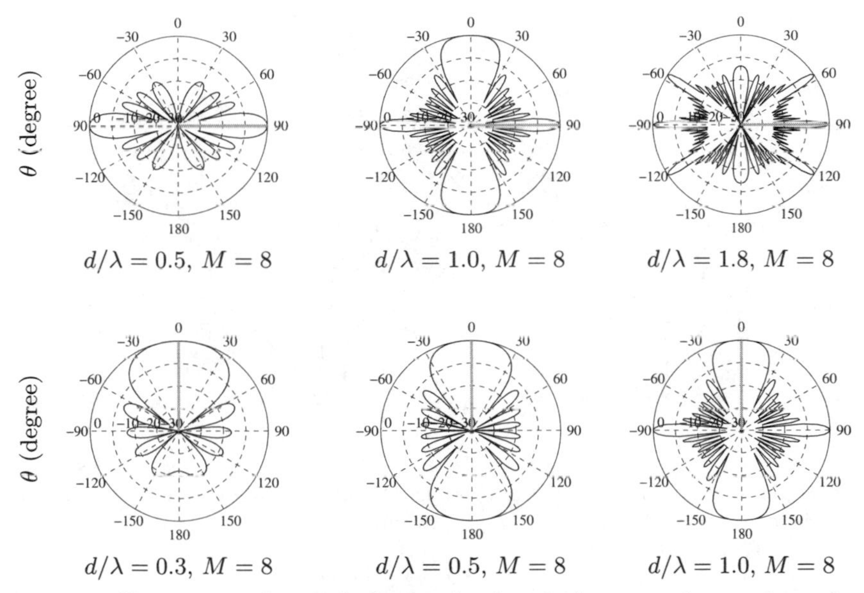

Fig. 4.6. Illustration of spatial aliasing for broadside arrays (upper figures) and endfire arrays (lower figures) with the power pattern $10 \log_{10} P(\omega; \theta, \phi)$ (in dB) of a beamformer with uniform sensor weighting using the uniformly spaced sensor array of Fig. 2.2b for various d/λ. The steering direction is indicated by the straight line

It may be noticed that the minimum sensor spacing d after (4.84) is an upper limit for all directions θ_{d} and θ in order to prevent spatial aliasing for far-field beamformers. Therefore, half-wavelength spacing is typically chosen for narrowband far-field beamformers using line arrays. For near-field line arrays, this half-wavelength sensor spacing is not sufficient, as shown in [AKW99, Abh99].

From this discussion, we conclude that sensor arrays can only be efficiently used for a given frequency range if spatial discrimination should be possible over the entire frequency range and if spatial aliasing should be prevented. A more detailed discussion can be found in, e.g., [Fla85a, Fla85b].

It was shown that the beampattern varies strongly with frequency ω. For realizing *frequency-invariant* beampatterns over a wide frequency range with equally spaced linear arrays, it is generally exploited that the steering vector $\mathbf{v}(\mathbf{k}, \omega)$ depends on the product ωd. (See (4.55).) The desired frequency range is divided into subbands with given center frequencies. For each subband, the sensor spacing d is adjusted such that the product of center frequency and sensor spacing is constant over all subbands. The size of the array thus scales inversely with frequency and leads to harmonically nested subarrays with specific beamformer design methods [GE93, Cho95, VSd96, WKW01].

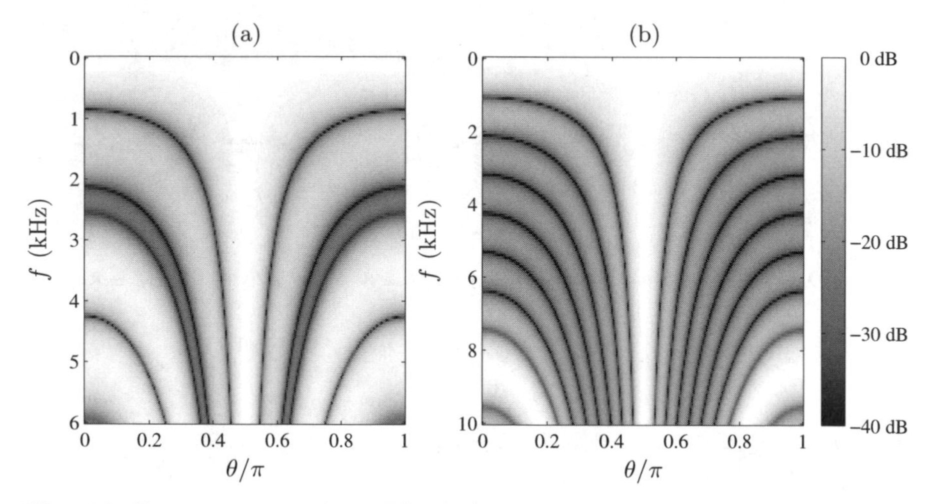

Fig. 4.7. Power pattern $10 \log_{10} P(\omega; \theta, \phi)$ as a function of frequency of a uniformly weighted beamformer ($\mathbf{w}_{\mathrm{f}}(\omega) = 1/M \cdot \mathbf{1}_{M \times 1}$) for the sensor array in App. B with **(a)** $M = 4$ and **(b)** $M = 8$

In our scenario, we generally use line arrays with $M \leq 16$ equally spaced sensors. For practical applications, we consider broadside-steered arrays with $M = 4$ sensors with the geometry defined in Fig. B.1 in App. B or with $M = 8$ uniformly-spaced sensors with spacing $d = 4$ cm. The aperture for both geometries is 28 cm such that the array can be mounted on or integrated in a laptop computer screen. For a uniformly weighted beamformer and for $M = 4$, the first spatial null appears at $f = 850$ Hz. Spatial aliasing occurs at $f \geq 3.5$ kHz (Fig. 4.7a). For $M = 8$, the first spatial null exists for $f \geq 1$ kHz,

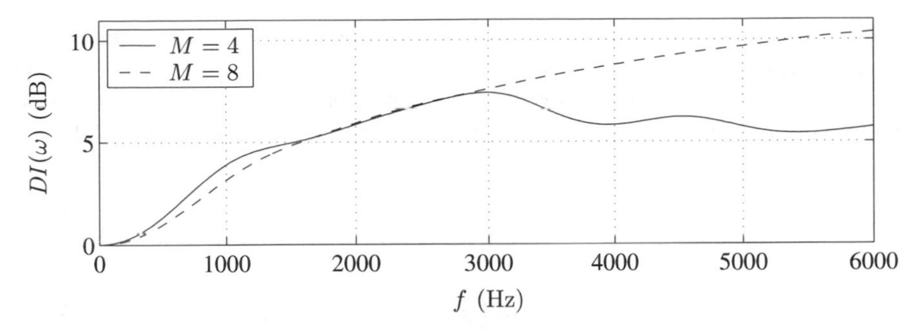

Fig. 4.8. Directivity index $DI(\omega)$ of a uniformly weighted beamformer ($\mathbf{w}_\mathrm{f}(\omega) = 1/M \cdot \mathbf{1}_{M\times 1}$) for the sensor array in App. B with $M = 4$ and $M = 8$

and spatial aliasing occurs for $f \geq 8.5\,\mathrm{kHz}$ (Fig. 4.7b). Both arrays are used in our experiments for the frequency range 200 Hz to 6 kHz at a sampling rate of $f_\mathrm{s} = 12\,\mathrm{kHz}$. At $f = 200\,\mathrm{Hz}$, the directivity index of the uniformly weighted beamformer is only 0.23 dB and 0.16 dB for $M = 4$ and for $M = 8$, respectively, so that it becomes more and more difficult to exploit the spatial selectivity of a broadside-steered uniformly weighted beamformer toward low frequencies (Fig. 4.8). Moreover, spatial aliasing is not avoided for the array with $M = 4$. We have chosen the large aperture for $M = 4$, in order to have maximum spatial selectivity at low frequencies for a given maximum array aperture of 28 cm.

4.3 Data-Independent Beamformer Design

For data-independent beamformers, the stacked weight vector $\mathbf{w}(k)$ is chosen to obtain a wavenumber-frequency response with specific properties. Usually, the beamformer response is shaped to pass signals arriving from a known position $\mathbf{p}$ with minimum distortion while suppressing undesired signals at known or unknown positions. With FIR filters in each sensor channel, we have spatial and temporal degrees of freedom to shape the spatio-temporal beamformer response. The beamformer designs are often specified for monochromatic plane waves with source positions in the far field of the array. For applying these beamformer designs to near-field sources, where spherical wave propagation must be considered, near-field/far-field reciprocity [KAW98, KWA99] can be exploited. For generalization to wideband signals, one may, first, apply the beamformer designs to a set of monochromatic waves which sample the desired frequency range and, second, use a conventional FIR filter design [PB87] to obtain the weight vectors.

For designing beamformers with the desired properties, generally all techniques which are known from temporal FIR filtering can be used due to a close relation between spatial and temporal filtering. For monochromatic plane waves with wavenumber vector $\mathbf{k}$ impinging on the linear uniformly-spaced sensor array of Fig. 2.2a with sensor weights $w_m = w_{0,m}$, $N = 1$, the correspondence between temporal filtering and beamforming becomes closest: The beampattern $B(\omega; \theta, \phi)$ (4.53), written with (4.45), (4.46), and (4.49) as

$$\begin{aligned} B(\omega; \theta, \phi) &= \sum_{m=0}^{M-1} w_m \exp(-j\omega\tau_m) \\ &= \sum_{m=0}^{M-1} w_m \exp\left(j\omega\left(\frac{M-1}{2} - m\right)\frac{d\cos\theta}{c}\right), \end{aligned} \quad (4.85)$$

is equivalent to the DTFT of the weight sequence w_m, $m = 0, 1, \ldots, M-1$, at frequency ω_0 except for a phase shift when identifying the product $\omega d \cos\theta / c$ with the normalized frequency $\omega_0 T_s$, $\omega_0 T_s = \omega d \cos\theta / c$. Consider as a special case a desired signal which arrives from broadside, i.e., $\theta_d = \pi/2$, and a uniformly weighted beamformer ($w_m = 1/M$). Then, the beamformer provides unity gain for the desired signal from $\theta_d = \pi/2$ in the same way as the corresponding temporal rectangular (uniform) window lets through signal components with frequency $\omega_0 = 0$, since $\omega_0 T_s := \omega d \cos\theta_d / c = 0$. Other directions θ are attenuated in the same way as other frequencies are attenuated by the corresponding temporal rectangular window.

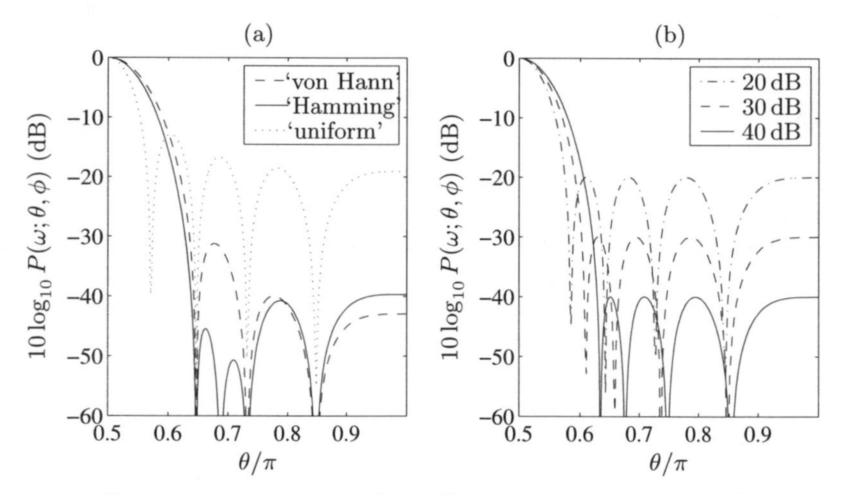

Fig. 4.9. Power pattern $10\log_{10} P(\omega; \theta, \phi)$ of a beamformer with **(a)** von Hann, Hamming, and uniform windowing, and **(b)** Dolph-Chebyshev windowing with different relative sidelobe levels; sensor array of Fig. 2.2a with $M = 9$, $d = \lambda/2$

For shaping the beamformer response for monochromatic plane waves of a given frequency ω, the amplitudes of the complex weighting $\mathbf{w}_f(\omega)$ may incorporate windowing (tapering) of the spatial aperture to trade the mainlobe width against relative sidelobe level. Besides classical windowing functions [Har78], Dolph-Chebyshev windowing [Dol46], which minimizes the mainlobe width for a given maximum sidelobe level, is widely used. Dolph-Chebyshev design gives a power pattern with equal sidelobe levels for a given frequency. In Fig. 4.9, power patterns for spatial windowing for monochromatic signals are depicted ($M = 9$, $d = \lambda/2$). Figure 4.9a shows power patterns for classical windows ('von Hann', 'Hamming' [Har78]) compared to uniform weighting. Figure 4.9b shows power patterns for Dolph-Chebyshev windows for relative sidelobe levels 20, 30, 40 dB.

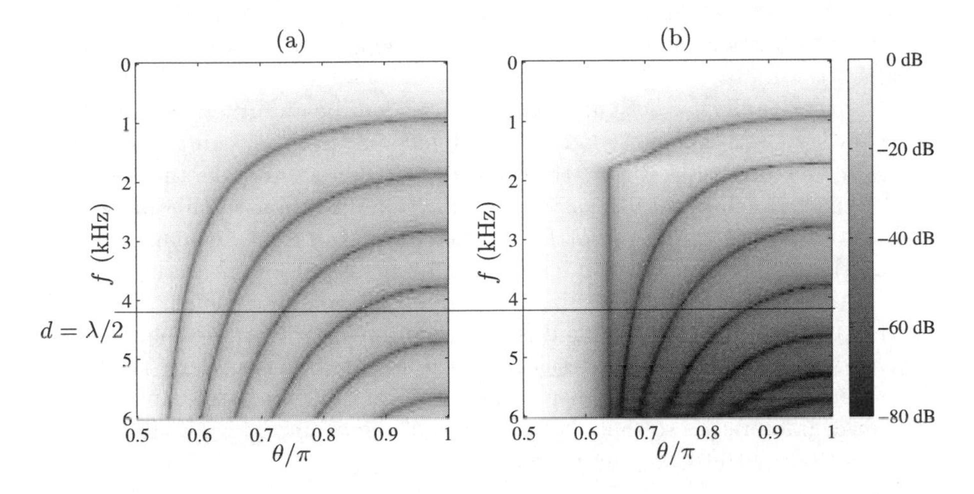

Fig. 4.10. Power pattern $10 \log_{10} P(\omega; \theta, \phi)$ of a wideband beamformer for the sensor array of Fig. 2.2a with $M = 9$, $d = 4$ cm with **(a)** uniform weighting, and **(b)** wideband Dolph-Chebyshev design ($N = 64$) with frequency-invariant peak-to-zero distance for frequencies $f > 1700$ Hz and uniform weighting for $f \leq 1700$ Hz. The line $d = \lambda/2$ relates Fig. 4.10a and Fig. 4.10b with Fig. 4.9a for uniform weighting and Fig. 4.9b for 40 dB sidelobe cancellation, respectively; $N = 64$

Figure 4.10 illustrates examples of power patterns for wideband signals ($M = 9$, $d = 4$ cm): In Fig. 4.10a, the power pattern of a wideband beamformer with uniform weighting is shown. We see that the peak-to-zero distance decreases with increasing frequency $f = \omega/(2\pi)$ according to (4.59). For high frequencies, the beamformer thus becomes sensitive to steering errors. For low frequencies, spatial separation of desired signal and interference becomes impossible [JD93, Chap. 3]. In Fig. 4.10b, the power pattern of a wideband beamformer using a Dolph-Chebyshev design is shown. The FIR filters $\mathbf{w}_m(k)$ are obtained by applying Dolph-Chebyshev windows to a set of discrete frequencies with a predefined frequency-invariant peak-to-zero distance of the power pattern. These frequency-dependent Dolph-Chebyshev windows are then fed into the Fourier approximation filter design [PB87] to determine the FIR filters $\mathbf{w}_m(k)$. Due to the limited spatial aperture of the sensor array, the Dolph-Chebyshev design is only used for $f \geq 1700\,\text{Hz}$. For $f < 1700\,\text{Hz}$, uniform weighting is applied. Dolph-Chebyshev windowing thus allows to specify an (almost) arbitrary mainlobe width above the minimum given by the uniform shading, which is frequency-independent over a wide range of frequencies, and which reduces the sensitivity due to steering errors.

For designing data-independent beamformers with more control about the beamformer response than simple windowing techniques, one may apply methods which are similar to the design of arbitrary FIR filter transfer functions [PB87] as, e.g., windowed Fourier approximation, least-squares approximation of the desired power pattern, or minimax design to control the maximum allowable variation of mainlobe level and sidelobe levels. (See, e.g., [Tre02, Chap. 3]).

Conventionally, for designing data-independent beamformers, it is assumed that the sensors are perfectly omni-directional and that they completely match in gain and phase. However, transducers are subject to tolerances in gain and phase, which generally vary over time. One solution to this problem is calibration of the sensors prior application [Syd94, TE01]. In many applications, the calibration procedure can only be done once so that the problem of the time-varying gain and phase mismatch cannot be resolved. Therefore, in [Doc03, DM03], a new method is proposed for improving the robustness of fixed wideband beamformer designs against such tolerances. Here, the beampattern of the fixed beamformer is designed for predefined tolerances of the sensor characteristics so that the variations of the beampattern relative to the desired beampattern are minimized.

4.4 Optimum Data-Dependent Beamformer Designs

In this section, we describe optimum data-dependent beamformer designs. Most often, optimum data-dependent beamformers are derived based on the narrowband assumption in the DTFT domain. These narrowband designs directly translate to at least WSS wideband signals if the wideband signals are decomposed into narrow frequency bins such that the narrowband assumption holds. For application to short-time stationary wideband signals, as, e.g., speech and audio signals, the DTFT can be approximated by the DFT, and the sensor data can be processed block-wise. However, data-dependent beamformers can also be derived in the time domain without the stationarity or short-time stationarity assumption.

For wideband signals, most data-dependent beamformers can be classified either as MMSE design or as linearly-constrained minimum variance (LCMV) design. Regardless of an implementation in the DFT domain or in the time domain, both approaches generally use time-averaging over a finite temporal aperture to estimate the relevant statistics of the sensor data. Due to the non-stationarity of speech and audio signals, the short-time DFTs [OS75] no longer produce stationary and mutually orthogonal frequency bin signals. Therefore, we use here a time-domain representation to rigorously derive optimum data-dependent beamformers. The non-stationarity of the sensor data and the finite temporal apertures are directly taken into account: Instead of using stochastic expectations, which must be replaced by time-averages for realizing the beamformer, we directly formulate the beamformers using LSE criteria over finite data blocks. We thus obtain the LSE and the *linearly-constrained least-squares error* (LCLSE) beamformer.

The LSE/MMSE design generally requires a reference of the desired signal. Often, a reference signal is not available and cannot be accurately estimated. The LCLSE/LCMV design avoids this problem by exploiting information about the location of the desired source. It is a spatially constrained design, where integrity of the desired signal is assured by constraining the beamformer response to unity gain for the position of the desired source.

In Sect. 4.4.1, the LSE/MMSE design is presented. In Sect. 4.4.2, the LCLSE/LCMV beamformer is derived, and it is shown how this beamformer can be efficiently realized in the so-called generalized sidelobe canceller (GSC) structure. The interpretation of LSE and LCLSE beamforming using eigenanalysis methods suggests the realization of optimum beamforming using eigenvalue decomposition, leading to eigenvector beamformers (Sect. 4.4.3). In Sect. 4.4.4, the problem of correlation between the desired signal and interference is discussed, which is of special interest in reverberant acoustic environments.

4.4.1 LSE/MMSE Design

In this section, the LSE beamformer for wideband non-stationary signals in the discrete time domain and the MMSE beamformer in the DTFT domain are presented. The LSE criterion allows to rigorously introduce the LSE/MMSE design with consideration of the length N of the beamformer weight vector and without making any assumptions about the frequency contents or the statistics of the sensor signals. As a consequence, the LSE description meets well the requirements of wideband non-stationary speech and audio signals. The MMSE design in the DTFT domain allows illustration of the fundamental characteristics of the LSE/MMSE design since many beamformer performance measures can be nicely illustrated in the DTFT domain.

This section is organized as follows: First, the optimum LSE processor is derived, and an eigenspace interpretation is given. Second, the MMSE beamformer in the DTFT domain is discussed. Third, the relation between the LSE design and the MMSE design is summarized. Finally, we outline applications of the LSE/MMSE beamformer for speech and audio signal processing.

LSE Beamforming

In this section, we first derive the optimum LSE beamformer. Second, the function of the LSE beamformer is interpreted in the eigenspace of interference-plus-noise.

Derivation of the Optimum Weight Vector

Following the vector-space interpretation of data-dependent beamformers in Sect. 4.2, the LSE design is directly related to an optimum MISO filter according to Fig. 3.2 with $P = 1$.

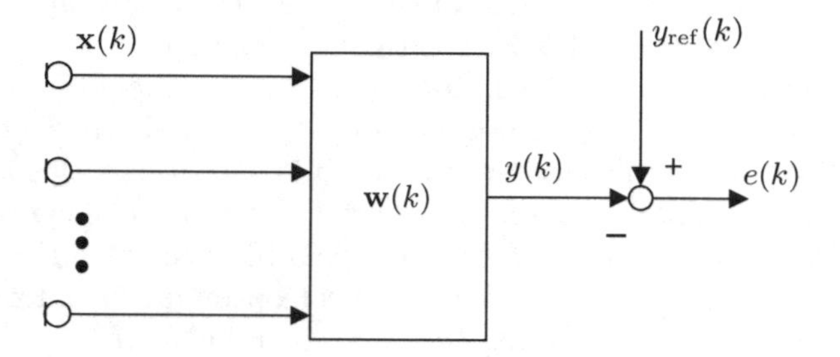

Fig. 4.11. Least-squares error (LSE) beamformer with the reference signal $y_{\mathrm{ref}}(k)$ as reference

The beamformer weight vector $\mathbf{w}(k)$ is designed such that the desired signal is contained in the space which is spanned by $\mathbf{w}(k)$ and such that the vector of interference-plus-noise is in the nullspace of $\mathbf{w}(k)$. This orthogonality w.r.t. interference-plus-noise corresponds to the principle of orthogonality of LS optimum linear filtering. Using in Fig. 3.2 a reference to the desired source signal $s_{\mathbf{p}_\mathrm{d}}(kT_\mathrm{s})$ as reference $y_{\mathrm{ref}}(k) := \mathbf{y}_{\mathrm{ref}}(k)$ (for $P = 1$), and identifying the MIMO filter $\mathbf{W}(k)$ with $\mathbf{w}(k)$, we obtain the LSE beamformer according to Fig. 4.11.

For formulating the LSE cost function, we define the estimation error $e(k)$ as the difference between the desired response $y_{\mathrm{ref}}(k)$ and the output signal of the beamformer

$$e(k) = y_{\mathrm{ref}}(k) - \mathbf{w}^T(k)\mathbf{x}(k) . \tag{4.86}$$

The cost function $\xi_{\mathrm{LSE}}(r)$ of the LSE beamformer is defined according to (3.7) with $P = 1$ as

$$\xi_{\mathrm{LSE}}(r) = \sum_{k=rR}^{rR+K-1} e^2(k) = \sum_{k=rR}^{rR+K-1} \left(y_{\mathrm{ref}}(k) - \mathbf{w}^T(rR)\mathbf{x}(k)\right)^2 . \tag{4.87}$$

Introducing the desired response vector $\mathbf{y}_\mathrm{r}(k)$ of size $K \times 1$

$$\mathbf{y}_\mathrm{r}(k) = \left(y_{\mathrm{ref}}(k),\, y_{\mathrm{ref}}(k+1),\, \ldots,\, y_{\mathrm{ref}}(k+K-1)\right)^T , \tag{4.88}$$

and using the $MN \times K$ data matrix $\mathbf{X}(k)$ defined in (3.11), then, the LSE cost function $\xi_{\text{LSE}}(r)$ may be written as:

$$\xi_{\text{LSE}}(r) = \left\| \mathbf{y}_{\text{r}}(rR) - \mathbf{X}^T(rR)\mathbf{w}(rR) \right\|_2^2 . \tag{4.89}$$

Since we assume full row rank of the data matrix $\mathbf{X}(rR)$, we use the solution for overdetermined systems of linear equations of Sect. 3.1.2 for solving (4.89) for $\mathbf{w}(rR)$. By differentiation of $\xi_{\text{LSE}}(r)$ w.r.t. $\mathbf{w}(rR)$ and by setting the derivative equal to zero, the minimum of $\xi_{\text{LSE}}(r)$ is obtained as

$$\mathbf{w}_{\text{LSE}}(rR) = \mathbf{\Phi}_{\mathbf{xx}}^{-1}(rR)\mathbf{X}(rR)\mathbf{y}_{\text{r}}(rR) , \tag{4.90}$$

with $\mathbf{\Phi}_{\mathbf{xx}}(k)$ given by (3.18). The product $\mathbf{\Phi}_{\mathbf{xx}}^{-1}(k)\mathbf{X}(k)$ is the pseudoinverse of $\mathbf{X}^T(k)$ for the overdetermined least-squares problem according to (3.28).

The product $\mathbf{X}(k)\mathbf{y}_{\text{r}}(k)$ can be interpreted as an estimate of the cross-correlation vector between the sensor signals $\mathbf{x}(k)$ and the desired response $y_{\text{ref}}(k)$ over a data block of length K. Assuming temporal orthogonality of the data matrix w.r.t. interference-plus-noise $\mathbf{N}(k)$ and the desired response vector $\mathbf{y}_{\text{r}}(k)$, i.e.,

$$\mathbf{N}(k)\mathbf{y}_{\text{r}}(k) = \mathbf{0}_{MN\times 1} , \tag{4.91}$$

yields with (4.6) and with (4.32)

$$\mathbf{w}_{\text{LSE}}(k) = \left(\mathbf{\Phi}_{\mathbf{nn}}(k) + \mathbf{D}(k)\mathbf{D}^T(k)\right)^{-1}\mathbf{D}(k)\mathbf{y}_{\text{r}}(k) . \tag{4.92}$$

Applying the matrix inversion lemma [Tre02]

$$\left(\mathbf{A}^{-1} + \mathbf{B}^H\mathbf{C}^{-1}\mathbf{B}\right)^{-1}\mathbf{B}^H\mathbf{C}^{-1} = \mathbf{A}\mathbf{B}^H\left(\mathbf{B}\mathbf{A}\mathbf{B}^H + \mathbf{C}\right)^{-1} \tag{4.93}$$

to $\left(\mathbf{\Phi}_{\mathbf{nn}}(k) + \mathbf{D}(k)\mathbf{D}^T(k)\right)^{-1}\mathbf{D}(k)$, we obtain the optimum weight vector as follows:

$$\mathbf{w}_{\text{LSE}}(k) = \mathbf{\Phi}_{\mathbf{nn}}^{-1}(k)\mathbf{D}(k)\left(\mathbf{\Lambda}_{\mathbf{D}}(k) + \mathbf{I}_{K\times K}\right)^{-1}\mathbf{y}_{\text{r}}(k) , \tag{4.94}$$

where $\mathbf{I}_{K\times K}$ is the identity matrix of size $K \times K$ and where

$$\mathbf{\Lambda}_{\mathbf{D}}(k) = \mathbf{D}^T(k)\mathbf{\Phi}_{\mathbf{nn}}^{-1}(k)\mathbf{D}(k) . \tag{4.95}$$

Introducing (4.94) into (4.89), the minimum of the LSE cost function $\xi_{\text{LSE}}(r)$ is obtained as

$$\xi_{\text{LSE,o}}(r) = \left\| \left(\mathbf{\Lambda}_{\mathbf{D}}(rR) + \mathbf{I}_{K\times K}\right)^{-1}\mathbf{y}_{\text{r}}(rR) \right\|_2^2 . \tag{4.96}$$

Eigenspace Interpretation

For interpreting LSE beamforming in the eigenspace of interference-plus-noise, we project the optimum weight vector of the LSE beamformer after (4.94) into the eigenspace of interference-plus-noise. This allows us to analyze the LSE beamforming operation in more detail.

According to the SVD of $\mathbf{X}(k)$ in (3.21), we introduce the SVD of the data matrix $\mathbf{N}(k)$ with the matrix of the left singular vectors $\mathbf{U}_{\mathbf{N}}(k)$, with the matrix of the right singular vectors $\mathbf{V}_{\mathbf{N}}(k)$, and with the matrix of the singular values $\mathbf{\Sigma}_{\mathbf{N}}(k)$. The matrix $\mathbf{\Sigma}_{\mathbf{N}}(k)$ is of size $MN \times MN$, since we assumed invertibility of $\mathbf{\Phi}_{\mathbf{nn}}(k)$. Using the SVD, we have $\mathbf{\Phi}_{\mathbf{nn}}(k)$ as

$$\mathbf{\Phi}_{\mathbf{nn}}(k) = \mathbf{U}_{\mathbf{N}}(k)\mathbf{\Sigma}_{\mathbf{N}}^2(k)\mathbf{U}_{\mathbf{N}}^T(k)\,, \tag{4.97}$$

and we can write the optimum weight vector of the LSE beamformer (4.94) as follows:

$$\mathbf{w}_{\mathrm{LSE}}(k) = \mathbf{U}_{\mathbf{N}}(k)\mathbf{\Sigma}_{\mathbf{N}}^{-2}(k)\mathbf{U}_{\mathbf{N}}^T(k)\mathbf{D}(k) \\ \left(\mathbf{D}^T(k)\mathbf{U}_{\mathbf{N}}(k)\mathbf{\Sigma}_{\mathbf{N}}^{-2}(k)\mathbf{U}_{\mathbf{N}}^T(k)\mathbf{D}(k) + \mathbf{I}_{K\times K}\right)^{-1}\mathbf{y}_{\mathrm{r}}(k)\,. \tag{4.98}$$

Then, we project the data matrices $\mathbf{D}(k)$ and $\mathbf{N}(k)$ into the space, which is spanned by $\mathbf{U}_{\mathbf{N}}(k)\mathbf{\Sigma}_{\mathbf{N}}^{-1}(k)$, i.e., onto the left singular vectors of $\mathbf{N}(k)$ weighted by the inverse of the singular values of $\mathbf{N}(k)$. We obtain

$$\widetilde{\mathbf{D}}(k) = \mathbf{\Sigma}_{\mathbf{N}}^{-1}(k)\mathbf{U}_{\mathbf{N}}^T(k)\mathbf{D}(k) \tag{4.99}$$

$$\widetilde{\mathbf{N}}(k) = \mathbf{\Sigma}_{\mathbf{N}}^{-1}(k)\mathbf{U}_{\mathbf{N}}^T(k)\mathbf{N}(k)\,. \tag{4.100}$$

We see from (4.100) that $\widetilde{\mathbf{N}}^T(k)\widetilde{\mathbf{N}}(k) = \mathbf{I}_{K\times K}$ and that $\widetilde{\mathbf{N}}(k)\widetilde{\mathbf{N}}^T(k) = \mathbf{I}_{MN\times MN}$. The combination of the projection onto $\mathbf{U}_{\mathbf{N}}(k)$ and the weighting by the inverse singular values $\mathbf{\Sigma}_{\mathbf{N}}^{-1}(k)$ is thus equivalent to joint whitening the spatially and temporally averaged auto-correlation matrix $\mathbf{N}^T(k)\mathbf{N}(k)$ and the sample spatio-temporal correlation matrix $\mathbf{\Phi}_{\mathbf{nn}}(k) = \mathbf{N}(k)\mathbf{N}^T(k)$. The projection of the data matrices $\mathbf{D}(k)$ and $\mathbf{N}(k)$ onto $\mathbf{U}_{\mathbf{N}}(k)\mathbf{\Sigma}_{\mathbf{N}}^{-1}(k)$ thus corresponds to a projection into the eigenspace of interference-plus-noise.[8] Introducing (4.99) into (4.98), we obtain the LSE beamformer as follows:

$$\mathbf{w}_{\mathrm{LSE}}(k) = \mathbf{U}_{\mathbf{N}}(k)\mathbf{\Sigma}_{\mathbf{N}}^{-1}(k)\left(\widetilde{\mathbf{D}}^T(k)\widetilde{\mathbf{D}}(k) + \mathbf{I}_{K\times K}\right)^{-1}\mathbf{y}_{\mathrm{r}}(k)\,. \tag{4.101}$$

The matrix $\widetilde{\mathbf{D}}^T(k)\widetilde{\mathbf{D}}(k)$ of size $K \times K$ may be interpreted as the (non-normalized) spatially and temporally averaged auto-correlation matrix at the sensors w.r.t. the desired signal in the eigenspace of interference-plus-noise. Defining finally the projection of the data vector $\mathbf{x}(k)$ onto $\mathbf{U}_{\mathbf{N}}(k)\mathbf{\Sigma}_{\mathbf{N}}^{-1}(k)$ as

$$\widetilde{\mathbf{x}}(k) = \mathbf{\Sigma}_{\mathbf{N}}^{-1}(k)\mathbf{U}_{\mathbf{N}}^T(k)\mathbf{x}(k) \tag{4.102}$$

$$= \mathbf{\Sigma}_{\mathbf{N}}^{-1}(k)\mathbf{U}_{\mathbf{N}}^T(k)[\mathbf{d}(k) + \mathbf{n}(k)] \tag{4.103}$$

$$= \widetilde{\mathbf{d}}(k) + \widetilde{\mathbf{n}}(k) \tag{4.104}$$

we can write the output signal of the LSE beamformer $y(k) = \mathbf{w}_{\mathrm{LSE}}^T(k)\mathbf{x}(k)$ as

[8] Note that $\widetilde{\mathbf{D}}(k) \neq \mathbf{0}_{MN\times K}$, since $\mathbf{\Phi}_{\mathbf{nn}}(k)$ is assumed to be invertible such that $\mathbf{U}_{\mathbf{N}}(k)\mathbf{\Sigma}_{\mathbf{N}}^{-1}(k)$ spans the entire estimation space.

$$\begin{aligned} y(k) &= \mathbf{y}_{\mathrm{r}}^T(k)\left(\widetilde{\mathbf{D}}^T(k)\widetilde{\mathbf{D}}(k) + \mathbf{I}_{K\times K}\right)^{-1}\widetilde{\mathbf{D}}^T(k)\widetilde{\mathbf{x}}(k) \\ &= \widetilde{\mathbf{w}}_{\mathrm{LSE}}^T(k)\widetilde{\mathbf{x}}(k)\,. \end{aligned} \tag{4.105}$$

The LSE beamforming operation (4.105) can be interpreted in the eigenspace of interference-plus-noise as follows: The matrix-vector product $\widetilde{\mathbf{D}}^T(k)\widetilde{\mathbf{x}}(k)$ projects the sensor data $\widetilde{\mathbf{x}}(k)$ into the space, which is spanned by $\widetilde{\mathbf{D}}(k)$. Interference-plus-noise components of $\widetilde{\mathbf{n}}(k)$, which are orthogonal to $\widetilde{\mathbf{D}}(k)$ are suppressed. Components of interference-plus-noise which are not orthogonal to $\widetilde{\mathbf{D}}(k)$, i.e., which are contained in the estimation space of the desired signal, cannot be completely cancelled. The attenuation factor depends on the inverse of the term in parentheses in (4.105). Considering first that the identity matrix $\mathbf{I}_{K\times K}$ in (4.105) can be neglected relative to the main diagonal of $\widetilde{\mathbf{D}}^T(k)\widetilde{\mathbf{D}}(k)$. This corresponds to the case where interference-plus-noise can be neglected relative to the desired signal in the estimation space of the desired signal. The resulting LSE beamformer

$$\widetilde{\mathbf{w}}_{\mathrm{LSE}}(k) = \widetilde{\mathbf{D}}(k)(\widetilde{\mathbf{D}}^T(k)\widetilde{\mathbf{D}}(k))^{-1}\mathbf{y}_{\mathrm{r}}(k) \tag{4.106}$$

in (4.105) can be identified with the solution of the underdetermined system of linear equations

$$\mathbf{y}_{\mathrm{r}}(k) = \widetilde{\mathbf{D}}^T(k)\widetilde{\mathbf{w}}_{\mathrm{LSE}}(k) \tag{4.107}$$

using (3.26) and (3.27). The reference $\mathbf{y}_{\mathrm{r}}(k)$ of the desired signal is thus contained in the space, which is spanned by the columns of $\widetilde{\mathbf{D}}^T(k)$. As a conclusion, this result means that the desired signal is not distorted by the LSE beamformer if interference-plus-noise that is contained in the estimation space of the desired signal can be neglected relative to the desired signal.

With presence of interference-plus-noise components in the estimation space of the desired signal that cannot be neglected relative to the desired signal, the augmentation of $\widetilde{\mathbf{D}}^T(k)\widetilde{\mathbf{D}}(k)$ by $\mathbf{I}_{K\times K}$ in (4.105) influences the output signal of the LSE beamformer: If a singular value in the matrix $\mathbf{\Sigma}_{\mathbf{N}}(k)$ which corresponds to interference-plus-noise components in the estimation space of the desired signal increases, the L_2-norm of the corresponding row of the data matrix $\widetilde{\mathbf{D}}(k)$ decreases according to (4.99). From (4.105), we see that this leads to an attenuation of the corresponding output signal components of the LSE beamformer, which affects both interference-plus-noise and the desired signal. It follows that, on the one hand, the suppression of interference-plus-noise increases, but that, on the other hand, the distortion of the desired signal increases, too.

From this eigenspace interpretation, we can summarize the behavior of the LSE beamformer as follows: Interference-plus-noise is completely cancelled and the desired signal is not distorted as long as the estimation space of interference-plus-noise at the sensors is orthogonal to the estimation space of the desired signal at the sensors. When the estimation spaces of interference-plus-noise and of the desired signal are non-orthogonal, the common subspace is neither completely cancelled nor completely preserved. The more interference-plus-noise components are contained in the common subspace of interference-plus-noise, the more the common subspace is attenuated, which means that interference-plus-noise is cancelled at the cost of distortion of the desired signal. Non-orthogonality may result from an estimation space $\mathbf{X}^T(k)$ with insufficient dimensions, i.e., spatial and temporal degrees of freedom, or from overlapping spatial and temporal characteristics of the desired signal and interference-plus-noise.

MMSE Beamforming

In this section, the MMSE beamformer design is introduced. First, the optimum weight vector is derived. Second, the characteristics of the MMSE beamformer are discussed.

Derivation of the Optimum Weight Vector

The statistical MMSE beamformer in the DTFT domain is obtained analogously to the linear MMSE MIMO system in Sect. 3.1.3 for $P = 1$ output channel. The MMSE cost function is given by

$$\xi_{\mathrm{MMSE}}(k) = E\left\{e^2(k)\right\} . \tag{4.108}$$

Minimization of $\xi_{\mathrm{MMSE}}(k)$ w.r.t. $\mathbf{w}(k)$ and transformation into the DTFT domain yields the optimum MMSE processor analogously to the optimum LSE processor (4.90) as follows:

$$\mathbf{w}_{\mathrm{f,MMSE}}(\omega) = \mathbf{S}_{\mathbf{xx}}^{-1}(\omega)\,\mathbf{s}_{\mathbf{x}y_{\mathrm{ref}}}(\omega) , \tag{4.109}$$

where $\mathbf{s}_{\mathbf{x}y_{\mathrm{ref}}}(\omega)$ is the CPSD vector of size $M \times 1$ between the sensor signals and the reference signal. Assuming a mutually uncorrelated reference signal $y_{\mathrm{ref}}(k)$ and interference-plus-noise $\mathbf{n}(k)$ and applying the matrix inversion lemma [Hay96], (4.109) can be written as a function of the PSD of the desired signal, $S_{dd}(\omega)$, of the vector $\mathbf{v}(\mathbf{k}_{\mathrm{d}}, \omega)$ after (4.50), and of the spatio-spectral correlation matrix $\mathbf{S}_{\mathbf{nn}}(\omega)$ of interference-plus-noise as

$$\mathbf{w}_{\mathrm{f,MMSE}}(\omega) = \frac{S_{dd}(\omega)}{S_{dd}(\omega) + \Lambda_{\mathbf{v}_{\mathrm{d}}}(\omega)} \Lambda_{\mathbf{v}_{\mathrm{d}}}(\omega)\mathbf{S}_{\mathbf{nn}}^{-1}(\omega)\mathbf{v}(\mathbf{k}_{\mathrm{d}}, \omega) , \tag{4.110}$$

where

$$\Lambda_{\mathbf{v}_{\mathrm{d}}}(\omega) = \left(\mathbf{v}^H(\mathbf{k}_{\mathrm{d}}, \omega)\mathbf{S}_{\mathbf{nn}}^{-1}(\omega)\mathbf{v}(\mathbf{k}_{\mathrm{d}}, \omega)\right)^{-1} \tag{4.111}$$

[EFK67, SBM01].

Interpretation

The MMSE beamformer can thus be interpreted as a beamformer

$$\mathbf{w}_{\mathrm{f,MVDR}}(\omega) = \Lambda_{\mathbf{v}_{\mathrm{d}}}(\omega)\mathbf{S}_{\mathbf{nn}}^{-1}(\omega)\mathbf{v}(\mathbf{k}_{\mathrm{d}}, \omega) \tag{4.112}$$

followed by the spectral weighting function $S_{dd}(\omega)/(S_{dd}(\omega)+\Lambda_{\mathbf{v}_{\mathrm{d}}}(\omega))$.[9] It can be easily verified that $\mathbf{w}_{\mathrm{f,MVDR}}(\omega)$ provides unity gain for the desired signal with steering vector $\mathbf{v}(\mathbf{k}_{\mathrm{d}},\omega)$, i.e., $\mathbf{w}_{\mathrm{f,MVDR}}^{H}(\omega)\mathbf{v}(\mathbf{k}_{\mathrm{d}},\omega) = 1$, by introducing (4.111) into (4.112). The term $\Lambda_{\mathbf{v}_{\mathrm{d}}}(\omega)$ is equivalent to the PSD of interference-plus-noise at the output of $\mathbf{w}_{\mathrm{f,MVDR}}(\omega)$, which is obtained by calculating the PSD of the output signal of $\mathbf{w}_{\mathrm{f,MVDR}}(\omega)$ for $\mathbf{S}_{\mathbf{xx}}(\omega) := \mathbf{S}_{\mathbf{nn}}(\omega)$ [EFK67, SBM01]. $\mathit{SINR}_{\mathrm{f,out}}(\omega)$ at the output of $\mathbf{w}_{\mathrm{f,MVDR}}(\omega)$ is thus obtained with (4.70) and with $\mathbf{S}_{\mathbf{dd}}(\omega) = S_{dd}(\omega)\mathbf{v}(\mathbf{k}_{\mathrm{d}},\omega)\mathbf{v}^{H}(\mathbf{k}_{\mathrm{d}},\omega)$ as

$$\mathit{SINR}_{\mathrm{f,out}}(\omega) = \frac{\mathbf{w}_{\mathrm{f,MVDR}}^{H}(\omega)\mathbf{S}_{\mathbf{dd}}(\omega)\mathbf{w}_{\mathrm{f,MVDR}}(\omega)}{\mathbf{w}_{\mathrm{f,MVDR}}^{H}(\omega)\mathbf{S}_{\mathbf{nn}}(\omega)\mathbf{w}_{\mathrm{f,MVDR}}(\omega)} = \frac{S_{dd}(\omega)}{\Lambda_{\mathbf{v}_{\mathrm{d}}}(\omega)} . \tag{4.113}$$

We obtain for (4.110)

$$\mathbf{w}_{\mathrm{f,MMSE}}(\omega) = \frac{1}{1+\mathit{SINR}_{\mathrm{f,out}}^{-1}(\omega)}\mathbf{w}_{\mathrm{f,MVDR}}(\omega) . \tag{4.114}$$

The MMSE processor after (4.114) is depicted in Fig. 4.12. The MMSE processor thus does not provide unity gain for the desired signal if the MVDR beamformer does not completely cancel interference-plus-noise ($\mathit{SINR}_{\mathrm{f,out}}(\omega) \to \infty$). With decreasing $\mathit{SINR}_{\mathrm{f,out}}(\omega)$, both the distortion of the desired signal and the interference rejection increase. The spectral weighting can be identified with a single-channel Wiener filter [LO79, Lim83]. It is often referred to as post-filter.

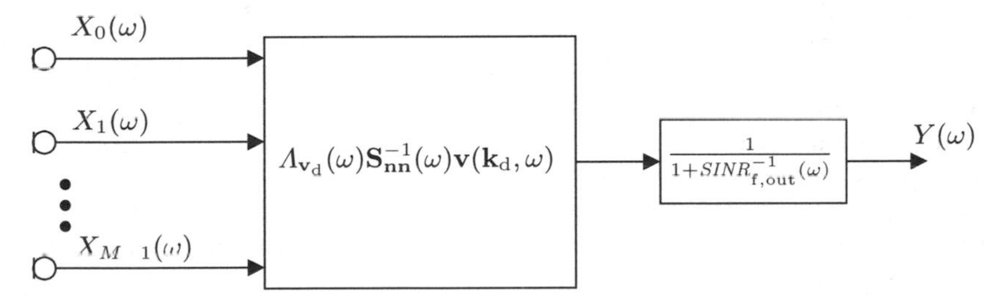

Fig. 4.12. Interpretation of the MMSE beamformer for monochromatic waves as a beamformer cascaded with a post-filter (after [Tre02, Chap. 6])

Figure 4.13a shows an example of a power pattern of a wideband MMSE beamformer. The desired signal arrives from broadside, $L = 14$ mutually uncorrelated and temporally white interferers arrive from directions $\theta_{\mathrm{n}}/\pi = 1/8 + l/36$, $l = 0, 1, \ldots, L-1$. Spatially and temporally white sensor noise is present. The PSDs $S_{dd}(\omega)$ and $S_{n_{\mathrm{c}}n_{\mathrm{c}}}(\omega)$, $S_{n_{\mathrm{w}}n_{\mathrm{w}}}(\omega)$ of the desired signal and of

[9] The weight vector $\mathbf{w}_{\mathrm{f,MVDR}}(\omega)$ is known as the minimum-variance distortionless response (MVDR) beamformer, which will be further discussed in Sect. 4.4.2 in the context of LCMV beamforming.

the interference, respectively, are depicted in Fig. 4.13b. The power pattern in Fig. 4.13a shows that the interference rejection is greater for directions with interference than for directions without interference presence. However, the interferers cannot be completely cancelled since the number of spatial degrees of freedom of the sensor array is not sufficient for placing spatial nulls in all directions where interference is present. In the frequency range $f \in [1;2]$ kHz, the suppression of the interference is greater than for other frequencies. Here, the value of the PSD $S_{dd}(\omega)$ of the desired signal is 25 dB less than the value of the PSD $S_{n_c n_c}(\omega)$ of the interference in contrast to 5 dB for the other frequencies (Fig. 4.13b). The MMSE beamformer attenuates the desired signal in order to minimize the mean-squared error (MSE) at the beamformer output, which leads to greater interference rejection for $f \in [1;2]$ kHz than for other frequencies.

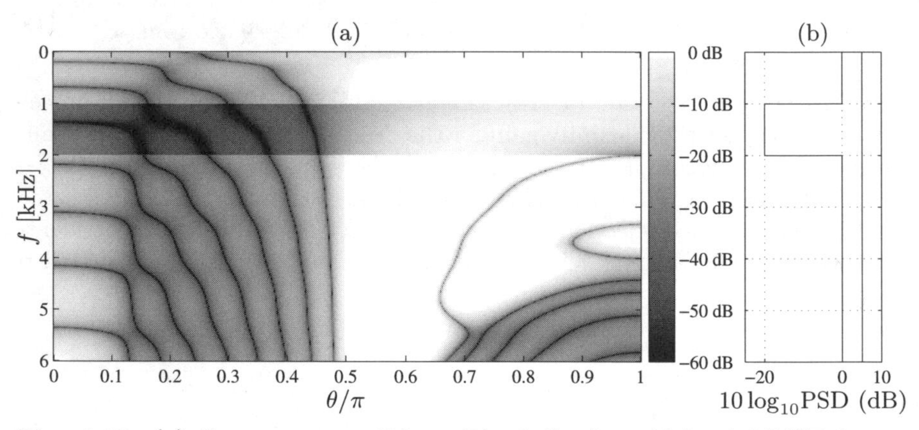

Fig. 4.13. **(a)** Power pattern $10 \log_{10} P(\omega;\theta,\phi)$ of a wideband MMSE beamformer with the desired signal arriving from broadside and with $L = 14$ mutually uncorrelated and temporally white interferers arriving from $\theta_{\mathrm{n}}/\pi = 1/8 + l/36$, $l = 0, 1, \ldots, L-1$, **(b)** PSDs $S_{dd}(\omega)$ and $S_{n_c n_c}(\omega)$ of the desired signal and of the interference, respectively; sensor array of Fig. 2.2a with $M = 9$, $d = 4$ cm; $10 \log_{10} \sigma^2_{n_{\mathrm{w}}} = -30$ dB

Assuming a spatially homogeneous wave field for interference-plus-noise and assuming matched conditions, the optimum MMSE beamformer after (4.110) can be reformulated as a function of $SINR_{\mathrm{f,in}}(\omega)$ and as a function of the spatial coherence matrix w.r.t. interference-plus-noise: According to (4.37), homogeneity allows to separate the spatio-spectral correlation matrix $\mathbf{S_{nn}}(\omega)$ into the product of the PSD of interference-plus-noise, $S_{nn}(\omega)$, and the spatial coherence matrix $\mathbf{\Gamma_{nn}}(\omega)$, i.e.,

$$\mathbf{S_{nn}}(\omega) = S_{nn}(\omega)\mathbf{\Gamma_{nn}}(\omega) .$$

$SINR_{\mathrm{f,in}}(\omega)$ after (4.69) thus simplifies with

$$\mathbf{S}_{\mathbf{dd}}(\omega) = S_{dd}(\omega)\mathbf{v}(\mathbf{k}_\mathrm{d}, \omega)\mathbf{v}^H(\mathbf{k}_\mathrm{d}, \omega)$$

to

$$SINR_{\mathrm{f,in}}(\omega) = S_{dd}(\omega)/S_{nn}(\omega)\,.$$

Substituting in (4.110) the term $\mathbf{S}_{\mathbf{nn}}^{-1}(\omega)$ by $S_{nn}^{-1}(\omega)\mathbf{\Gamma}_{\mathbf{nn}}^{-1}(\omega)$, and dividing in (4.110) the numerator and the denominator of the spectral weighting (post-filter) function by $S_{nn}(\omega)$, we obtain with $SINR_{\mathrm{f,in}}(\omega) = S_{dd}(\omega)/S_{nn}(\omega)$ the expression

$$\mathbf{w}_{\mathrm{f,MMSE}}(\omega) = \frac{SINR_{\mathrm{f,in}}(\omega)}{SINR_{\mathrm{f,in}}(\omega) + \Lambda'_{\mathbf{v}_\mathrm{d}}(\omega)} \Lambda'_{\mathbf{v}_\mathrm{d}}(\omega)\mathbf{\Gamma}_{\mathbf{nn}}^{-1}(\omega)\mathbf{v}(\mathbf{k}_\mathrm{d}, \omega)\,, \tag{4.115}$$

where

$$\Lambda'_{\mathbf{v}_\mathrm{d}}(\omega) = \left(\mathbf{v}^H(\mathbf{k}_\mathrm{d}, \omega)\mathbf{\Gamma}_{\mathbf{nn}}^{-1}(\omega)\mathbf{v}(\mathbf{k}_\mathrm{d}, \omega)\right)^{-1}\,. \tag{4.116}$$

The optimum MMSE beamformer for spatially homogeneous interference-plus-noise thus only depends on the density ratio $SINR_{\mathrm{f,in}}(\omega)$, on the DOA of the desired signal and on the spatial coherence matrix w.r.t. interference-plus-noise. This is advantageous for implementing the MMSE beamformer, as we will see later.

Comparison Between the Descriptions of the LSE and MMSE Beamformer

The derivations of the optimum LSE beamformer and of the optimum MMSE beamformer show that an analog description can be obtained for the LSE beamformer in the time domain (4.90) and for the MMSE beamformer in the DTFT domain (4.109). However, the LSE formulation explicitly takes the length N of the beamformer weight vector into account. Application of the matrix inversion lemma to (4.90) and to (4.109) leads to descriptions (4.94) and (4.110) of the optimum LSE beamformer and of the optimum MMSE beamformer, respectively, which allow to interpret the function of the LSE/MMSE beamformer. For interpretation, the wideband LSE beamformer requires a projection into the eigenspace of interference-plus-noise for diagonalizing the sample spatio-temporal correlation matrix of interference-plus-noise in (4.94). The MMSE beamformer in the DTFT domain does not require this additional transformation of (4.110). The eigenspace interpretation of the LSE beamformer is more detailed than the interpretation of the MMSE beamformer: While the eigenspace interpretation of the LSE beamformer takes into account spatial and temporal degrees of freedom of the beamformer, i.e., the number of sensors M and the length N of the beamformer weight vectors, the interpretation of the MMSE beamformer in the DTFT domain only considers spatial degrees of freedom.

Application to Audio Signal Processing

LSE/MMSE beamformers for audio signal acquisition are usually realized in the DFT domain based on (4.110) with the assumption of short-time stationary desired signals [FM01, SBM01, Mar01b, MB02]. Three methods are currently known for addressing the problem of estimating the PSD $S_{dd}(\omega)$ of the desired signal and the spatio-spectral correlation matrix $\mathbf{S_{nn}}(\omega)$ w.r.t. interference-plus-noise separately:

1. Stationary, or – relative to the desired signal – slowly time-varying interference is assumed. Then, the PSD of interference-plus-noise can be estimated during absence of the desired signal. An estimate of the PSD of the desired signal is given by the difference of the magnitude spectra between the estimated PSD of interference-plus-noise and a reference signal, which contains desired signal and interference-plus-noise. For estimating the PSD of the desired signal and/or of interference-plus-noise, methods which are known from single-channel noise reduction can be generalized to the multi-channel case (see, e.g., [EM85, Mar01a, Coh02, CB02]). LSE/MMSE beamformers using weighting functions similar to single-channel spectral subtraction are presented in [FM01, AHBK03].
2. Assumptions about the spatio-spectral correlation matrix w.r.t. interference and noise $\mathbf{S_{nn}}(k)$ at the sensors are made in order to improve the estimate of the desired signal at the output of the beamformer (4.112). This approach is often realized as a beamformer-plus-post-filter structure, similar to (4.110). The LSE/MMSE criterion is usually not fulfilled by this class of beamformers [SBM01, Mar01b, MB02].
3. A new method for estimating the SINR at the sensors and spatial coherence functions w.r.t. to the desired signal and w.r.t. interference-plus-noise at the sensors is presented in App. A and summarized in [HTK03]. It can be applied to realize beamformers as in (4.115). This approach explicitly relaxes the assumption of slowly time-varying PSD of interference-plus-noise relative to the PSD of the desired signal, and allows strong time-variance for both of them.

The discussion of LSE/MMSE beamforming has shown that the minimization of the LSE/MMSE introduces unsupervised distortion into the desired signal at the output of the beamformer depending on the properties of the acoustic wave fields. For controlling the distortion of the desired signal and for optimally adjusting the LSE/MMSE techniques to the human auditory system for human listeners, psycho-acoustic masking effects can be exploited. This allows to reduce distortion ogf the desired signal in periods or at frequencies where the distortion is noticeable. This idea was first used in single-channel noise-reduction, e.g., in [Vir95, TMK97, GJV98, Vir99, JC01b, JC02, LHA02], and is extended to sensor arrays in [RBB03].

4.4.2 Linearly-Constrained Least-Squares Error (LCLSE) and Linearly-Constrained Minimum Variance (LCMV) Design

Application of LSE/MMSE beamforming to speech and audio signal acquisition is limited, since – potentially unacceptable – distortion of the desired signal is often introduced because of inaccurate estimation of the desired signal or its PSD. However, if additional information about the desired source is available, this information can be introduced into the beamformer cost function to form a constrained optimization problem. The resulting linearly-constrained beamformer using a least-squares formulation (LCLSE) is investigated in this section.

This section is organized as follows: First, we outline the LCLSE design using the direct beamformer structure of Fig. 4.2. The relation of LCLSE beamforming with LCMV beamforming is studied. The MVDR beamformer that we encountered in Sect. 4.4.1 is shown to be a special case of LCMV beamforming. Second, we give the LCLSE/LCMV design in the generalized sidelobe canceller structure, which allows more efficient realization of LCLSE/LCMV beamformers. The generalized sidelobe canceller will be used in Chap. 5 for deriving a practical audio acquisition system.

Direct LCLSE/LCMV Design

The optimum LCLSE beamformer is derived first. Second, commonly used constraints that assure a distortionless beamformer response for the desired signal are reviewed. Third, we interpret LCLSE beamforming using a vector-space representation. For mismatched conditions, the constraints which assure a distortionless response for the desired signal are not sufficient for preventing distortion of the desired signal. For this purpose, we consider additional robustness constraints which may be introduced into the optimum weight vector. In a next step, we show the relation of wideband LCLSE beamforming with wideband LSE beamforming and with narrowband LCMV beamforming. Finally, we summarize applications of the direct LCLSE/LCMV design to audio signal processing.

Optimum LCLSE Beamformer

Since an accurate estimate of the desired signal is usually not available, it is desirable to introduce additional information about the desired source into the beamformer cost function. This transforms the unconstrained optimization problem into a constrained form as follows:

$$\xi_{\mathrm{LC}}(r) = \sum_{k=rR}^{rR+K-1} y^2(k) = \left\|\mathbf{X}^T(rR)\mathbf{w}(rR)\right\|_2^2 \rightarrow \min \qquad (4.117)$$

subject to C constraints

$$\mathbf{C}^T(rR)\mathbf{w}(rR) = \mathbf{c}(rR). \tag{4.118}$$

The matrix $\mathbf{C}(k)$ of size $MN \times CN$ is the *constraint matrix* with linearly independent columns. The vector $\mathbf{c}(k)$ of size $CN \times 1$ is the *constraint vector.*[10] The constraint matrix $\mathbf{C}(k)$ defines the directions where constraints should be put on. The constraint vector $\mathbf{c}(k)$ specifies the beamformer response for the constrained directions.

C spatial constraints require C spatial degrees of freedom of the weight vector $\mathbf{w}(k)$. Thus, only $M-C$ spatial degrees of freedom are available for minimization of $\xi_{\mathrm{LC}}(r)$. The optimum LCLSE beamformer ('Frost beamformer') is found by minimization of the constrained cost function using Lagrange multipliers [Fro72, JD93] as

$$\mathbf{w}_{\mathrm{LC}}(rR) = \mathbf{\Phi}_{\mathbf{xx}}^{-1}(rR)\mathbf{C}(rR)\left(\mathbf{C}^T(rR)\mathbf{\Phi}_{\mathbf{xx}}^{-1}(rR)\mathbf{C}(rR)\right)^{-1}\mathbf{c}(rR), \tag{4.119}$$

where $\mathbf{\Phi}_{\mathbf{xx}}(k)$ is defined in (3.18).

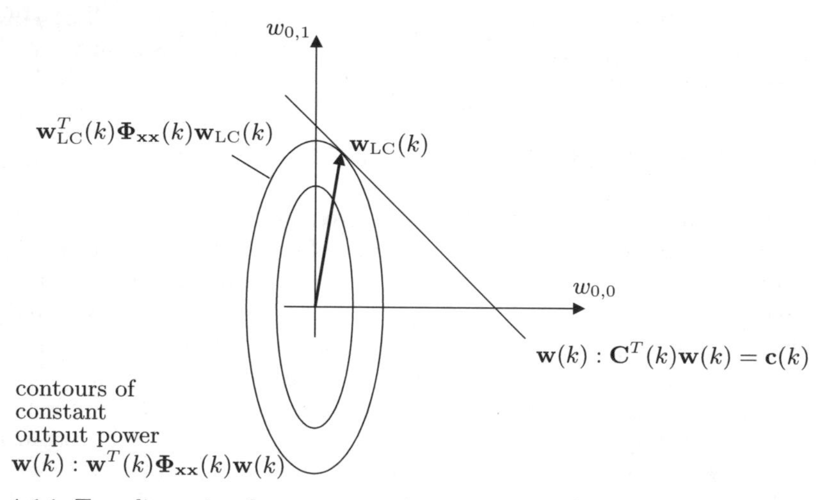

Fig. 4.14. Two-dimensional vector-space interpretation of the LCLSE after [Fro72, GJ82, Buc86] for $M = 2$, $N = 1$, i.e., $\mathbf{w}(k) = (w_{0,0}(k),\, w_{0,1}(k))^T$

Multiplication of (4.119) with $\mathbf{C}^T(rR)$ from the left shows that the constraints (4.118) are fulfilled. The optimum processor $\mathbf{w}_{\mathrm{LC}}(rR)$ yields the minimum of $\xi_{\mathrm{LC}}(r)$:

$$\xi_{\mathrm{LC,o}}(r) = \mathbf{w}_{\mathrm{LC}}^T(rR)\mathbf{\Phi}_{\mathbf{xx}}(rR)\mathbf{w}_{\mathrm{LC}}(rR) \tag{4.120}$$

$$= \mathbf{c}^T(rR)\left(\mathbf{C}^T(rR)\mathbf{\Phi}_{\mathbf{xx}}^{-1}(rR)\mathbf{C}(rR)\right)^{-1}\mathbf{c}(rR). \tag{4.121}$$

[10] In the constrained optimization problem (4.117), (4.118), the constraints (4.118) are only evaluated once every R samples and not for each sample k. It must thus be assumed that the constraints for times rR meet the requirements for the entire r-th block of input data $\mathbf{X}(rR)$.

The LCLSE beamformer may be illustrated in the vector space geometrically according to Fig. 4.14 [Fro72, GJ82, Buc86]: The constraint equation (4.118) defines an CN-dimensional constraint plane $\mathbf{w}(k) : \mathbf{C}^T(k)\mathbf{w}(k) = \mathbf{c}(k)$. The optimum weight vector $\mathbf{w}_{\mathrm{LC}}(rR)$, which fulfills the constraints, ends on this plane. The optimum weight vector $\mathbf{w}_{\mathrm{LC}}(rR)$ is not necessarily orthogonal to the constraint plane. The quadratic form $\mathbf{w}(k) : \mathbf{w}^T(k)\mathbf{\Phi}_{\mathbf{xx}}(k)\mathbf{w}(k)$ with positive definite $\mathbf{\Phi}_{\mathbf{xx}}(k)$ may be interpreted as an $MN-1$-dimensional ellipsoid and represents according to (4.120) the contours of constant energy $\xi_{\mathrm{LC}}(r)$. The minimum of the cost function $\xi_{\mathrm{LC}}(r)$ is given by the ellipsoid, which is tangent to the constraint plane.

Spatial Constraints

Generally, the constraints are designed to assure a distortionless response for the desired signal. Beamformer constraints can be formulated in various ways:

1. *Directional constraints*, also referred to as distortionless response criteria, require knowledge about the true position of the desired source. If several desired sources are present, multiple directional constraints may be used. Directional constraints often lead to cancellation of the desired signal if the source position is not known exactly [Zah72, Cox73a, Vur77, Vur79, Com80, Com82].
2. For a better robustness against such look-direction errors, *derivative constraints* are often used. Thereby, the derivatives up to a given order of the beampattern $B(\omega;\theta,\phi)$ w.r.t. (θ, ϕ) for the array steering direction must be zero. Derivative constraints thus increase the angular range of the directional constraints. The number of derivatives trades maximum allowable uncertainty of the position of the desired sources against the number of spatial degrees of freedom for suppression of interferers. For sufficient suppression of interferers, this technique is typically limited to small orders [Cox73a, AC76, Vur77, Vur79, Ste83, EC83, BG86, Tse92, TG92, TCL95, ZT02].
3. A greater flexibility is offered by *eigenvector constraints*. They influence the beamformer response over a specified region in the wavenumber-frequency space while minimizing the necessary number of degrees of freedom. The number of degrees of freedom for shaping the desired response is controlled by selecting a specified set of eigenvectors of the constrained wavenumber-frequency space for representing the desired response [EC85, Buc87].
4. By *quiescent pattern constraints*, a desired power pattern ('quiescent pattern') is specified over the entire wavenumber-frequency space. The cost function $\xi_{\mathrm{LC}}(r)$ is minimized by simultaneous approximation of the quiescent pattern in a least-squares sense [AC76, GB87, Vee90, Tse92, SE96].

In Fig. 4.15, the magnitudes of the beamformer responses $G(\mathbf{k},\omega)$ for an LCLSE beamformer with directional constraints, 2-nd-order derivative constraints, and 4-th-order derivative constraints are compared for varying DOA

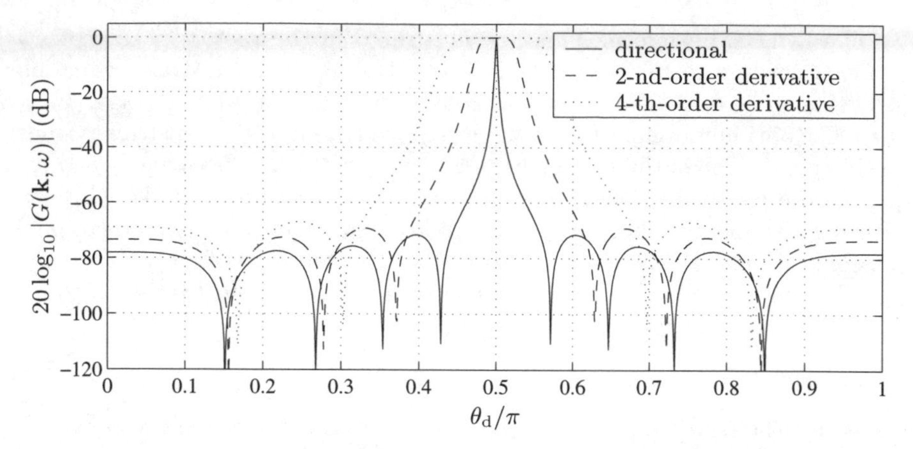

Fig. 4.15. Squared magnitude $20\log_{10}|G(\mathbf{k},\omega)|$ of the beamformer response for an LCLSE beamformer with directional constraint (solid line), 2-nd-order derivative constraints (dashed line), and 4-th-order derivative constraints (dotted line), with the position of the desired source varying in $\theta_\mathrm{d} \in [0, \pi]$ with presence of sensor noise; SNR= 20 dB, sensor array of Fig. 2.2a with $M = 9$, $d = \lambda/2$

θ_d of the desired source. The directional constraint assures a distortionless response for broadside direction. The 2-nd-order derivative constraints and the 4-th-order derivative constraints additionally set the derivatives of the beampattern w.r.t. θ up to 2-nd-order and up to 4-th-order equal to zero, respectively. The desired source is assumed to be a monochromatic plane wave with wavelength λ and with the DOA varying in $\theta_\mathrm{d} \in [0, \pi]$. Sensor noise with SNR=20 dB is present. The sensor array of Fig. 2.2a with $M = 9$ and $d = \lambda/2$ is used. For each θ_d a new weight vector $\mathbf{w}_\mathrm{LC}(k)$ is determined. The beamformer responses are normalized to the beamformer response for broadside direction. For the directional constraint, even small deviations from $\theta_\mathrm{d} = \pi/2$ lead to significant cancellation of the desired signal. With increasing order of the derivative constraints, on the one hand, cancellation of the desired signal can be better prevented for DOAs around $\theta_\mathrm{d} = \pi/2$. On the other hand, suppression of undesired signals arriving from other directions is reduced due to the reduced number of degrees of freedom and due to the widening of the pattern around $\theta = \pi/2$.

Note that for realistic application of LCLSE/LCMV beamforming, not only mismatch of the steering direction of the array but also mismatch of the positions and of the complex gain of the sensors has to be considered for the design of the constraints corresponding to the design of data-independent beamformers (Sect. 4.3). Otherwise, as for mismatched steering direction, the desired signal might be seriously distorted at the output of the beamformer.

Eigenspace Interpretation

For illustration, the LCLSE beamformer after (4.119) may be interpreted using the eigenvalue decomposition of $\mathbf{\Phi}_{\mathbf{xx}}(k)$. Projection of the LCLSE optimization criterion into the eigenspace of $\mathbf{\Phi}_{\mathbf{xx}}^{-1}(k)$ using the eigenvalue decomposition after (3.23) yields with

$$\widetilde{\mathbf{C}}(k) = \mathbf{\Sigma}_{\mathbf{X}}^{-1}(k)\mathbf{U}_{\mathbf{X}}^{T}(k)\mathbf{C}(k)\,, \tag{4.122}$$

$$\widetilde{\mathbf{w}}(k) = \mathbf{\Sigma}_{\mathbf{X}}^{-1}(k)\mathbf{U}_{\mathbf{X}}^{T}(k)\mathbf{w}(k)\,, \tag{4.123}$$

$$\widetilde{\mathbf{x}}(k) = \mathbf{\Sigma}_{\mathbf{X}}^{-1}(k)\mathbf{U}_{\mathbf{X}}^{T}(k)\mathbf{x}(k)\,, \tag{4.124}$$

the optimum LCLSE weight vector

$$\widetilde{\mathbf{w}}_{\mathrm{LC}}(rR) = \widetilde{\mathbf{C}}(rR)\left(\widetilde{\mathbf{C}}^{T}(rR)\widetilde{\mathbf{C}}(rR)\right)^{-1}\mathbf{c}(rR)\,. \tag{4.125}$$

Using the eigenspace description (4.124) and (4.125), we can write the beamformer output signal $y(k) = \mathbf{w}_{\mathrm{LC}}^{T}(k)\mathbf{x}(k)$ as:

$$y(k) = \mathbf{c}^{T}(k)\left(\widetilde{\mathbf{C}}^{T}(k)\widetilde{\mathbf{C}}(k)\right)^{-1}\widetilde{\mathbf{C}}^{T}(k)\widetilde{\mathbf{x}}(k)\,. \tag{4.126}$$

Equation (4.122) may be interpreted as a projection of the constraint matrix $\mathbf{C}(k)$ into the eigenspace of the spatio-temporal correlation matrix of the sensor data. Constraints which are in the null space of $\mathbf{X}(k)$ cancel out. Accordingly, (4.123) and (4.124) project the weight vector $\mathbf{w}(k)$ and the sensor data vector $\mathbf{x}(k)$ into the eigenspace of $\mathbf{\Phi}_{\mathbf{xx}}^{-1}(k)$, respectively. We obtain the beamformer formulation in the eigenspace of $\mathbf{\Phi}_{\mathbf{xx}}^{-1}(k)$ after (4.126): The matrix-vector product $\widetilde{\mathbf{C}}^{T}(k)\widetilde{\mathbf{x}}(k)$ projects the sensor data onto the constraints, and, thus, cancels signal components which are not contained in the constrained subspace ('interference suppression'). Sensor data, which is contained in the constrained subspace is normalized by the matrix product $\widetilde{\mathbf{C}}^{T}(k)\widetilde{\mathbf{C}}(k)$ and weighted by the vector $\mathbf{c}(k)$ such that the constraints (4.118) are fulfilled.

This eigenspace interpretation is illustrated geometrically in Fig. 4.16. The optimum LCLSE weight vector $\widetilde{\mathbf{w}}_{\mathrm{LC}}(rR)$ in the eigenspace of the sensor signals fulfills the constraint equation

$$\widetilde{\mathbf{C}}^{T}(rR)\widetilde{\mathbf{w}}(k) = \mathbf{c}(rR)\,. \tag{4.127}$$

Equivalently, the constraint equation can be interpreted as an underdetermined system of linear equations (4.127) with $\widetilde{\mathbf{w}}_{\mathrm{LC}}(rR)$ being the solution with the smallest L_2-norm. (See (3.27) in Sect. 3.1.2.) Since the weight vector $\widetilde{\mathbf{w}}_{\mathrm{LC}}(rR)$ fulfills the constraint equation (4.127), the weight vector $\widetilde{\mathbf{w}}_{\mathrm{LC}}(rR)$ ends on the constraint plane (4.127). The weight vector $\widetilde{\mathbf{w}}_{\mathrm{LC}}(rR)$ is normal to the constraint plane (4.127), since $\widetilde{\mathbf{w}}_{\mathrm{LC}}(rR)$ is the solution of (4.127) with the smallest L_2-norm [Fle65, Chap. 4][Fro72]. Finally, when determining the contours of constant output power $\xi_{\mathrm{LC}}(rR)$ by introducing (4.126) into (4.120), we find that the contours of constant output power transform to circles due to the diagonalization of the sample spatio-temporal correlation matrix $\mathbf{\Phi}_{\mathbf{xx}}(k)$. That is, $\widetilde{\mathbf{w}}(k) : \|\widetilde{\mathbf{w}}(k)\|_2^2$.

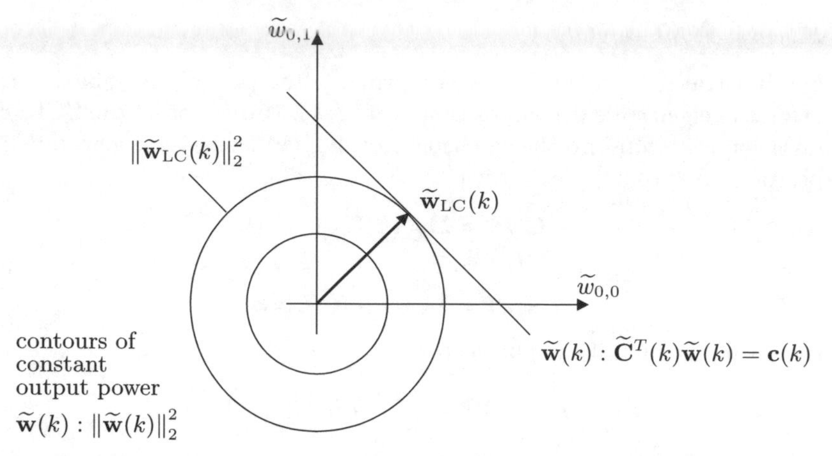

Fig. 4.16. Eigenspace interpretation of the LCLSE beamformer for $M = 2$, $N = 1$, i.e., $\widetilde{\mathbf{w}}(k) = (\widetilde{w}_{0,0}(k), \widetilde{w}_{0,1}(k))^T$

Robustness Improvement

Robust beamformers are beamformers whose performance degrades only smoothly in the presence of mismatched conditions. Typical problems are mismatched distortionless response constraints where the position of the desired source and the array look-direction do not match, and array imperfections like random errors in position, and amplitude and phase differences of sensors. (See, e.g., [God86, Jab86a, YU94].) For improving robustness of optimum beamformers, two techniques were developed which are both useful for unreliable spatial constraints and array perturbations:

1. *Diagonal loading* increases non-coherent sensor signal components relative to coherent signal components by augmenting $\mathbf{\Phi}_{\mathbf{xx}}(k)$ by a diagonal matrix $\sigma_{\mathrm{dl}}^2(k)\mathbf{I}_{MN\times MN}$,

$$\mathbf{\Phi}_{\mathbf{xx},\mathrm{dl}}(k) = \mathbf{\Phi}_{\mathbf{xx}}(k) + \sigma_{\mathrm{dl}}^2(k)\mathbf{I}_{MN\times MN} \,. \tag{4.128}$$

   It can be shown that this puts an upper limit $T_0(k)$ to the array sensitivity $T(k)$ against uncorrelated errors (4.63), $T(k) = 2\pi/T_{\mathrm{s}}\mathbf{w}^T(k)\mathbf{w}(k) \leq T_0(k)$, where $T_0(k)$ depends on the parameter $\sigma_{\mathrm{dl}}^2(k)$ [GM55, US56, CZK86, CZK87].
   An example for the influence of diagonal loading on the beamformer response $G(\mathbf{k}, \omega)$ is depicted in Fig. 4.17. Figure 4.17 shows $|G(\mathbf{k}, \omega)|^2$ on a logarithmic scale for directional constraints, 2-nd-order and 4-th-order derivative constraints for $M = 9$ and $d = \lambda/2$ for varying DOA θ_{d} with fixed broadside steering. Compared to Fig. 4.15, it may be seen that the diagonal loading reduces the suppression of the desired signal if the steering direction ($\theta = \pi/2$) does not match the DOA of the desired source ($|\theta_{\mathrm{d}} - \pi/2| \approx 0$), which improves robustness against cancellation of the

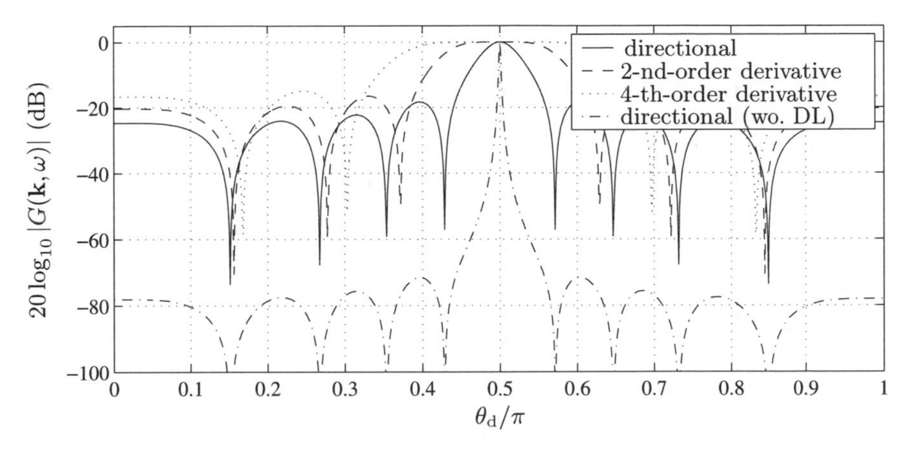

Fig. 4.17. Squared magnitude $20\log_{10}|G(\mathbf{k},\omega)|$ of the beamformer response for an LCLSE beamformer with diagonal loading (DL) for directional and for derivative constraints; $M=9$, $d=\lambda/2$, $\sigma_{\mathrm{dl}}^2=10$

desired signal. However, the interference rejection ($|\theta_d - \pi/2| \ll 0$) is also reduced.

2. We can perform the beamforming in the vector space, which is spanned by interference-plus-noise [Cox73a] instead of in the vector space spanned by $\mathbf{\Phi}_{\mathbf{xx}}(rR)$. This thus replaces $\mathbf{\Phi}_{\mathbf{xx}}(rR)$ in (4.119) by $\mathbf{\Phi}_{\mathbf{nn}}(rR)$,

$$\mathbf{w}_{\mathrm{LC}}(rR) = \mathbf{\Phi}_{\mathbf{nn}}^{-1}(rR)\mathbf{C}(rR)\left(\mathbf{C}^T(rR)\mathbf{\Phi}_{\mathbf{nn}}^{-1}(rR)\mathbf{C}(rR)\right)^{-1}\mathbf{c}(rR)\,. \quad (4.129)$$

This can be understood when replacing in (4.122)–(4.124) the eigenvalue decomposition of $\mathbf{\Phi}_{\mathbf{xx}}(rR)$ by the eigenvalue decomposition of $\mathbf{\Phi}_{\mathbf{nn}}(rR)$. Essentially, this means that the desired signal has no influence on the beamformer optimization. For mismatched conditions, the beamformer does not try to eliminate parts of the desired signal that are assumed to be interference or noise, which improves robustness against cancellation of the desired signal.

An example is illustrated in Fig. 4.18. Figure 4.18 shows the squared beamformer response $|G(\mathbf{k},\omega)|^2$ for an LCLSE beamformer with $M=9$, $d=\lambda/2$ for presence of two interferers with $\theta_n/\pi = 0.1$ and $\theta_n/\pi = 0.4$ for varying θ_d with fixed broadside steering. The beamformer response is equal to the beamformer response without mismatched steering direction. The cancellation of the desired signal depends on the positions of the interferers.

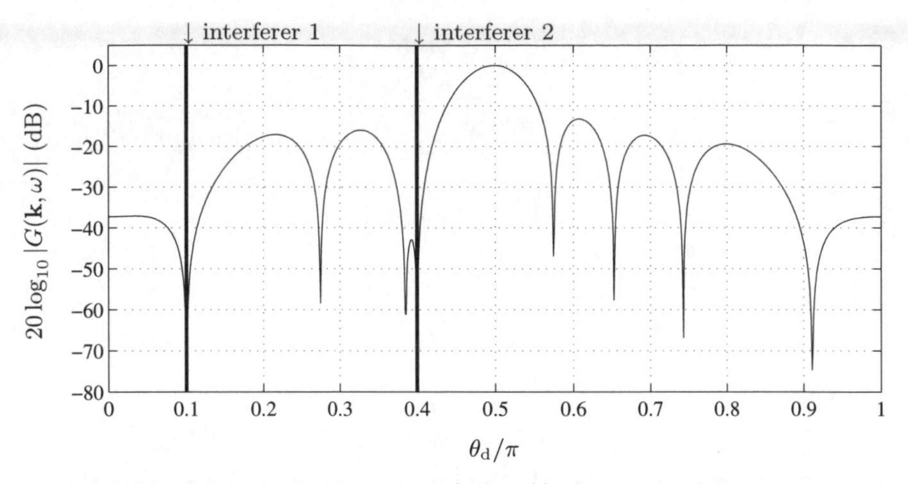

Fig. 4.18. Squared magnitude $20\log_{10}|G(\mathbf{k},\omega)|$ of the beamformer response for an LCLSE beamformer with exlusion of the desired signal from the beamformer optimization; $M = 9$, $d = \lambda/2$

Relation with Other Beamforming Techniques

Here, the relation of the LCLSE beamformer with the LSE beamformer, with the statistical LCMV beamformer, and with the minimum variance distortionless responses (MVDR) beamformer is established.

(1) Relation between LCLSE and LSE beamforming. The relation of LCLSE beamforming to LSE beamforming is established using the special constraints

$$\mathbf{D}^T(rR)\mathbf{w}(rR) = \mathbf{y}_\mathrm{r}(rR)\,, \tag{4.130}$$

and by identifying (4.130) with (4.118). For the optimum LCLSE weight vector after (4.119), we then obtain with (4.95)

$$\mathbf{w}_\mathrm{LC}(rR) = \mathbf{\Phi}_{\mathbf{xx}}^{-1}(rR)\mathbf{X}_\mathbf{d}^T(rR)\mathbf{\Lambda}_\mathbf{D}^{-1}(rR)\mathbf{y}_\mathrm{r}(rR)\,. \tag{4.131}$$

The robustness-improved version of (4.131),

$$\mathbf{w}_\mathrm{LC}(rR) = \mathbf{\Phi}_{\mathbf{nn}}^{-1}(rR)\mathbf{X}_\mathbf{d}^T(rR)\mathbf{\Lambda}_\mathbf{D}^{-1}(rR)\mathbf{y}_\mathrm{r}(rR)\,, \tag{4.132}$$

is equivalent to the optimum LSE beamformer (4.94) except for the inverse of $\mathbf{\Lambda}_\mathbf{D}(rR)$, which is not augmented by the identity matrix here. The augmentation by the identity matrix results in distortion of the desired signal, which depends on the interference characteristics. (See eigenspace interpretation of the LSE beamformer in Sect. 4.4.1.)

(2) Relation between LCLSE and LCMV beamforming. We replace in the cost function $\xi_{\mathrm{LC}}(r)$ of the LCLSE beamformer after (4.117) the temporal averaging by statistical expectation:

$$\xi_{\mathrm{LCMV}}(k) = E\{y^2(k)\} = E\left\{\left(\mathbf{w}^T(k)\mathbf{x}(k)\right)^2\right\} . \tag{4.133}$$

Minimization of $\xi_{\mathrm{LCMV}}(k)$ subject to the constraints (4.118) using Lagrange multipliers (see, e.g., [Tre02, Chap. 6]), the assumption of WSS sensor signals and time-invariant constraints, and transformation into the DTFT domain yields the optimum LCMV beamformer in the DTFT domain as follows:

$$\mathbf{w}_{\mathrm{f,LCMV}}(\omega) = \mathbf{S}_{\mathbf{xx}}^{-1}(\omega)\mathbf{C}_{\mathrm{f}}(\omega)\left(\mathbf{C}_{\mathrm{f}}^H(\omega)\mathbf{S}_{\mathbf{xx}}^{-1}(\omega)\mathbf{C}_{\mathrm{f}}(\omega)\right)^{-1}\mathbf{c}_{\mathrm{f}}(\omega) . \tag{4.134}$$

Equation (4.134) fulfills the constraint equation

$$\mathbf{C}_{\mathrm{f}}^H(\omega)\mathbf{w}_{\mathrm{f}}(\omega) = \mathbf{c}_{\mathrm{f}}(\omega) , \tag{4.135}$$

where $\mathbf{C}_{\mathrm{f}}(\omega)$ and $\mathbf{c}_{\mathrm{f}}(\omega)$ are the constraint matrix of size $M \times C$ and the constraint vector of size $C \times 1$ in the DTFT domain, respectively.

(3) Relation between LCLSE and MVDR beamforming. The statistical version of the special LCMV beamformer given by (4.131), is referred to as the *minimum variance distortionless response* beamformer. The MVDR processor in the DTFT domain is obtained by assuming a desired signal with steering vector $\mathbf{v}(\mathbf{k}_{\mathrm{d}}, \omega)$ and by choosing the constraints

$$\mathbf{C}_{\mathrm{f}}(\omega) = \mathbf{v}(\mathbf{k}_{\mathrm{d}}, \omega) , \tag{4.136}$$

$$\mathbf{c}_{\mathrm{f}}(\omega) = 1 . \tag{4.137}$$

The constraint matrix thus selects the DOA of the desired source signal, and the constraint vector sets the beamformer response for this direction equal to 1. Using (4.136) and (4.137) in (4.134), we obtain the MVDR beamformer in the DTFT domain as

$$\mathbf{w}_{\mathrm{f,MVDR}}(\omega) = \Lambda_{\mathbf{v}_{\mathrm{d}}}(\omega)\mathbf{S}_{\mathbf{xx}}^{-1}(\omega)\mathbf{v}(\mathbf{k}_{\mathrm{d}}, \omega) , \tag{4.138}$$

which we already encountered in the context of MMSE beamforming in (4.112) in a robustness-improved version. In Fig. 4.20, a power pattern of an MVDR beamformer for wideband signals with wideband interference arriving from $\theta_{\mathrm{n}} = 0.3\pi$ is illustrated.

Application to Audio Signal Processing

In this section, we have illustrated (a) the basic concepts of constrained optimum beamforming and (b) the relation of constrained optimum beamforming to LSE/MMSE beamforming. Separate estimates of sample spatio-temporal correlation matrices for desired signal and for interference-plus-noise

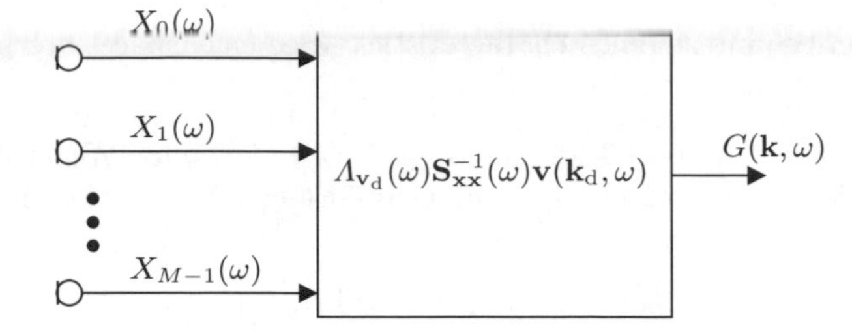

Fig. 4.19. Minimum variance distortionless response (MVDR) beamformer

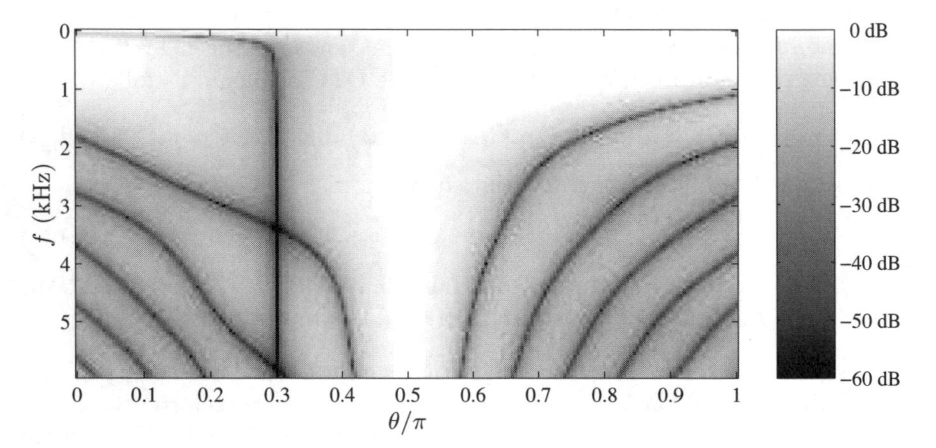

Fig. 4.20. Power pattern $10\log_{10} P(\omega;\theta,\phi)$ of a wideband MVDR beamformer with interference arriving from $\theta_\mathrm{n} = 0.3\pi$; sensor array of Fig. 2.2a with $M = 9$, $d = 4\,\mathrm{cm}$

– as required for LSE/MMSE processors – are usually difficult to obtain. In LCLSE/LCMV beamforming, these separate estimates are therefore replaced by the sample spatio-temporal correlation matrix of the sensor signals in combination with constraints for preserving a distortionless response for the desired signal. This is especially important for non-stationary source signals, such as speech and audio signals, since separate estimates of sample spatio-temporal correlation matrices for desired signal and for interference-plus-noise are generally difficult to obtain.

For audio signal processing, positions of the desired source are generally fluctuating. Therefore, we introduced constraints which explicitly depend on time k. For additional robustness against mismatched conditions and robustness against distortion of the desired signal, a combination of three methods is possible: (a) widening the spatial constraints, (b) diagonal loading of the sample spatio-temporal correlation matrix $\mathbf{\Phi}_{\mathbf{xx}}(k)$, and (c) by excluding the desired signal from the optimization of the LCLSE beamformer by replacing $\mathbf{\Phi}_{\mathbf{xx}}(k)$ by $\mathbf{\Phi}_{\mathbf{nn}}(k)$ in the optimum LCLSE processor. In this case, of course, it must be possible to estimate $\mathbf{\Phi}_{\mathbf{nn}}(k)$.

LCLSE/LCMV beamforming has been extensively studied for audio signal processing for general applications in [BS01], for hearing aids in [GZ01], and for speech recognition in [OMS01, MS01]. For array apertures, which are much smaller than the signal wavelength, LCLSE/LCMV beamformers can be realized as differential ('super-directive') arrays [Elk96, Elk00].

Generalized Sidelobe Canceller (GSC)

An efficient realization of the LCLSE/LCMV beamformer is the *generalized sidelobe canceller* [GJ82] in Fig. 4.21. The GSC is especially advantageous for adaptive realizations of LCLSE/LCMV beamformers, since the constrained optimization problem is transformed into an unconstrained one. In the following, we first derive the GSC in the time domain. Second, we introduce the GSC in the DTFT domain.

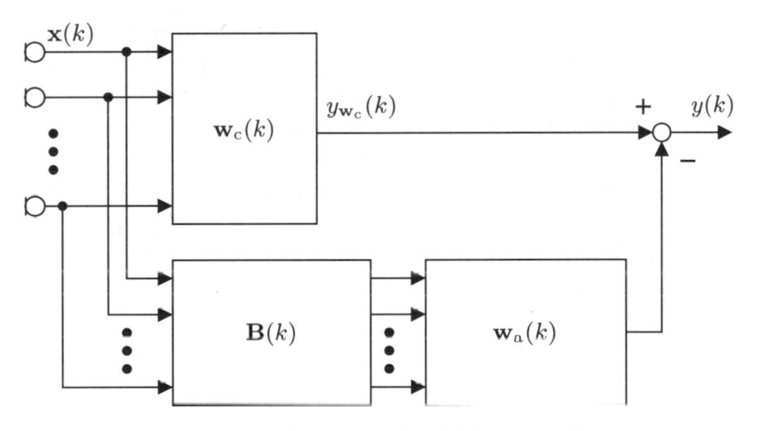

Fig. 4.21. Generalized sidelobe canceller

The LCLSE Beamformer in GSC Structure

The GSC splits the LCLSE beamformer into two orthogonal subspaces: The first subspace satisfies the constraints, and, thus, ideally contains undistorted desired signal and filtered interference. The first subspace is given by the vector space of the upper processor in Fig. 4.21. This upper signal path is referred to as *reference path.* The $MN \times 1$ weight vector of the reference path is defined as

$$\mathbf{w}_c(k) = (\mathbf{w}_{c,0}(k), \mathbf{w}_{c,1}(k), \ldots, \mathbf{w}_{c,M-1}(k))^T . \tag{4.139}$$

The output signal of the weight vector $\mathbf{w}_c(k)$ is defined analogously to (4.42) as

$$y_{\mathbf{w}_c}(k) = \mathbf{w}_c^T(k)\mathbf{x}(k) . \tag{4.140}$$

The vector $\mathbf{w}_c(rR)$ is often referred to as *quiescent weight vector*. For fixed quiescent weight vectors $\mathbf{w}_c =: \mathbf{w}_c(k)$ the dependency on the discrete time k is dropped. Equation (4.139) has to fulfill the constraint equation (4.118) in order to provide an undistorted desired signal at the output of the GSC. Therefore, the weight vector $\mathbf{w}_c(k)$ is chosen such that the constraint equation (4.118) is fulfilled:

$$\mathbf{C}^T(rR)\mathbf{w}_c(rR) \stackrel{!}{=} \mathbf{c}(rR)\,. \tag{4.141}$$

The vector $\mathbf{w}_c(rR)$ is obtained by projecting the optimum LCLSE weight vector $\mathbf{w}_{\mathrm{LC}}(rR)$ orthogonally onto the constraint matrix $\mathbf{C}(rR)$:

$$\begin{aligned}\mathbf{w}_c(rR) &= \mathbf{C}(rR)\left(\mathbf{C}^T(rR)\mathbf{C}(rR)\right)^{-1}\mathbf{C}^T(rR)\mathbf{w}_{\mathrm{LC}}(rR)\\ &= \mathbf{C}(rR)\left(\mathbf{C}^T(rR)\mathbf{C}(rR)\right)^{-1}\mathbf{c}(rR)\,.\end{aligned} \tag{4.142}$$

Note that this projection onto the constraint matrix $\mathbf{C}(rR)$ is equivalent to solving the underdetermined system of linear equations (4.141). The matrix $\mathbf{C}(rR)\left(\mathbf{C}^T(rR)\mathbf{C}(rR)\right)^{-1}$ in (4.142) is equivalent to the pseudoinverse of the constraint matrix $\mathbf{C}^T(k)$ for an underdetermined systems of linear equations. (See (3.27).) If the quiescent weight vector $\mathbf{w}_c(rR)$ fulfills (4.142), $\mathbf{w}_c(rR)$ is the shortest vector which ends on the constraint plane, since the solution (4.142) to the underdetermined system of linear equations (4.141) is according to Sect. 3.1.2 the solution with the smallest L_2-norm. The quiescent weight vector $\mathbf{w}_c(rR)$ is thus normal to the constraint plane $\mathbf{C}^T(rR)\mathbf{w}(rR) = \mathbf{c}(rR)$ [Fle65, Chap. 4][Fro72]. (See Fig. 4.22.)

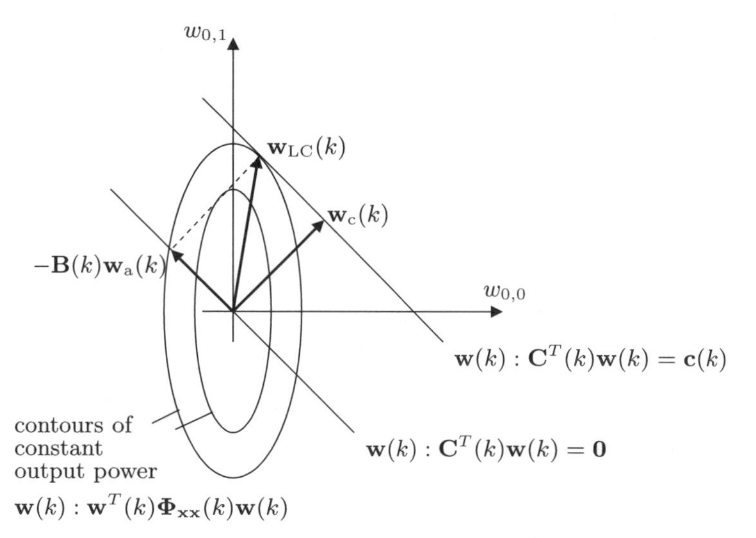

Fig. 4.22. Vector-space interpretation of the GSC after [Fro72, GJ82, Buc86] for $M = 2$, $N = 1$, i.e., $\mathbf{w}(k) = (w_{0,0}(k), w_{0,1}(k))^T$

The second subspace is given by the lower signal path of Fig. 4.21. It is often called *sidelobe cancelling path.* This second subspace is orthogonal to $\mathbf{w}_c(k)$. Orthogonality is assured by an $MN \times BN$ matrix $\mathbf{B}(k)$, which is orthogonal to each column of $\mathbf{C}(k)$:

$$\mathbf{C}^T(k)\mathbf{B}(k) = \mathbf{0}\,. \tag{4.143}$$

Equation (4.143) is the homogeneous form of the constraint equation (4.118). The matrix $\mathbf{B}(k)$ is called *blocking matrix* (BM), since signals which are orthogonal to $\mathbf{B}(k)$ (or, equivalently, in the vector space of the constraints) are rejected. Ideally, the output of $\mathbf{B}(k)$ does not contain desired signal components, and, thus, is a reference for interference-plus-noise. The value B is equivalent to the number of output signals of the blocking matrix. Generally, it is assumed that the columns of the blocking matrix $\mathbf{B}(k)$ are linearly independent (e.g. [GJ82]). However, in our context, we do not necessarily assume orthogonality of the columns of $\mathbf{B}(k)$ for generality. This implies that $B \geq M - C$, with equality for orthogonality of the columns of $\mathbf{B}(k)$. Note that the number of spatial degrees of freedom after the blocking matrix is equal to $M - C$. This will be used in Chap. 5.

The remaining $(M-C)N$ degrees of freedom of the blocking matrix output are used to minimize the squared GSC output signal $y(k)$. With the tap-stacked weight vector

$$\mathbf{w}_a(k) = \left(\mathbf{w}_{a,0}^T(k),\, \mathbf{w}_{a,1}^T(k),\, \dots,\, \mathbf{w}_{a,B-1}^T(k)\right)^T\,, \tag{4.144}$$

where the B filters $\mathbf{w}_{a,m}(k)$ of length N are given by

$$\mathbf{w}_{a,m}(k) = \left(w_{a,0,m}(k),\, w_{a,1,m}(k),\, \dots,\, w_{a,N-1,m}(k)\right)^T\,, \tag{4.145}$$

we write

$$\begin{aligned} \xi_{LC}(r) &= \sum_{k=rR}^{rR+K-1} y^2(k) \\ &= \left\| \left(\mathbf{w}_c^T(rR) - \mathbf{w}_a^T(rR)\mathbf{B}^T(rR)\right)\mathbf{X}(rR)\right\|_2^2 \to \min\,. \end{aligned} \tag{4.146}$$

For simplicity, the same number of filter taps N is assumed for the FIR filters in $\mathbf{w}_c(k)$ and in $\mathbf{w}_a(k)$. The constrained minimization problem (4.117) is thus transformed to an unconstrained one, since the weight vector $\mathbf{w}_a(k)$ is optimized by unconstrained minimization of (4.146). The constraints (4.118) are fulfilled by the quiescent weight vector $\mathbf{w}_c(k)$ after (4.142) and by the blocking matrix $\mathbf{B}(k)$ fulfilling (4.143). The optimum solution $\mathbf{w}_{a,o}(rR)$ is thus found by unconstrained LSE optimization according to Sect. 3.1.2:

$$\mathbf{w}_{a,o}(rR) = \left(\mathbf{B}^T(rR)\mathbf{\Phi}_{\mathbf{xx}}(rR)\mathbf{B}(rR)\right)^+ \mathbf{B}^T(rR)\mathbf{\Phi}_{\mathbf{xx}}(rR)\mathbf{w}_c(rR)\,, \tag{4.147}$$

where $(\cdot)^+$ is the pseudoinverse of the matrix in parentheses. Recalling (3.18), $\mathbf{\Phi}_{\mathbf{xx}}(k) = \mathbf{X}(k)\mathbf{X}^T(k)$, it may be seen that the inverted term in parentheses is the time-averaged estimate of the spatio-temporal correlation matrix of the blocking matrix output signals at time rR. Furthermore, the right term is the time-averaged estimate of the cross-correlation vector between the blocking matrix output signals $\mathbf{B}^T(k)\mathbf{x}(k)$ and the output signal of the quiescent weight vector $y_{\mathbf{w}_\mathrm{c}}(k) = \mathbf{w}_\mathrm{c}^T(k)\mathbf{x}(k)$. The weight vector $\mathbf{w}_{\mathrm{a,o}}(rR)$ thus corresponds to an LSE processor with the output of the blocking matrix as input signal and the output of the quiescent weight vector as reference signal. (See Sect. 3.1.1 and (4.90).) As the blocking matrix output does not contain desired signal components (for carefully designed constraints), it may be easily verified that the LSE-type beamformer $\mathbf{w}_{\mathrm{a,o}}(rR)$ does not distort the desired signal. Since the weight vector $\mathbf{w}_{\mathrm{a,o}}(rR)$ assures suppression of interference-plus-noise, the weight vector $\mathbf{w}_\mathrm{a}(k)$ is referred to as *interference canceller* (IC). We use the pseudoinverse of the spatio-temporal correlation matrix w.r.t. the output signals of the blocking matrix in (4.147) since we do not necessarily assume invertibility of the term in brackets. While we assume presence of spatially and temporally white sensor noise, we do not assume full rank of the blocking matrix $\mathbf{B}(k)$. From the vector space illustration of the GSC according to Fig. 4.22, we can draw the following conclusion concerning the quiescent weight vector: The weight vector $\mathbf{w}_\mathrm{c}(k)$ is optimum in the LCLSE sense for the given constraints if $\mathbf{w}_\mathrm{c}(k) = \mathbf{w}_\mathrm{LC}(k)$, i.e., if $\mathbf{w}_\mathrm{LC}(k)$ is normal to the constraint plane $\mathbf{w}(k) : \mathbf{C}^T(k)\mathbf{w}(k) = \mathbf{c}(k)$. From Fig. 4.16, we know that the optimum LCLSE weight vector $\widetilde{\mathbf{w}}_\mathrm{LC}(k)$ is normal to the constraint plane in the eigenspace of the sample spatio-temporal correlation matrix of the sensor signals, i.e., for diagonalized sample spatio-temporal correlation matrix. Therefore, $\mathbf{w}_\mathrm{LC}(k)$ is normal to the constraint plane if the sample spatio-temporal correlation matrix of the sensor signals is diagonal, which is equivalent to presence of spatially and temporally white noise only. For presence of only spatially and temporally white noise, the interference canceller $\mathbf{w}_{\mathrm{a,o}}(rR)$ is thus inefficient, i.e., $\mathbf{w}_{\mathrm{a,o}}(rR) = \mathbf{0}_{BN\times 1}$.

GSC in the DTFT Domain

The description of the GSC in the DTFT domain is obtained when writing $\mathbf{w}_\mathrm{f}(\omega)$ as

$$\mathbf{w}_\mathrm{f}(\omega) = \mathbf{w}_{\mathrm{f,c}}(\omega) - \mathbf{B}_\mathrm{f}(\omega)\mathbf{w}_{\mathrm{f,a}}(\omega)\,, \tag{4.148}$$

corresponding to (4.146). The $M \times 1$-vector $\mathbf{w}_{\mathrm{f,c}}(\omega)$ is defined according to (4.46). The vector $\mathbf{w}_{\mathrm{f,c}}(\omega)$ captures the complex conjugate transfer functions $W_{\mathrm{c},m}^*(\omega)$ of the fixed weight vectors $\mathbf{w}_{\mathrm{c},m} := \mathbf{w}_{\mathrm{c},m}(k)$ as follows:

$$\mathbf{w}_{\mathrm{f,c}}(\omega) = \left(W_{\mathrm{c},0}(\omega),\, W_{\mathrm{c},1}(\omega),\, \ldots,\, W_{\mathrm{c},M-1}(\omega)\right)^H . \tag{4.149}$$

The blocking matrix $\mathbf{B}_\mathrm{f}(\omega)$ in the DTFT domain fulfills the constraints

$$\mathbf{C}_\mathrm{f}^H(\omega)\mathbf{B}_\mathrm{f}^*(\omega) = \mathbf{0}_{C\times M} \tag{4.150}$$

in analogy to (4.143). The optimum interference canceller $\mathbf{w}_{\mathrm{f,a,o}}(\omega)$ is found by unconstrained minimization (MMSE optimization) of (4.133) with $\mathbf{w}_\mathrm{f}(\omega)$

replaced by (4.148). We obtain the optimum weight vector according to Sect. 3.1.3 as

$$\mathbf{w}_{\mathrm{f,a,o}}(\omega) = \left(\mathbf{B}_{\mathrm{f}}^{H}(\omega)\mathbf{S}_{\mathbf{xx}}(\omega)\mathbf{B}_{\mathrm{f}}(\omega)\right)^{+}\mathbf{B}_{\mathrm{f}}^{H}(\omega)\mathbf{S}_{\mathbf{xx}}(\omega)\mathbf{w}_{\mathrm{f,c}}(\omega) \tag{4.151}$$

in analogy to (4.147).

Consider for example a desired plane wave from broadside direction, i.e., $\mathbf{v}(\mathbf{k}_{\mathrm{d}},\omega) = \mathbf{1}_{M\times 1}$, and an MVDR beamformer with the constraints according to (4.136) and (4.137). The equivalent GSC realization is obtained by designing, first, a quiescent weight vector by using (4.142), which gives the uniformly weighted beamformer $\mathbf{w}_{\mathrm{f,c}}(\omega) = 1/M \cdot \mathbf{1}_{M\times 1}$. Second, a blocking matrix, which fulfills (4.150), is, e.g., given by

$$\mathbf{B}_{\mathrm{f}}^{H}(\omega) = \begin{pmatrix} 1 & -1 & 0 & \cdots & 0 \\ 0 & 1 & -1 & \cdots & 0 \\ \vdots & & \ddots & \ddots & \vdots \\ 0 & 0 & \cdots & 1 & -1 \end{pmatrix} \tag{4.152}$$

[GJ82]. The blocking matrix $\mathbf{B}_{\mathrm{f}}(\omega)$ after (4.152) cancels signal components from broadside direction, i.e., with steering vector $\mathbf{v}(\mathbf{k}_{\mathrm{d}},\omega) := \mathbf{1}_{M\times 1}$, since $\mathbf{B}_{\mathrm{f}}^{H}(\omega)\mathbf{1}_{M\times 1} = \mathbf{0}_{M-1\times 1}$. The weight vector $\mathbf{w}_{\mathrm{f,a,o}}(\omega)$ depends on the spatio-spectral correlation matrix of interference-plus-noise and is optimized for the present wave-field according to (4.151).

Equivalence of the GSC with LCLSE/LCMV beamformers is shown in [Jim77, Buc86]. Detailed studies of the performance of the GSC can be found in, e.g., [Jab86a, Jab86b, NCE92, KCY92]. The GSC structure has been applied to audio signal processing in, e.g., [Com90, NCB93, AG97, HSH99, NFB01, GBW01, HK02b, HBK03a].

4.4.3 Eigenvector Beamformers

Eigenvector beamformers reduce the optimization space of data-dependent beamformers for improving the reliability of the estimates of the spatio-temporal correlation matrix $\mathbf{\Phi}_{\mathbf{xx}}(k)$. Recall that a decreasing number of filter weights MN improves the reliability of $\mathbf{\Phi}_{\mathbf{xx}}(k)$ for a given observation interval K. (See Sect. 2.2.2 and [RMB74][Tre02, Chap. 7].) For adaptive realizations of data-dependent beamformers, the convergence speed of the adaptive weight vectors is increased due to the reduced number of dimensions of the estimation space.

The eigenspace interpretation of LCLSE beamforming suggests to concentrate on the suppression of the large, most disturbing eigenvalues of $\mathbf{\Phi}_{\mathbf{nn}}(k)$ in order to exploit the advantages of spatio-temporal correlation matrices with less dimensions. For deriving the LCLSE beamformer in the eigenspace of desired-signal-plus-interference, we assume that the V' largest eigenvalues of $\mathbf{\Phi}_{\mathbf{xx}}(k)$ belong to the subspace of desired-signal-plus-interference. The remaining $MN - V'$ small eigenvalues of $\mathbf{\Phi}_{\mathbf{xx}}(k)$ belong to the subspace with sensor noise only, which should not be taken into account by the eigenspace LCLSE beamformer. The matrix with eigenvectors of $\mathbf{\Phi}_{\mathbf{xx}}(k)$ which belong to the V' largest eigenvalues is $\mathbf{U}_{\mathbf{X},V'}(k)$. The matrix $\mathbf{\Sigma}^2_{\mathbf{X},V'}(k)$ is the diagonal matrix with the V' largest eigenvalues. Introducing the constraint matrix $\mathbf{C}(rR)$ in the eigenspace of desired-signal-plus-interference as $\widetilde{\mathbf{C}}_{V'}(k) = \mathbf{U}^T_{\mathbf{X},V'}(k)\mathbf{C}(k)$, the eigenspace optimum beamformer (*principal-component beamformer*) can be obtained as:

$$\mathbf{w}_{V',\mathrm{o}}(rR) = \mathbf{U}_{\mathbf{X},V'}(rR)\mathbf{\Sigma}^{-2}_{\mathbf{X},V'}(rR)\widetilde{\mathbf{C}}_{V'}(rR)\mathbf{\Lambda}^{-1}_{\widetilde{\mathbf{C}}_{V'}}(rR)\mathbf{c}(rR)\,, \quad (4.153)$$

$$\mathbf{\Lambda}_{\widetilde{\mathbf{C}}_{V'}}(rR) = \widetilde{\mathbf{C}}^T_{V'}(rR)\mathbf{\Sigma}^{-2}_{\mathbf{X},V'}(rR)\widetilde{\mathbf{C}}_{V'}(rR)\,. \quad (4.154)$$

We notice that the projection $\mathbf{U}^T_{\mathbf{X},V'}(k)\mathbf{C}(k)$ of the constraint matrix $\mathbf{C}(rR)$ into the eigenspace of desired-signal-plus-interference yields a diagonalization of the sample spatio-temporal correlation matrix w.r.t. desired-signal-plus-interference. The principal component beamformer (4.153) can thus be interpreted as a beamformer which processes the sensor data in the vector space of $\mathbf{U}_{\mathbf{X},V'}(rR)$. See Fig. 4.23.

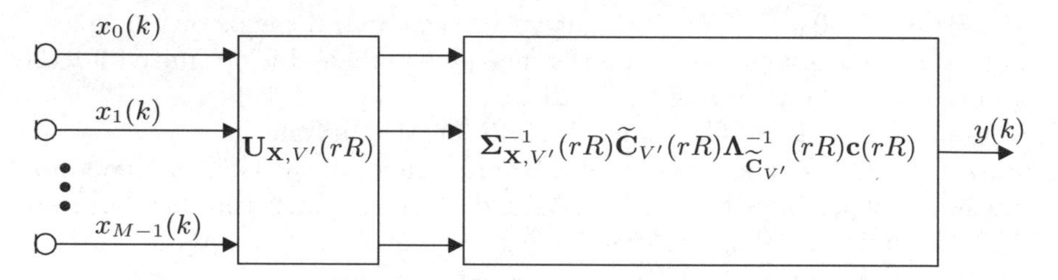

Fig. 4.23. Interpretation of the eigenvector beamformer (after [Tre02, Chap. 6])

Signal components which correspond to the $MN - V'$ small eigenvalues – ideally only sensor noise – are orthogonal to the eigenspace of desired-signal-plus-interference. These signal components are thus suppressed without 'seeing' the beamformer constraints. Signal components which correspond to the V' large eigenvalues are suppressed according to the eigenvector realization of the LCLSE beamformer (4.153). Eigenvector beamformers are investigated in more detail in, e.g., [HT83, CK84, Gab86, Fri88, KU89, Hai91, CY92, YU94, YY95, Hai96, YL96, MG97, LL97, LL00]. Eigenvector beamformers using the GSC structure are discussed in, e.g., [Ows85, Vee88, SM94, YY95].

In [Wur03], DFT-domain realizations of eigenvector beamformers are analyzed for usage in speech and audio signal processing. The results show that in realistic applications only little improvement of interference rejection can be obtained compared to a standard MVDR beamformer due to the necessity to use diagonal loading in order to prevent singularity of the estimates of the spatio-spectral correlation matrices.

For speech and audio signal acquisition, present eigenspace techniques are based on LSE/MMSE optimization criteria. They were first used for reducing artifacts of spectral subtraction in single-channel speech enhancement. There, the eigenspace of the interference is estimated during silence of the desired source. Interference components, which are orthogonal to the eigenspace of the desired signal are suppressed. Interference components in the eigenspace of desired-signal-plus-interference are attenuated, while constraining the noise level at the processor output for limiting the distortion of the desired signal. Application is restricted to situations with slowly time-varying interference, i.e., with quasi-stationary background noise [DBC91, ET95, JHHS95, MP00, AHYN00, RG01, JC02, KK02a]. More recently, the eigenspace methods have been extended to the multi-channel case [JC01a, HL01, DM01, Doc03]. Again, desired signal distortion is minimized in the desired-signal-plus-interference eigenspace subject to a constraint on the maximum allowed interference level at the beamformer output.

4.4.4 Suppression of Correlated Interference

The signal model, which has been described in Sect. 4.1, treats multi-path propagation of the desired source signal as interference and only the direct signal path is considered to be the desired signal. This model is desirable, as, ideally, it allows to dereverberate the desired signal. However, in combination with data-dependent beamformers as discussed so far, this model leads to cancellation of the desired signal due to the high temporal correlation of reflected signal components with the direct path signal [Ows85, CG80, WDGN82, Duv83] resulting in annoying signal distortion.

For preventing cancellation of the desired signal, various methods have been presented for applications others than audio. They can be classified into two categories: one class performs a spatio-spectral decorrelation step prior to conventional beamforming. The other class integrates decorrelation with the beamforming.

1. Two techniques are considered, which decorrelate the sensor signal prior to the beamforming. First, spatial averaging [SK85, SSW86, TKY86, TK87, PYC88, PK89] carries out the temporal correlation reduction by spatially averaging spatio-spectral correlation matrices over a bank of subarrays. This effects a simultaneous reduction of desired signal distortion and interference suppression capability, since the effective number of sensors and the effective spatial aperture is only that of the subarrays [RPK87, YCP89]. Second, frequency-domain averaging [YK90] reduces the temporal correlation by averaging over frequency. This method is only useful if the inverse of the signal bandwidth is less than twice the time delays between the desired signal and the correlated interference. Estimates of interferer positions are required [YK90].
2. Integration of temporal decorrelation and beamforming include the split polarity transformation [LH93], a special partially adaptive beamformer

[QV95a], and (generalized) quadratic constraints [QV95b]. These methods require knowledge (or models) of the temporal correlation between the desired signal and the temporally correlated interferers at the sensors. For reverberated wideband signals, the temporal correlation is generally described by room impulse responses, which are difficult to estimate in realistic scenarios. Therefore, these techniques can hardly be used for speech and audio signal processing.

All these techniques have not yet been studied for speech and audio signal acquisition so far. Most of the methods require estimates of the second-order characteristics of the correlated interference. In highly non-stationary acoustic environments, estimation of these parameters is difficult. If any, spatial averaging seems to be the most promising approach. However, in highly correlated environments, a large number of subarrays may be necessary which conflicts with common limitations on spatial aperture and number of sensors.

Being unable to separate desired signal and reverberation in most promising approaches, reverberation is considered as desired signal for the constraint design to avoid cancellation of the desired signal. Thereby, on the one hand, dereverberation capability is reduced to increasing the power ratio of direct path to reverberation. On the other hand, robustness against cancellation of the desired signal is assured [HSH99, GBW01, HK02b]. (See also Chap. 5.)

4.5 Discussion

This chapter provided an overview of basic properties of beamforming for microphone arrays. Particularly, the concept of optimum adaptive sensor arrays was reviewed, and it was shown how these methods may be applied to meet the requirements of audio signal processing. We derived optimum beamformers using time-domain LSE criteria. Thereby, we were not constrained by the stationarity assumption, and we presented a more general representation of various beamforming aspects which are relevant to our application. For illustration, the relationship between DTFT-domain beamforming and time-domain beamforming was studied.

Starting from a space-time signal model, the fundamental concept of beamforming was introduced as a spatio-temporal filtering problem. From interference-independent beamformer performance measures, we derived dependencies of beamformers on the geometry of the sensor array. For wideband audio signal processing, a trade-off has to be found between maximum array aperture, small sensor spacing, and the number of microphones, while considering product design constraints. On the one hand, the aperture of the sensor array should be as large as possible in order to allow efficient separation of desired signal and interference at low frequencies. On the other hand, large sensor spacings lead to spatial aliasing at high frequencies, which limits the array aperture with a limited number of sensors.

Table 4.1. Overview of optimum data-dependent beamformers for audio signal processing (1)

Type	LSE/MMSE	LCLSE/LCMV
	Non-stationary wideband signals (discrete time domain)	
	LSE	LCLSE
Cost function	$\xi_{\mathrm{LSE}}(r) = \Vert \mathbf{y}_{\mathrm{r}}(rR) - \mathbf{X}^T(rR)\mathbf{w}(rR)\Vert_2^2$	$\xi_{\mathrm{LC}}(r) = \Vert \mathbf{X}^T(rR)\mathbf{w}(rR)\Vert_2^2$ subject to $\mathbf{C}(rR)^T\mathbf{w}(rR) = \mathbf{c}(rR)$
Optimum filters	$\mathbf{w}_{\mathrm{LSE}}(rR) = \mathbf{\Phi}_{\mathbf{xx}}^{-1}(rR)\mathbf{X}(rR)\mathbf{y}_{\mathrm{r}}(rR)$	$\mathbf{w}_{\mathrm{LC}}(rR) = \mathbf{\Phi}_{\mathbf{xx}}^{-1}(rR)\mathbf{C}(rR)\mathbf{\Lambda}_{\mathbf{C}}^{-1}(rR)\mathbf{c}(rR)$ with $\mathbf{\Lambda}_{\mathbf{C}}(rR) = \mathbf{C}^T(rR)\mathbf{\Phi}_{\mathbf{xx}}^{-1}(rR)\mathbf{C}(rR)$
	Stationary wideband signals (DTFT domain)	
	MMSE	LCMV
Optimum filters	$\mathbf{w}_{\mathrm{f,MMSE}}(\omega) = \frac{S_{dd}(\omega)}{S_{dd}(\omega)+\Lambda_{\mathbf{v}_{\mathrm{d}}}(\omega)}\mathbf{w}_{\mathrm{f,MVDR}}(\omega)$ with $\Lambda_{\mathbf{v}_{\mathrm{d}}}(\omega) = \left(\mathbf{v}^H(\mathbf{k}_{\mathrm{d}},\omega)\mathbf{S}_{\mathbf{nn}}^{-1}(\omega)\mathbf{v}(\mathbf{k}_{\mathrm{d}},\omega)\right)^{-1}$	$\mathbf{w}_{\mathrm{f,LCMV}}(\omega) = \mathbf{S}_{\mathbf{xx}}^{-1}(\omega)\mathbf{C}_{\mathrm{f}}(\omega)\mathbf{\Lambda}_{\mathbf{C}_{\mathrm{f}}}(\omega)\mathbf{c}_{\mathrm{f}}(\omega)$ with $\mathbf{\Lambda}_{\mathbf{C}_{\mathrm{f}}}(\omega) = \left(\mathbf{C}_{\mathrm{f}}^T(\omega)\mathbf{S}_{\mathbf{xx}}^{-1}(\omega)\mathbf{C}_{\mathrm{f}}(\omega)\right)^{-1}$ Special case MVDR: $\mathbf{w}_{\mathrm{f,MVDR}}(\omega) = \Lambda_{\mathbf{v}_{\mathrm{d}}}(\omega)\mathbf{S}_{\mathbf{nn}}^{-1}(\omega)\mathbf{v}(\mathbf{k}_{\mathrm{d}},\omega)$

For a practical application, the importance of these opposite requirements depends on the frequency contents of the sensor signals. In our scenario, we, generally, deal with signals with a strong low-frequency content. Therefore, due to the limited number of sensors, we tend to choose the array aperture as large as the application allows while possibly violating the spatial sampling theorem at high frequencies (not considering nested arrays).

For measuring the performance of data-dependent beamformers, inter-fe-rence-de-pen-dent beamformer performance measures are more appropriate than interference-independent ones. Therefore, we generally use the cancellation of the desired signal for measuring the distortion of the desired signal and the interference rejection for measuring the suppression of interference-plus-noise.

Table 4.2. Overview of optimum data-dependent beamformers for audio signal processing (2)

Type	LSE/MMSE	LCLSE/LCMV
Advantages/Disadvantages:		
	+ High interference rejection,	
	+ Exploitation of eigenvector beamforming,	
	− Dereverberation capability limited,	
	+ Positions of desired source and of sensors not required (robustness against array perturbations), but	+ Robustness against array perturbations by appropriate constraint design, and
	− reference of the desired signal necessary,	+ no of reference of the desired signal necessary,
	+ No sensitivity to reverberation of the desired signal,	− Sensitivity to reverberation of the desired signal,
	+ Higher interference rejection than LCLSE/LCMV design at cost of	− Less interference rejection than LSE/MMSE design, but
	− distortion of the desired signal,	+ no distortion of desired signal,
		− High complexity, but
		+ efficient realization as GSC.
	+ Can be realized as MVDR beamformer plus post-filter → Combined advantages of LCMV and MMSE beamforming.	

Although we focus in this work on data-dependent beamformers because of the high interference rejection with a limited number of sensors, we briefly outlined methods of data-independent beamforming. Data-independent beamformer designs, will be used, e.g., as a fixed beamformer for realizing the practical audio acquisition system in Chap. 5. Data-dependent beamformers can be categorized into LSE/MMSE beamforming and LCLSE/LCMV beamforming, as summarized in Table 4.1 and in Table 4.2. The LSE/MMSE approach can be seen as the more general approach, since information about the positions of the desired source and of the sensors is not required. The LSE/MMSE beamformer yields higher interference rejection than the LCLSE/ LCMV beamformer. However, for the LSE/MMSE beamformer, second-order statistics of the desired signal must be known. Often, it is difficult to estimate the second-order statistics of the desired signal accurately, as e.g., for mixtures of non-stationary desired signals and non-stationary interferers. Inaccurate estimation of the second-order statistics of the desired signal introduces artifacts into the output signal of the LSE/MMSE beamformer. In the extreme case, speech intelligibility is reduced relative to the sensor signals or interference is distorted in an unacceptable way for human listeners. Therefore, the LCLSE/LCMV beamformer may be the better choice since information about the position of the desired source relative to the sensor positions replaces the knowledge about the waveform or the spatio-spectral correlation matrix of the desired signal. Since spatial information is used for the weight vector of the beamformer, LCLSE/LCMV beamformers are less robust against mismatched conditions, i.e., assumed geometry of the sensors, assumed position of the desired source, variations of gain and phase of the sensors, or multi-path propagation of the desired source signal. The LCLSE/LCMV beamformer can be efficiently realized in the structure of the GSC.

For our practical audio acquisition system, we will use LCLSE/LCMV beamforming in the structure of the GSC as a basic architecture for three reasons:

1. It allows computationally efficient realizations of LCLSE/LCMV beamformers by transforming the constrained optimization problem into an unconstrained one, which is important for real-time implementations.
2. In contrast to the LSE/MMSE beamformer, LCLSE/LCMV beamforming does not require a reference of the desired source, but 'only' knowledge of the position of the desired source relative to the sensors.
3. Since the LSE/MMSE beamformer can be realized as an LCLSE/LCMV beamformer plus post-filter, the advantage of higher interference rejection of LSE/MMSE beamforming can be exploited using the GSC plus a cascaded post-filter. Post-filtering is not studied within our context.

Two crucial drawbacks of traditional LCLSE/LCMV beamforming in the context of speech audio signal acquisition are the sensitivity to reverberation of the desired signal and the sensitivity to steering errors and to array perturbations, which may lead to cancellation of the desired signal at the beamformer output. For exploiting the advantages of the GSC, these problems have to be resolved first. Therefore, we derive in the next chapter a realization of the GSC, which is robust against steering mismatch, array perturbations, and reverberation w.r.t. the desired signal without impairing the advantages of the GSC.

5

A Practical Audio Acquisition System Using a Robust GSC (RGSC)

In the preceding chapter, we have seen that data-dependent beamformers can be efficiently realized in GSC structures. However, GSCs, or more general LCLSE/LCMV beamformers, are sensitive to steering errors, to array perturbations and to reverberation w.r.t. the desired signal. Therefore, the performance of LCLSE/LCMV beamformers may decrease in realistic implementations: The desired signal may be distorted, and/or interference rejection may be reduced. In order to overcome these problems and to fully profit from the advantages of the GSC, we derive in this chapter a robust version of the GSC (RGSC). The problems of LCLSE/LCMV beamformers are resolved by using time-varying spatio-temporal constraints instead of purely spatial constraints [HK02a, HK03]. In contrast to spatial constraints, these spatio-temporal constraints explicitly take multi-path propagation w.r.t. the desired source into account by modeling the propagation by room impulse responses, such that the direct signal path and the secondary signal paths are cancelled at the output of the blocking matrix of the RGSC. Therefore, the RGSC is robust against cancellation of the desired signal in reverberant acoustic environments. In Chap. 2, it was shown that not only the propagation in a reflective environment can be modeled by impulse responses, but that non-perfect transducer characteristics, i.e., directional-dependent and time-varying tolerances of the complex gain of the transducers, can be captured in impulse responses, too. Therefore, the RGSC is robust against tolerances of the characteristics or positions of the transducers without requiring explicit calibration. Time-varying spatio-temporal constraints take into account the temporal variations of the impulse responses between the desired source and the sensor signals due to speaker movements, changes in the acoustic environment, or time-varying complex gains of the sensors.

The original structure of the RGSC was first introduced in [HSH99] as a GSC for improved robustness against mismatched conditions. In [HK02a], the relation of the RGSC with optimum LCMV beamforming in the DTFT domain using the narrowband assumption was established. The link of the RGSC with wideband LCLSE beamforming in the time domain using spatio-temporal constraints is described in [HK03]. In this chapter, we extend the analysis of the RGSC by identifying blocking matrix and interference canceller with MIMO optimum filters. This gives new insights into the characteristics of the wideband RGSC in FIR filter realizations with a limited number of filter coefficients.

This chapter is organized as follows: The spatio-temporal constraints are introduced in Sect. 5.1. The RGSC as a wideband LCLSE beamformer using spatio-temporal constraints is derived in Sect. 5.2. The special case of WSS signals is considered in Sect. 5.3. In Sect. 5.4, the relationship of the blocking matrix and of the interference canceller with MIMO optimum filters is shown, and consequences on the performance of the RGSC for real applications are studied. These properties are verified in Sect. 5.5 experimentally. In Sect. 5.6, the strategy is presented for determining the optimum blocking matrix and the optimum interference canceller. This strategy is indispensable for obtaining maximum robustness against mismatched conditions in time-varying environments with non-stationary wideband signals. The relation with alternative GSC realizations is summarized in Sect. 5.7. Finally, the results of this chapter are discussed in Sect. 5.8.

5.1 Spatio-temporal Constraints

The spatial constraints (4.118) are generally not appropriate for audio signals for two reasons: First, multi-path propagation of the desired signal cannot be included by placing distortionless constraints in all DOAs of the signal of the desired source, since the number of reflections is much greater than the number of spatial degrees of freedom. However, the multi-path propagation of the desired source signal can be modeled by temporal impulse responses $h(\mathbf{p}_m, \mathbf{p}_\mathrm{d}; t, \tau)$ according to (2.10) between the position of the desired source $\mathbf{p}_\mathrm{d}$ and the position of the sensors $\mathbf{p}_m$. This temporal correlation cannot be incorporated into the spatial constraints (4.118). Second, temporal information about the temporally highly auto-correlated desired source signal cannot be incorporated into the spatial constraints either. As a consequence, the spatial constraints (4.118) do not sufficiently meet the requirements of speech and audio signals.

It is rather desirable to provide constraints which depend on both spatial and temporal characteristics of the desired signal, and to extend the spatial constraints to *spatio-temporal constraints.* We rewrite (4.118) as

$$\mathbf{C}_\mathrm{s}^T(k)\mathbf{w}_\mathrm{s}(k) = \mathbf{c}_\mathrm{s}(k)\,, \tag{5.1}$$

with the $MN^2 \times 1$ beamformer weight vector $\mathbf{w}_\mathrm{s}(k)$, with the $MN^2 \times CN^2$ constraint matrix $\mathbf{C}_\mathrm{s}(k)$, and with the $CN^2 \times 1$ constraint vector $\mathbf{c}_\mathrm{s}(k)$,

$$\mathbf{w}_\mathrm{s}(k) = \left(\mathbf{w}^T(k)\,, \mathbf{w}^T(k-1)\,, \dots,\, \mathbf{w}^T(k-N+1)\right)^T\,, \tag{5.2}$$

$$\mathbf{C}_{\mathrm{s}}(k) = \mathrm{diag}\left\{\mathbf{C}(k),\, \mathbf{C}(k-1),\, \ldots,\, \mathbf{C}(k-N+1)\right\}, \tag{5.3}$$

$$\mathbf{c}_{\mathrm{s}}(k) = \left(\mathbf{c}^T(k),\, \mathbf{c}^T(k-1),\, \ldots,\, \mathbf{c}^T(k-N+1)\right)^T, \tag{5.4}$$

respectively. The constraint equation (5.1) repeats the spatial constraints $\mathbf{C}^T(k-n)\mathbf{w}(k-n) = \mathbf{c}(k-n)$ for N successive time instants $n = 0, 1, \ldots, N-1$. For simplicity, we use the same block length N as for the length of the filters of a MIMO system. Translating the constraints (5.1) to the LCLSE beamformer, the constraints at time k and $N-1$ past constraints must be fulfilled simultaneously by the beamformer weight vector. Therefore, the constraint vectors $\mathbf{c}(k-n)$ can be chosen in order to specify a desired waveform with the spatial characteristics defined in the constraint matrices $\mathbf{C}(k-n)$. Defining a stacked $MN^2 \times K$ matrix

$$\mathbf{X}_{\mathrm{s}}(k) = \left(\mathbf{X}^T(k),\, \mathbf{X}^T(k-1),\, \ldots,\, \mathbf{X}^T(k-N+1)\right)^T, \tag{5.5}$$

where $\mathbf{X}(k)$ is given by (3.11), we can rewrite the optimization criterion (4.117), (4.118) as

$$\xi_{\mathrm{LC}}^{(\mathrm{s})}(r) = \left\|\mathbf{w}_{\mathrm{s}}^T(rR)\mathbf{X}_{\mathrm{s}}(rR)\right\|_2^2 \tag{5.6}$$

subject to the constraints (5.1) evaluated at time $k = rR$. With

$$\mathbf{\Phi}_{\mathbf{xx},\mathrm{s}}(k) = \mathbf{X}_{\mathrm{s}}(k)\mathbf{X}_{\mathrm{s}}^T(k), \tag{5.7}$$

the solution of the constrained optimization problem is given by

$$\mathbf{w}_{\mathrm{LC,o}}^{(\mathrm{s})}(rR) = \mathbf{\Phi}_{\mathbf{xx},\mathrm{s}}^{-1}(rR)\mathbf{C}_{\mathrm{s}}(rR)\left(\mathbf{C}_{\mathrm{s}}^T(rR)\mathbf{\Phi}_{\mathbf{xx},\mathrm{s}}^{-1}(rR)\mathbf{C}_{\mathrm{s}}(rR)\right)^{-1}\mathbf{c}_{\mathrm{s}}(rR), \tag{5.8}$$

which is simply an extended version of (4.119). We notice that the computation of the optimum weight vector in this direct form requires the inversion of large matrices, which is not practical for real-time realizations. However, with the GSC structure, these matrix inversions can be avoided, so that practical implementations are possible as we will see later in Chap. 7.

5.2 RGSC as an LCLSE Beamformer with Spatio-temporal Constraints

In this section, we rigorously derive the RGSC after [HSH99] as a solution to the LCLSE optimization criterion with spatio-temporal constraints. The derivation illustrates the advantages of the RGSC for audio signal acquisition in realistic reverberant environments.

This section is organized as follows: In Sect. 5.2.1, the distortionless response criterion for the RGSC is given. In Sect. 5.2.2, it is illustrated how these constraints can be efficiently implemented as a blocking matrix of the GSC. Finally, in Sect. 5.2.3, a closed-form solution of the optimum weight vector of the interference canceller is derived.

5.2.1 Quiescent Weight Vector

We specialize the spatio-temporal distortionless response criterion given by (5.1) as follows: We demand that the desired signal processed by a quiescent weight vector $\mathbf{w}_c(k)$ is not distorted at the output of the LCLSE beamformer over a data block of N successive samples. With the vector $\mathbf{d}(k)$ of the desired signal at the sensors after (4.4), such a constraint can be put into the form of (5.1) with $C = 1$ as follows:

$$\mathbf{C}_s(k) = \operatorname{diag}\{(\mathbf{d}(k), \mathbf{d}(k-1), \ldots, \mathbf{d}(k-N+1))\} , \tag{5.9}$$

$$\mathbf{c}_s(k) = \mathbf{C}_s^T(k)\mathbf{w}_{c,s}(k) , \tag{5.10}$$

where

$$\mathbf{w}_{c,s}(k) = \left(\mathbf{w}_c^T(k), \mathbf{w}_c^T(k-1), \ldots, \mathbf{w}_c^T(k-N+1)\right)^T . \tag{5.11}$$

Note that the constraint vector $\mathbf{c}_s(k)$ depends on the quiescent weight vector $\mathbf{w}_c(k)$ so that the design of $\mathbf{c}_s(k)$ is replaced by the appropriate choice of the weight vector $\mathbf{w}_{c,s}(k)$. This has the advantage that any beamformer design can be used to specify the constraints, or, equivalently, to specify the beamformer response w.r.t. the desired signal. The quiescent weight vector may be, e.g., realized as a fixed beamformer, if the position of the desired source only varies little within a given interval of DOAs. Typical applications are, e.g., sensor arrays for human speakers in front of computer screens or for drivers of vehicles. When designing the fixed beamformer, one has to assure that the mainlobe width of the fixed beamformer is sufficiently wide for not distorting the desired signal. Robustness against array perturbations can be obtained by using appropriate designs of fixed beamformers. (See also Sect. 4.3.) The optimum weight vector of the LCLSE beamformer is obtained by using the constraints (5.9)–(5.11) in (5.8).

For realizing the LCLSE beamformer as a GSC, the quiescent weight vector of the RGSC must fulfill the constraint equation (5.1). Since the stacked constraint vector $\mathbf{c}_s(k)$ after (5.10) is equal to the left side of (5.1) with $\mathbf{w}_s(k)$ replaced by $\mathbf{w}_{c,s}(k)$, the constraint equation (5.1) is always met if the quiescent weight vector of the RGSC is equal to $\mathbf{w}_{c,s}(k)$.

5.2.2 Blocking Matrix

Furthermore, for realizing the LCLSE beamformer as a GSC, the columns of $\mathbf{C}_s(k)$ must be pairwise orthogonal to the columns of a *spatio-temporal blocking matrix* $\mathbf{B}_s(k)$ of size $MN^2 \times M$, according to (4.143). This means, with the spatio-temporal constraints (5.9) and (5.10), that the reference path in Fig. 5.1 has to be orthogonal to the sidelobe cancelling path w.r.t. the desired signal for $\mathbf{d}(k-n)$, $n = 0, 1, \ldots, N-1$. It follows that desired signal components must be suppressed in the sidelobe cancelling path by using the information about the desired signal from an observation window of length N. This can be achieved by introducing FIR filters with coefficient vectors $\mathbf{b}_m(k)$ between the output of the quiescent beamformer $\mathbf{w}_c(k)$ and each of the M channels of the sidelobe cancelling path. This realization of the blocking matrix is depicted in Fig. 5.1.

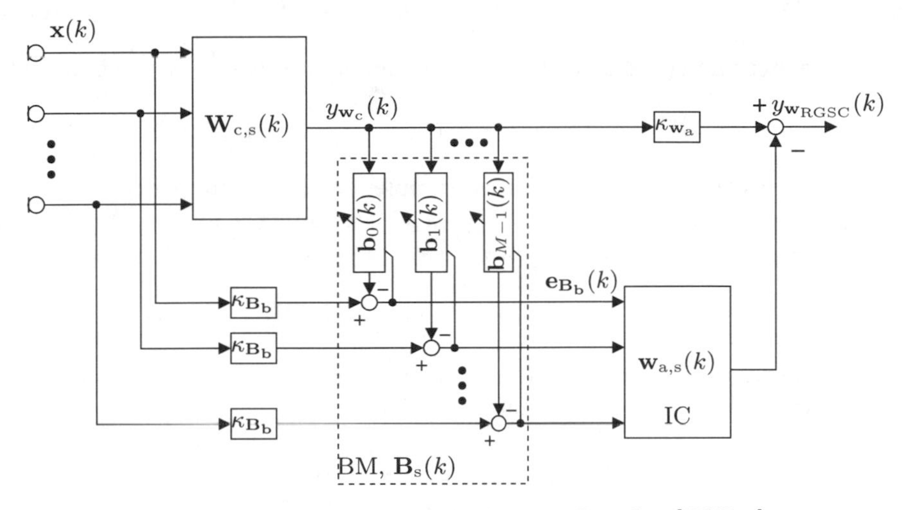

Fig. 5.1. Robust generalized sidelobe canceller after [HSH99]

For simplicity we use $N \times 1$ vectors $\mathbf{b}_m(k)$ with elements $b_{n,m}(k)$, which are captured in an $N \times M$ matrix $\mathbf{B}_\mathrm{b}(k)$ as:

$$\mathbf{B}_\mathrm{b}(k) = (\mathbf{b}_0(k),\, \mathbf{b}_1(k),\, \ldots,\, \mathbf{b}_{M-1}(k)) \,. \tag{5.12}$$

Delays $\kappa_{\mathbf{B}_\mathrm{b}}$ between each of the sensor signals $x_m(k)$ and the input of the blocking matrix $y_{\mathbf{w}_\mathrm{c}}(k)$ are required for causality and for synchronization of the output signal of the quiescent weight vector and the sensor signals. For simplifying the analysis, these delays $\kappa_{\mathbf{B}_\mathrm{b}}$ are set to zero for now.

For suppressing desired signal components in the sidelobe cancelling path, the output signal of the quiescent weight vector $\mathbf{w}_{\mathrm{c,s}}(k)$ has to be orthogonal to each channel of the sidelobe cancelling path for the desired signal. This can be achieved by determining $\mathbf{B}_\mathrm{b}(k)$ such that the principle of orthogonality for a data vector of length N is fulfilled for the desired signal. We can thus identify the filters $\mathbf{B}_\mathrm{b}(k)$ with an optimum MIMO system with one input channel and M output channels (Fig. 3.2). We introduce a column vector $\mathbf{e}_{\mathbf{B}_\mathrm{b}}(k)$ of length M, which captures one sample of each of the M output signals of the blocking matrix:

$$\mathbf{e}_{\mathbf{B}_\mathrm{b}}(k) = (e_{\mathbf{B}_\mathrm{b},0}(k),\, e_{\mathbf{B}_\mathrm{b},1}(k),\, \ldots,\, e_{\mathbf{B}_\mathrm{b},M-1}(k))^T \,. \tag{5.13}$$

Analogously to (3.9), the column vectors $\mathbf{e}_{\mathbf{B}_\mathrm{b}}(k+n)$ are then captured in a matrix $\mathbf{E}_{\mathbf{B}_\mathrm{b}}(k)$ of size $K \times M$ for $n = 0, 1, \ldots, K-1$ as:

$$\mathbf{E}_{\mathbf{B}_\mathrm{b}}(k) = (\mathbf{e}_{\mathbf{B}_\mathrm{b}}(k), \mathbf{e}_{\mathbf{B}_\mathrm{b}}(k+1), \ldots, \mathbf{e}_{\mathbf{B}_\mathrm{b}}(k+K-1))^T . \tag{5.14}$$

We introduce a block-diagonal matrix $\mathbf{W}_{\mathrm{c,s}}(k)$ of size $MN^2 \times N$ with weight vectors $\mathbf{w}_\mathrm{c}(k-n)$, $n = 0, 1, \ldots, N-1$, on the main diagonal,

$$\mathbf{W}_{\mathrm{c,s}}(k) = \mathrm{diag}\left\{(\mathbf{w}_\mathrm{c}(k), \mathbf{w}_\mathrm{c}(k-1), \ldots, \mathbf{w}_\mathrm{c}(k-N+1))\right\} , \tag{5.15}$$

and we capture N data vectors $\mathbf{x}(k-n)$ for $n = 0, 1, \ldots, N-1$ in a vector $\mathbf{x}_\mathrm{s}(k)$ as

$$\mathbf{x}_\mathrm{s}(k) = \left(\mathbf{x}^T(k), \mathbf{x}^T(k-1), \ldots, \mathbf{x}^T(k-N+1)\right)^T . \tag{5.16}$$

This allows to write N consecutive samples of the output signal of the quiescent beamformer $\mathbf{w}_\mathrm{c}(k)$ as $\mathbf{W}_{\mathrm{c,s}}^T(k)\mathbf{x}_\mathrm{s}(k)$. Forming a matrix $\mathbf{D}_\mathrm{s}(k)$ of the matrices $\mathbf{D}(k-n)$ after (4.5) for $n = 0, 1, \ldots, N-1$,

$$\mathbf{D}_\mathrm{s}(k) = \left(\mathbf{D}^T(k), \mathbf{D}^T(k-1), \ldots, \mathbf{D}^T(k-N+1)\right)^T \tag{5.17}$$

analogously to (5.5), we obtain the principle of orthogonality (3.30) w.r.t. the desired signal at the blocking matrix output as

$$\mathbf{W}_{\mathrm{c,s}}^T(k)\mathbf{D}_\mathrm{s}(k)\mathbf{E}_{\mathbf{B}_\mathrm{b}}(k) \stackrel{!}{=} \mathbf{0}_{N\times M} . \tag{5.18}$$

We rewrite next the matrix $\mathbf{E}_{\mathbf{B}_\mathrm{b}}(k)$ of the blocking matrix output signals as a function of the data matrix $\mathbf{X}_\mathrm{s}(k)$. We get

$$\mathbf{E}_{\mathbf{B}_\mathrm{b}}(k) = \mathbf{X}_\mathrm{s}^T(k)\left(\mathbf{J}_{MN^2\times M}^{(N)} - \mathbf{W}_{\mathrm{c,s}}(k)\mathbf{B}_\mathrm{b}(k)\right) , \tag{5.19}$$

where the term in parentheses is the spatio-temporal blocking matrix $\mathbf{B}_\mathrm{s}(k)$ of size $MN^2 \times M$, which fulfills the spatio-temporal constraints (5.9), (5.10):

$$\mathbf{B}_\mathrm{s}(k) = \mathbf{J}_{MN^2\times M}^{(N)} - \mathbf{W}_{\mathrm{c,s}}(k)\mathbf{B}_\mathrm{b}(k) . \tag{5.20}$$

The $MN^2 \times M$ matrix $\mathbf{J}_{MN^2\times M}^{(N)}$ is given by[1]

$$\mathbf{J}_{MN^2\times M}^{(N)} = \left(\mathbf{1}_{MN^2\times 1}^{(0)}, \mathbf{1}_{MN^2\times 1}^{(N)}, \ldots, \mathbf{1}_{MN^2\times 1}^{(MN-N)}\right) , \tag{5.21}$$

where the vector $\mathbf{1}_{MN\times 1}^{(n)}$ is a vector with zeroes and with the n-th element equal to one. The matrix $\mathbf{J}_{MN^2\times M}^{(N)}$ selects entries of $\mathbf{x}_\mathrm{s}(k)$ to match with the matrix $\mathbf{E}_{\mathbf{B}_\mathrm{b}}(k)$.

The optimum coefficient matrix $\mathbf{B}_{\mathrm{b,o}}(k)$ is obtained as follows: We first replace in (5.18) the error matrix $\mathbf{E}_{\mathbf{B}_\mathrm{b}}(k)$ by (5.19). Then, we solve (5.18) for $\mathbf{B}_\mathrm{b}(k)$, and we replace the data matrix $\mathbf{X}_\mathrm{s}(k)$ by

[1]The index n of $\mathbf{J}_{MN^2\times M}^{(n)}$ means that the non-zero entry of the columns of $\mathbf{J}_{MN^2\times M}^{(n)}$ is shifted by n entries from one column to the next.

$$\mathbf{X}_{\mathrm{s}}(k) = \mathbf{D}_{\mathrm{s}}(k) + \mathbf{N}_{\mathrm{s}}(k), \tag{5.22}$$

where the stacked data matrix $\mathbf{N}_{\mathrm{s}}(k)$ of interference-plus-noise at the sensors is defined as

$$\mathbf{N}_{\mathrm{s}}(k) = \left(\mathbf{N}^T(k),\, \mathbf{N}^T(k-1),\, \ldots,\, \mathbf{N}^T(k-N+1)\right)^T, \tag{5.23}$$

analogously to (5.17). Assuming orthogonality of $\mathbf{D}_{\mathrm{s}}^T(k)$ and $\mathbf{N}_{\mathrm{s}}^T(k)$, i.e., $\mathbf{D}_{\mathrm{s}}(k)\mathbf{N}_{\mathrm{s}}^T(k) = \mathbf{0}_{MN^2 \times MN^2}$, and defining the sample spatio-temporal correlation matrix $\mathbf{\Phi}_{\mathbf{dd},\mathrm{s}}(k)$ of size $MN^2 \times MN^2$ as

$$\mathbf{\Phi}_{\mathbf{dd},\mathrm{s}}(k) = \mathbf{D}_{\mathrm{s}}(k)\mathbf{D}_{\mathrm{s}}^T(k), \tag{5.24}$$

we finally obtain the optimum coefficient matrix $\mathbf{B}_{\mathbf{b},\mathrm{o}}(k)$ as

$$\mathbf{B}_{\mathbf{b},\mathrm{o}}(k) = \left(\mathbf{\Lambda}_{\mathbf{W}_{\mathrm{c,s}}}(k)\right)^+ \mathbf{W}_{\mathrm{c,s}}^T(k)\mathbf{\Phi}_{\mathbf{dd},\mathrm{s}}(k)\mathbf{J}_{MN^2 \times M}^{(N)}, \tag{5.25}$$

where $\left(\mathbf{\Lambda}_{\mathbf{W}_{\mathrm{c,s}}}(k)\right)^+$ is the pseudoinverse of the matrix $\mathbf{\Lambda}_{\mathbf{W}_{\mathrm{c,s}}}(k)$,

$$\mathbf{\Lambda}_{\mathbf{W}_{\mathrm{c,s}}}(k) = \mathbf{W}_{\mathrm{c,s}}^T(k)\mathbf{\Phi}_{\mathbf{dd},\mathrm{s}}(k)\mathbf{W}_{\mathrm{c,s}}(k). \tag{5.26}$$

The matrix $\mathbf{W}_{\mathrm{c,s}}^T(k)\mathbf{\Phi}_{\mathbf{dd},\mathrm{s}}(k)\mathbf{J}_{MN^2 \times M}^{(N)}$ is a matrix of the latest N non-normalized estimates of the correlation functions between the output of the beamformer $\mathbf{w}_{\mathrm{c}}(k)$ and each of the sensor signals w.r.t. the desired signal. Accordingly, the matrix $\mathbf{\Lambda}_{\mathbf{W}_{\mathrm{c,s}}}(k)$ is a matrix of the latest N estimates of the auto-correlation function of the output signal of $\mathbf{w}_{\mathrm{c}}(k)$.

The matrix $\mathbf{\Lambda}_{\mathbf{W}_{\mathrm{c,s}}}(k)$ is generally not invertible: The matrix $\mathbf{W}_{\mathrm{c,s}}(k)$ is of full column rank for practical applications. However, (a) the sample spatio-temporal correlation matrix $\mathbf{\Phi}_{\mathbf{dd},\mathrm{s}}(k)$ is generally not of full rank. (b) Some columns of $\mathbf{\Phi}_{\mathbf{dd},\mathrm{s}}(k)$ may be orthogonal to the space which is spanned by the columns of $\mathbf{W}_{\mathrm{c,s}}(k)$ due to suppression of the direct signal path because of steering errors ('cancellation of the desired signal'), or due to suppression of reflections ('dereverberation'). Hence, we use the pseudoinverse of $\mathbf{\Lambda}_{\mathbf{W}_{\mathrm{c,s}}}(k)$, $\left(\mathbf{\Lambda}_{\mathbf{W}_{\mathrm{c,s}}}(k)\right)^+$, to solve the optimization problem for the optimum coefficient matrix $\mathbf{B}_{\mathbf{b},\mathrm{o}}(k)$. Using the SVD of $\mathbf{W}_{\mathrm{c,s}}^T(k)\mathbf{D}_{\mathrm{s}}(k)$, (5.25) may be simplified to

$$\mathbf{B}_{\mathbf{b},\mathrm{o}}(k) = \left(\mathbf{D}_{\mathrm{s}}^T(k)\mathbf{W}_{\mathrm{c,s}}(k)\right)^+ \mathbf{D}_{\mathrm{s}}^T(k)\mathbf{J}_{MN^2 \times M}^{(N)}. \tag{5.27}$$

The optimum blocking matrix $\mathbf{B}_{\mathrm{s,o}}(k)$, which is obtained by replacing in (5.20) the coefficient matrix $\mathbf{B}_{\mathbf{b}}(k)$ by the optimum coefficient matrix $\mathbf{B}_{\mathbf{b},\mathrm{o}}(k)$, does not have full column rank: With $\mathbf{B}_{\mathrm{s,o}}(k)$, we rewrite the principle of orthogonality (5.18) with $\mathbf{E}_{\mathbf{B}_{\mathbf{b}}}(k) = \mathbf{X}_{\mathrm{s}}^T(k)\mathbf{B}_{\mathrm{s,o}}(k)$, with (5.24), and with $\mathbf{D}_{\mathrm{s}}(k)\mathbf{N}_{\mathrm{s}}^T(k) = \mathbf{0}_{MN \times MN}$ as

$$\mathbf{W}_{\mathrm{c,s}}^T(k)\mathbf{D}_{\mathrm{s}}(k)\mathbf{D}_{\mathrm{s}}^T(k)\mathbf{B}_{\mathrm{s,o}}(k) = \mathbf{0}_{N \times M}. \tag{5.28}$$

We see that the matrix $\mathbf{W}_{c,s}(k)$ is orthogonal to the blocking matrix $\mathbf{B}_{s,o}(k)$ in the space which is spanned by spatio-temporal correlation matrix $\mathbf{\Phi}_{\mathbf{dd},s}(k) = \mathbf{D}_s(k)\mathbf{D}_s^T(k)$ w.r.t. the desired signal, denoted by span$\{\mathbf{\Phi}_{\mathbf{dd},s}(k)\}$. This is in contrast to the GSC with spatial constraints, where orthogonality of $\mathbf{w}_c(k)$ and $\mathbf{B}(k)$ is required in the entire signal space. (See Sect. 4.4.2.) This means that the dimension of the orthogonal complement of span$\{\mathbf{B}_{s,o}^T(k)\}$, which is denoted by dim$\{(\mathbf{B}_{s,o}^T(k))^\perp\}$, increases with decreasing dim$\{\mathbf{\Phi}_{\mathbf{dd},s}(k)\}$. Since span$\{(\mathbf{B}_{s,o}^T(k))^\perp\}$ is equivalent to the space in which suppression of interference-plus-noise is possible, the number of degrees of freedom for suppressing interference-plus-noise increases with decreasing dim$\{\mathbf{\Phi}_{\mathbf{dd},s}(k)\}$.

We conclude that the RGSC using spatio-temporal constraints better meets the requirements of wideband non-stationary signals than the GSC using basic spatial constraints. On the one hand, ideally, the constraints match the characteristics of the desired signal at the sensors. All propagation paths of the desired source signal are taken into account by the constraints so that – in contrast to the original spatial blocking matrix – the direct signal path and reverberation is suppressed by the blocking matrix. Distortion of the desired signal by the RGSC can thus be efficiently prevented in reverberant acoustic environments. On the other hand, since the influence of the constraints on interference-plus-noise is minimized in the sidelobe cancelling path, interference rejection by the RGSC is maximized by the spatio-temporal constraints.

5.2.3 Interference Canceller

We now formulate the output signal $y_{\mathbf{w}_{\mathrm{RGSC}}}(k)$ of the GSC using the spatio-temporal blocking matrix according to $y(k)$ in (4.146). This allows to determine the interference canceller for the RGSC in the same way as $\mathbf{w}_{a,o}(k)$ in (4.147). Delays $\kappa_{\mathbf{w}_a}$, which are required for synchronization of the output signal $y_{\mathbf{w}_c}(k)$ of the quiescent weight vector and the output signals $\mathbf{e}_{\mathbf{B}_b}(k)$ of the blocking matrix are set to zero for simplicity. (See Fig. 5.1.) The output signal of the RGSC can be written as

$$\begin{aligned} y_{\mathbf{w}_{\mathrm{RGSC}}}(k) &= \left(\mathbf{w}_c^T(k)\mathbf{J}_{MN^3\times MN}^{(1)T} - \mathbf{w}_{a,s}^T(k)\mathbf{B}_{ss}^T(k)\right)\mathbf{x}_{ss}(k) \\ &= \mathbf{w}_{\mathrm{RGSC}}^T(k)\mathbf{x}_{ss}(k)\,, \end{aligned} \tag{5.29}$$

where

$$\mathbf{B}_{ss}(k) = \mathrm{diag}\left\{(\mathbf{B}_s(k),\, \mathbf{B}_s(k-1),\, \dots,\, \mathbf{B}_s(k-N+1))\right\}\,, \tag{5.30}$$

$$\mathbf{x}_{ss}(k) = \left(\mathbf{x}_s^T(k),\, \mathbf{x}_s^T(k-1),\, \dots,\, \mathbf{x}_s^T(k-N+1)\right)^T\,, \tag{5.31}$$

$$\mathbf{J}_{MN^3\times MN}^{(1)} = \left(\mathbf{1}_{MN^3\times 1}^{(0)},\, \mathbf{1}_{MN^3\times 1}^{(1)},\, \dots,\, \mathbf{1}_{MN^3\times 1}^{(MN-1)}\right)\,, \tag{5.32}$$

and where $\mathbf{w}_{a,s}(k)$ is arranged as

$$\mathbf{w}_{\mathrm{a,s}}(k) = \begin{pmatrix} w_{\mathrm{a},0,0}(k) \\ w_{\mathrm{a},0,1}(k) \\ \vdots \\ w_{\mathrm{a},0,M-1}(k) \\ w_{\mathrm{a},1,0}(k) \\ \vdots \\ w_{\mathrm{a},N-1,M-1}(k) \end{pmatrix} . \tag{5.33}$$

The same number of filter coefficients N are assumed for the weight vectors in $\mathbf{w}_{\mathrm{c}}(k)$, $\mathbf{B}_{\mathbf{b}}(k)$, and $\mathbf{w}_{\mathrm{a,s}}(k)$ for simplicity. Identification of $y_{\mathbf{w}_{\mathrm{RGSC}}}(k)$ after (5.29) with $y(k)$ in (4.146) yields with

$$\mathbf{X}_{\mathrm{ss}}(k) = \left(\mathbf{X}_{\mathrm{s}}^T(k),\, \mathbf{X}_{\mathrm{s}}^T(k-1),\, \ldots,\, \mathbf{X}_{\mathrm{s}}^T(k-N+1)\right)^T , \tag{5.34}$$

$$\mathbf{\Phi}_{\mathbf{xx},\mathrm{ss}}(k) = \mathbf{X}_{\mathrm{ss}}(k)\mathbf{X}_{\mathrm{ss}}^T(k) , \tag{5.35}$$

the optimum weights $\mathbf{w}_{\mathrm{a,s,o}}(k)$ of the interference canceller as

$$\mathbf{w}_{\mathrm{a,s,o}}(k) = \left(\mathbf{B}_{\mathrm{ss}}^T(k)\mathbf{\Phi}_{\mathbf{xx},\mathrm{ss}}(k)\mathbf{B}_{\mathrm{ss}}(k)\right)^+ \mathbf{B}_{\mathrm{ss}}^T(k)\mathbf{\Phi}_{\mathbf{xx},\mathrm{ss}}(k)\mathbf{J}_{MN^3\times MN}^{(1)}\mathbf{w}_{\mathrm{c}}(k) , \tag{5.36}$$

analogously to (4.147). Here, we replaced the matrix inverse with the pseudoinverse, since the spatio-temporal correlation matrix $\mathbf{B}_{\mathrm{ss}}^T(k)\mathbf{\Phi}_{\mathbf{xx},\mathrm{ss}}(k)\mathbf{B}_{\mathrm{ss}}(k)$ w.r.t. the output signals of the blocking matrix is generally not invertible: While we assumed full rank of $\mathbf{\Phi}_{\mathbf{xx},\mathrm{ss}}(k)$, the block-diagonal matrix $\mathbf{B}_{\mathrm{ss}}(k)$ does generally not have full column rank, depending on the rank of $\mathbf{B}_{\mathrm{s}}(k-n)$ with $n = 0, 1, \ldots, N-1$. (See Sect. 5.2.2.)

The formulation of the optimum weight vector of the interference canceller completes the derivation of the RGSC. The derivation shows the formal equivalence of the RGSC with the GSC after [GJ82] for the case where the spatial constraints are extended to spatio-temporal constraints. The RGSC meets an LCLSE optimization criterion.

5.3 RGSC in the DTFT Domain

In this section, we consider the RGSC in the DTFT domain for WSS signals. We thus obtain a closed-form solution for the filters of the blocking matrix. This closed-form solution (a) illustrates the capability of the RGSC to efficiently deal with reverberant environments, (b) shows the flexibility of the spatio-temporal constraints to optimally benefit from colored desired signals, and (c) allows to relate the RGSC to alternative GSC realizations.

We assume that the desired source is located at a position $\mathbf{p}_{\mathrm{d}}$. The wave field in the reverberant environment is modeled by room transfer functions. According to (2.14), the signal of the desired source, which is received at the m-th sensor is thus given by

$$D_m(\omega) = S_{\mathbf{p}_\mathrm{d}}(\omega) H(\mathbf{p}_m, \mathbf{p}_\mathrm{d}; \omega) \,, \tag{5.37}$$

where $S_{\mathbf{p}_\mathrm{d}}(\omega)$ is the DTFT of the desired source signal $s_{\mathbf{p}_\mathrm{d}}(kT_\mathrm{s})$. The transfer functions between the position of the desired source and the M sensors are captured in an $M \times 1$ vector according to (2.52) as follows:

$$\mathbf{h}_\mathrm{f}(\mathbf{p}_\mathrm{d}, \omega) = \left(H(\mathbf{p}_0, \mathbf{p}_\mathrm{d}; \omega),\, H(\mathbf{p}_1, \mathbf{p}_\mathrm{d}; \omega),\, \ldots,\, H(\mathbf{p}_{M-1}, \mathbf{p}_\mathrm{d}; \omega)\right)^T . \tag{5.38}$$

The constraints which are given by (5.9) and by (5.10) can thus be formulated in the DTFT domain as

$$\mathbf{C}_\mathrm{f}(\omega) = S_{\mathbf{p}_\mathrm{d}}(\omega) \mathbf{h}_\mathrm{f}(\mathbf{p}_\mathrm{d}, \omega) \,, \tag{5.39}$$

$$\mathbf{c}_\mathrm{f}(\omega) = S^*_{\mathbf{p}_\mathrm{d}}(\omega) \mathbf{h}_\mathrm{f}^H(\mathbf{p}_\mathrm{d}, \omega) \mathbf{w}_{\mathrm{f,c}}(\omega) \,. \tag{5.40}$$

The constraints depend on the spectrum of the desired signal. For wideband signals, the constraints are only active at frequencies with desired signal activity. At frequencies where the desired signal is not active, the constrained optimization problem reduces to an unconstrained optimization problem. Hence, the optimum is found by unconstrained minimization of $\xi_{\mathrm{f,LCMV}}(\omega)$ w.r.t. $\mathbf{w}_\mathrm{f}(\omega)$ after (4.133). This yields the trivial weight vector

$$\mathbf{w}_{\mathrm{f,RGSC}}(\omega) := \mathbf{0}_{M \times 1} \,, \tag{5.41}$$

for the RGSC, which means that the optimum weight vector of the interference canceller is equivalent to the quiescent weight vector,

$$\mathbf{w}_{\mathrm{f,a,o}}(\omega) = \mathbf{w}_{\mathrm{f,c}}(\omega) \,. \tag{5.42}$$

At frequencies with activity of the desired signal, the RGSC is equivalent to the LCMV beamformer with spatial constraints (4.134), i.e.,

$$\mathbf{w}_{\mathrm{f,RGSC}}(\omega) := \mathbf{w}_{\mathrm{f,LCMV}}(\omega) \,. \tag{5.43}$$

The transformation of the weight vectors $\mathbf{b}_m(k)$ of the blocking matrix into the DTFT domain yields $B_m(\omega)$. The weights $B_m(\omega)$, $m = 0, 1, \ldots, M-1$, are captured in an $M \times 1$ vector as

$$\mathbf{b}_\mathrm{f}(\omega) = \left(B_0(\omega),\, B_1(\omega),\, \ldots,\, B_{M-1}(\omega)\right)^H . \tag{5.44}$$

The optimum weight vector $\mathbf{b}_\mathrm{f}(\omega)$ is found by applying the principle of orthogonality in the DTFT domain. It yields with $\mathbf{S}_{\mathbf{dd}}(\omega) \neq \mathbf{0}_{M \times M}$

$$\mathbf{b}^*_{\mathrm{f,o}}(\omega) = \left(\mathbf{w}^H_{\mathrm{f,c}}(\omega) \mathbf{S}_{\mathbf{dd}}(\omega) \mathbf{w}_{\mathrm{f,c}}(\omega)\right)^{-1} \mathbf{S}_{\mathbf{dd}}(\omega) \mathbf{w}_{\mathrm{f,c}}(\omega) \,, \tag{5.45}$$

where the term in parentheses is the PSD of the desired signal at the output of the quiescent weight vector and where the term $\mathbf{S}_{\mathbf{dd}}(\omega)\mathbf{w}_{\mathrm{f,c}}(\omega)$ is the cross-power spectral density between the output of $\mathbf{w}_{\mathrm{f,c}}(\omega)$ and the desired signal at the sensors. We further substitute in (5.45) $\mathbf{S}_{\mathbf{dd}}(\omega)$ by the spatio-spectral correlation matrix of a point source in a reverberant environment according to (2.53)

$$\mathbf{S}_{\mathbf{dd}}(\omega) := S_{dd}(\omega)\mathbf{h}_{\mathrm{f}}(\mathbf{p}_{\mathrm{d}},\omega)\mathbf{h}_{\mathrm{f}}^{H}(\mathbf{p}_{\mathrm{d}},\omega)\,. \tag{5.46}$$

We thus obtain after some rearrangements for $\mathbf{w}_{\mathrm{f,c}}^{H}(\omega)\mathbf{h}_{\mathrm{f}}(\mathbf{p}_{\mathrm{d}},\omega) \neq 0$ and for $\mathbf{S}_{\mathbf{dd}}(\omega) \neq 0$ the optimum filters of the blocking matrix as follows:

$$\mathbf{b}_{\mathrm{f,o}}^{*}(\omega) = \frac{\mathbf{h}_{\mathrm{f}}(\mathbf{p}_{\mathrm{d}},\omega)}{\mathbf{w}_{\mathrm{f,c}}^{H}(\omega)\mathbf{h}_{\mathrm{f}}(\mathbf{p}_{\mathrm{d}},\omega)}\,. \tag{5.47}$$

Note that the optimum filters have to inversely model a weighted sum of the room transfer functions between the desired source and the sensors. The relation between the blocking matrix $\mathbf{B}_{\mathrm{f}}(\omega)$ and the weight vector $\mathbf{b}_{\mathrm{f}}(\omega)$ is determined according to (5.20) as follows:

$$\mathbf{B}_{\mathrm{f}}^{H}(\omega) = \mathbf{I}_{M\times M} - \mathbf{b}_{\mathrm{f}}^{*}(\omega)\mathbf{w}_{\mathrm{f,c}}^{H}(\omega)\,. \tag{5.48}$$

If we use the weight vector $\mathbf{b}_{\mathrm{f,o}}^{*}(\omega)$ in (5.48), a closed-form expression for the optimum blocking matrix can be computed as:

$$\mathbf{B}_{\mathrm{f,o}}^{H} = \frac{1}{\mathbf{w}_{\mathrm{f,c}}^{H}\mathbf{h}_{\mathrm{f}}(\mathbf{p}_{\mathrm{d}})}\times \tag{5.49}$$
$$\times \begin{pmatrix} \sum\limits_{m=1}^{M-1} H(\mathbf{p}_m,\mathbf{p}_{\mathrm{d}})W_{\mathrm{c},m} & -H(\mathbf{p}_0,\mathbf{p}_{\mathrm{d}})W_{\mathrm{c},1} & \cdots & -H(\mathbf{p}_0,\mathbf{p}_{\mathrm{d}})W_{\mathrm{c},M-1} \\ -H(\mathbf{p}_1,\mathbf{p}_{\mathrm{d}})W_{\mathrm{c},0} & \sum\limits_{\substack{m=0\\ m\neq 2}}^{M-1} H(\mathbf{p}_m,\mathbf{p}_{\mathrm{d}})W_{\mathrm{c},m} & \cdots & -H(\mathbf{p}_1,\mathbf{p}_{\mathrm{d}})W_{\mathrm{c},M-1} \\ \vdots & & \ddots & \\ -H(\mathbf{p}_{M-1},\mathbf{p}_{\mathrm{d}})W_{\mathrm{c},0} & & \cdots & \sum\limits_{m=0}^{M-2} H(\mathbf{p}_m,\mathbf{p}_{\mathrm{d}})W_{\mathrm{c},m} \end{pmatrix},$$

where we neglected the dependency on ω for a better reading. It can be easily verified that multi-path propagation of the desired source signal is effectively cancelled by $\mathbf{B}_{\mathrm{f,o}}(\omega)$, i.e., $\mathbf{B}_{\mathrm{f,o}}^{H}(\omega)\mathbf{h}_{\mathrm{f}}(\mathbf{p}_{\mathrm{d}},\omega) = \mathbf{0}_{M\times 1}$. In analogy to the constraints defined by (5.39) and by (5.40), the blocking matrix is only active at frequencies where the desired signal is present at the output of the quiescent weight vector. With activity of the desired signal, the blocking matrix $\mathbf{B}_{\mathrm{f,o}}(\omega)$ does not have full rank, i.e., $\mathrm{rk}\,\{\mathbf{B}_{\mathrm{f,o}}(\omega)\} = M-1$, according to one spatial constraint ($C = 1$). Without activity of the desired signal, i.e., $\mathbf{w}_{\mathrm{f,c}}^{H}(\omega)\mathbf{h}_{\mathrm{f}}(\mathbf{p}_{\mathrm{d}},\omega)S_{\mathbf{p}_{\mathrm{d}}}(\omega) = 0$, we have $\mathbf{b}_{\mathrm{f,o}}^{*}(\omega) = \mathbf{0}_{M\times 1}$, which yields

$$\mathbf{B}_{\mathrm{f,o}}(\omega) = \mathbf{I}_{M\times M}\,. \tag{5.50}$$

The blocking matrix is the identity matrix, which means that $\mathrm{rk}\,\{\mathbf{B}_{\mathrm{f,o}}(\omega)\} = M$. In contrast to 'conventional' LCMV beamformers, for $\mathbf{B}_{\mathrm{f,o}}(\omega) = \mathbf{I}_{M\times M}$, the suppression of interference is not limited by the number of spatial degrees of freedom, since the optimum weight vector of the RGSC nulls the beamformer output signal according to (5.41). For $\mathbf{B}_{\mathrm{f,o}}(\omega) \neq \mathbf{I}_{M\times M}$, we have equivalence with 'conventional' LCMV beamforming with the same interference rejection. This illustrates that, ideally, the constraints are only active if the desired signal is present. This effect can also be viewed as a spectral weighting of the output of the RGSC according to the MMSE weighting function, with the exception that the weighting is constrained such that the desired signal is not distorted.

The blocking matrix (5.49) can be related to the blocking matrix for a plane-wave desired signal from broadside after (4.152), when assuming a uniformly weighted beamformer, $\mathbf{w}_{\mathrm{f}}(\omega) = 1/M \cdot \mathbf{1}_{M\times 1}$. The optimum blocking matrix $\mathbf{B}_{\mathrm{f,o}}(\omega)$ simplifies to

$$\mathbf{B}_{\mathrm{f,o}}^{H}(\omega) = \frac{1}{M}\begin{pmatrix} M-1 & -1 & \cdots & -1 \\ -1 & M-1 & \cdots & -1 \\ \vdots & & \ddots & \\ -1 & -1 & \cdots & M-1 \end{pmatrix}. \tag{5.51}$$

Both blocking matrices are equivalent, except that the blocking matrix $\mathbf{B}_{\mathrm{f}}(\omega)$ after (4.152) is constructed to have linearly independent columns, while the blocking matrix $\mathbf{B}_{\mathrm{f,o}}(\omega)$ after (5.51) has one linearly dependent column.

5.4 RGSC Viewed from Inverse Modeling and from System Identification

In the four classes of applications of MIMO optimum filtering after [Hay96], data-dependent optimum beamforming is generally classified as interference cancellation. However, due to the close relation of the different classes of optimum filtering, data-dependent optimum beamforming can also be interpreted as inverse modeling. (See Sect. 3.2.) In this section, we will see that this interpretation gives new insights into the behavior of the RGSC and yields consequences for practical realizations. The results are valid for other data-dependent optimum beamformers, too.

In Sect. 5.4.1, the blocking matrix is examined. Section 5.4.2 studies the interference canceller. The discussion shows the fundamental differences between both units. The dependency of their performance on the number of sensors, on the filter lengths, and on the number of interferers is studied.

5.4.1 Blocking Matrix

In this section, the spatio-temporal blocking matrix is analyzed. First, the blocking matrix is identified with an optimum single-input multiple-output (SIMO) system, and its characteristics are discussed for stationary conditions. Second, we translate these properties to the transient case for practical applications. Interference and noise is assumed to be inactive.

Stationary Conditions

The filters of the blocking matrix $\mathbf{B_b}(k)$ can be seen as the SIMO system that is depicted in Fig. 5.2.

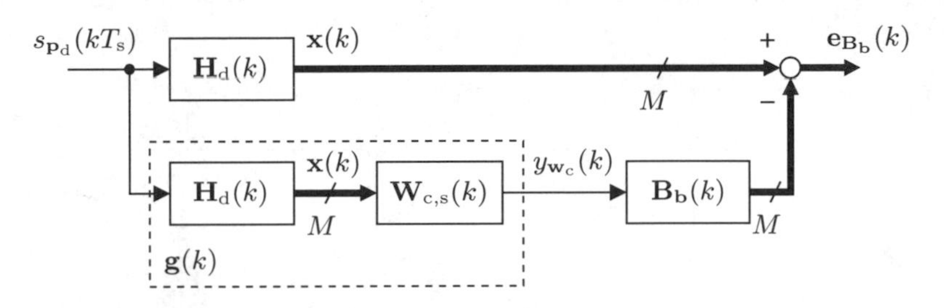

Fig. 5.2. Blocking matrix filters $\mathbf{B_b}(k)$ as a SIMO system for inverse modeling

The desired source signal $s_{\mathbf{p}_\mathrm{d}}(kT_\mathrm{s})$ is filtered by the room impulse responses $h(\mathbf{p}_m, \mathbf{p}_\mathrm{d}; kT_\mathrm{s}, nT_\mathrm{s})$ of length $N_{\mathbf{h}_\mathrm{d}}$ between the position of the desired source and the sensors,

$$\mathbf{h}_{\mathrm{d},m}(k) = \begin{pmatrix} h(\mathbf{p}_m, \mathbf{p}_\mathrm{d}; kT_\mathrm{s}, 0) \\ h(\mathbf{p}_m, \mathbf{p}_\mathrm{d}; kT_\mathrm{s}, T_\mathrm{s}) \\ \vdots \\ h(\mathbf{p}_m, \mathbf{p}_\mathrm{d}; kT_\mathrm{s}, (N_{\mathbf{h}_\mathrm{d}} - 1)T_\mathrm{s}) \end{pmatrix}, \tag{5.52}$$

which are captured in an $N_{\mathbf{h}_\mathrm{d}} \times M$ matrix $\mathbf{H}_\mathrm{d}(k)$ as

$$\mathbf{H}_\mathrm{d}(k) = (\mathbf{h}_{\mathrm{d},0}(k), \mathbf{h}_{\mathrm{d},1}(k), \ldots, \mathbf{h}_{\mathrm{d},M-1}(k))^T . \tag{5.53}$$

The output signal $y_{\mathbf{w}_\mathrm{c}}(k)$ of the quiescent weight vector $\mathbf{W}_{\mathrm{c,s}}(k)$ forms the single-channel input for the filters of the blocking matrix. The reference signals are given by the sensor signals $\mathbf{x}(k)$.

From an inverse modeling perspective, the filters $\mathbf{B_b}(k)$ have to model the combined system of the inverse of the SISO system $\mathbf{g}(k)$, which consists of the cascade of $\mathbf{H}_\mathrm{d}(k)$ and of $\mathbf{W}_{\mathrm{c,s}}(k)$ in the lower signal path of Fig. 5.2, and of the system $\mathbf{H}_\mathrm{d}(k)$ in the upper signal path.

Room impulse responses for acoustic environments are generally non-minimum phase with most of the zeroes distributed in the z-plane close to the unit circle. The filters $\mathbf{B_b}(k)$ have to model the inverse of room impulse responses as indicated in (5.47). As a consequence of the characteristics of the room impulse responses, the filters in $\mathbf{B_b}(k)$ must be of infinite length for complete suppression of the desired signal at the blocking matrix output. The MINT after Sect. 3.2.2 for modeling the inverse of an arbitrary system using a finite number of filter coefficients does not apply, since the system between $s_{\mathbf{p}_\mathrm{d}}(kT_\mathrm{s})$ and the input signal of $\mathbf{B_b}(k)$ is a SISO system (see (3.43)). Moreover, the optimum filters $\mathbf{B}_{\mathbf{b},\mathrm{o}}(k)$ are not necessarily causal, which makes the synchronization delays $\kappa_{\mathbf{B}_\mathbf{b}}$ in Fig. 5.1 necessary. In Fig. 5.3a, a typical example of optimum filter coefficients $b_{\mathrm{o},n,0}(k)$ for the environment with $T_{60} = 400\,\mathrm{ms}$ (see App. B) for $\kappa_{\mathbf{B}_\mathbf{b}} = N/2$ is shown. Note the symmetry of $b_{\mathrm{o},n,0}(k)$ w.r.t. $n := \kappa_{\mathbf{B}_\mathbf{b}}$, which motivates this (typical) choice of $\kappa_{\mathbf{B}_\mathbf{b}} = N/2$.

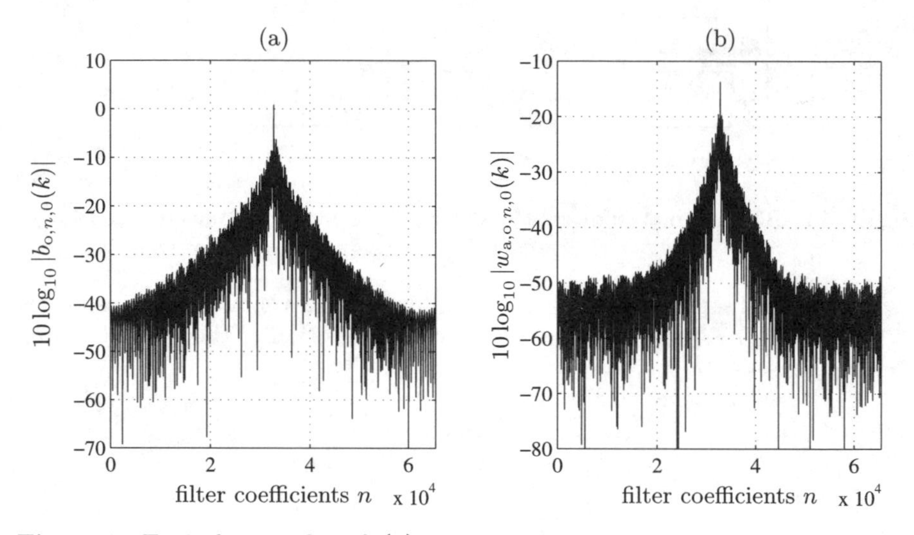

Fig. 5.3. Typical examples of **(a)** optimum blocking matrix filter coefficients $b_{\mathrm{o},n,0}(k)$, and **(b)** optimum interference canceller filter coefficients $w_{\mathrm{a,o},n,0}(k)$ for $\kappa_{\mathbf{B}_\mathbf{b}} = N/2$ and $\kappa_{\mathbf{w}_\mathrm{a}} = \kappa_{\mathbf{B}_\mathbf{b}} + N/2$ computed with the real-time realization of the RGSC described in Chap. 7; sensor array with $M = 8$ in the multimedia room with $T_{60} = 400\,\mathrm{ms}$ (see App. B), uniform weight vector $\mathbf{w}_\mathrm{c}(k)$, desired signal from $\theta_\mathrm{d} = \pi/3$, interferer from $\theta_\mathrm{n} = 0$

Assuming now that the room impulse responses between the position of the desired source and the sensors are known, then, the quiescent weight vector $\mathbf{W}_{\mathrm{c,s}}(k)$ can be chosen according to the MINT to invert the room impulse responses and to recover the signal $s_{\mathbf{p}_\mathrm{d}}(kT_\mathrm{s})$ of the desired source at the output of $\mathbf{W}_{\mathrm{c,s}}(k)$. The determination of $\mathbf{B}_\mathbf{b}(k)$ is then a system identification problem. The weight vectors of the blocking matrix reduce to the room impulse responses between the position of the desired source and the sensors, i.e., $b_{n,m}(k) = h(\mathbf{p}_m, \mathbf{p}_\mathrm{d}; kT_\mathrm{s}, nT_\mathrm{s})$.

The MINT does also apply if the input signals $\mathbf{x}_\mathrm{s}(k)$ of the quiescent weight vector are considered as input signals of a MIMO system, which replaces the cascade of $\mathbf{W}_{\mathrm{c,s}}(k)$ and $\mathbf{B}_\mathbf{b}(k)$. However, since the sensor signals are used as input of $\mathbf{B}_\mathbf{b}(k)$, interference is not attenuated by $\mathbf{W}_{\mathrm{c,s}}(k)$, and robustness against cancellation of interference by the blocking matrix reduces. Obviously, a quiescent weight vector is still required as reference for the determination of $\mathbf{w}_{a,s}(k)$.

Transient Conditions

In our realization of the RGSC in Chap. 7, we use adaptive filters for implementing the blocking matrix. The adaptive filters have to meet two competing criteria: On the one hand, the length of the adaptive filters must be large in order to minimize leakage of the desired signal through the blocking matrix for stationary conditions. On the other hand, the adaptive filters must be kept short for maximum speed of convergence. As a result, the choice of the filter length should always be chosen with regard to the present operating conditions and of the speed of convergence of the adaptation algorithm. In a practical system, the number of filter taps of the blocking matrix is generally between $N = 64 \dots 512$ depending on the speed of convergence of the adaptation algorithm and depending on the time-variance of the source signals and of the acoustic environment. For the realization, it must therefore be considered that the desired signal cannot be completely cancelled by the blocking matrix. The residual components of the desired signal at the input of the interference canceller may lead to cancellation of the desired signal at the output of the RGSC. This may require measures of robustness improvement according to Sect. 4.4.2.

5.4.2 Interference Canceller

In this section, the interference canceller is studied. First, we identify the interference canceller with an optimum MISO system, and discusses its characteristics for stationary conditions. Second, we extend the discussion to the transient case for practical systems. The desired signal is assumed to be inactive.

Stationary Conditions

The weight vector $\mathbf{w}_{a,s}(k)$ can be interpreted as a MISO system, as shown in Fig. 5.4.

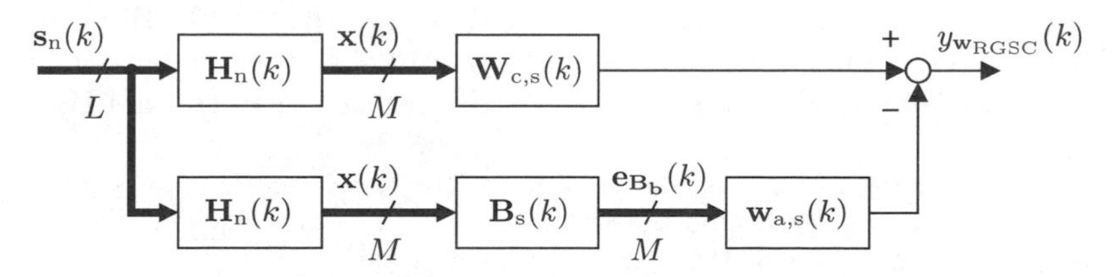

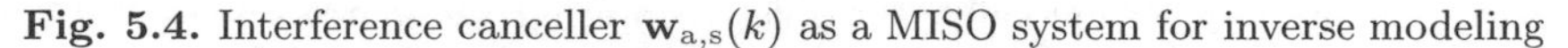
Fig. 5.4. Interference canceller $\mathbf{w}_{a,s}(k)$ as a MISO system for inverse modeling

We assume presence of L mutually uncorrelated interferers, which are modeled as point sources at positions $\mathbf{p}_{\mathrm{n},l}$, i.e., $s_{\mathrm{n},l}(\mathbf{p},t) := s_{\mathbf{p}_{\mathrm{n},l}}(t)$ in a reverberant environment. The room characteristics between the source positions and the sensors are modeled by impulse responses $h(\mathbf{p}_m, \mathbf{p}_{\mathrm{n},l}; kT_\mathrm{s}, nT_\mathrm{s})$ of length $N_{\mathbf{h}_\mathrm{n}}$, which are captured in an $LN_{\mathbf{h}_\mathrm{n}} \times M$ matrix $\mathbf{H}_\mathrm{n}(k)$ as[2]

$$\mathbf{H}_\mathrm{n}(k) = \left(\mathbf{h}_{\mathrm{n},0}(k),\, \mathbf{h}_{\mathrm{n},1}(k),\, \dots,\, \mathbf{h}_{\mathrm{n},M-1}(k)\right)^T , \tag{5.54}$$

$$\mathbf{h}_{\mathrm{n},m}(k) = \left(\mathbf{h}^T_{\mathrm{n},0,m}(k),\, \mathbf{h}^T_{\mathrm{n},1,m}(k),\, \dots,\, \mathbf{h}^T_{\mathrm{n},L-1,m}(k)\right)^T , \tag{5.55}$$

where the vectors $\mathbf{h}_{\mathrm{n},l,m}(k)$ of length $N_{\mathbf{h}_\mathrm{n}}$ capture the room impulse responses between the position of the l-th interferer and the m-th sensor analogously to (5.52). The signals $s_{\mathbf{p}_{\mathrm{n},l}}(kT_\mathrm{s})$ are captured in a stacked column vector $\mathbf{s}_\mathrm{n}(k)$ of length $LN_{\mathbf{h}_\mathrm{n}}$ according to the stacked vector of sensor signals, $\mathbf{x}(k)$, after (2.29) and (2.30):

$$\mathbf{s}_{\mathrm{n},l}(k) = \left(s_{\mathbf{p}_{\mathrm{n},l}}(kT_\mathrm{s}),\, s_{\mathbf{p}_{\mathrm{n},l}}((k-1)T_\mathrm{s}),\, \dots,\, s_{\mathbf{p}_{\mathrm{n},l}}((k-N_{\mathbf{h}_\mathrm{n}}+1)T_\mathrm{s})\right)^T \tag{5.56}$$

$$\mathbf{s}_\mathrm{n}(k) = \left(\mathbf{s}_{\mathrm{n},0}(k),\, \mathbf{s}_{\mathrm{n},1}(k),\, \dots,\, \mathbf{s}_{\mathrm{n},L-1}(k)\right)^T . \tag{5.57}$$

The sensor signals are determined by the matrix-vector product $\mathbf{H}_\mathrm{n}^T(k)\mathbf{s}_\mathrm{n}(k)$ according to (3.2). For minimizing the contribution of the interferers to the output signal $y_{\mathbf{w}_\mathrm{RGSC}}(k)$ of the RGSC, the weight vector $\mathbf{w}_{\mathrm{a,s}}(k)$ models the combined system of the inverse of the cascade of $\mathbf{H}_\mathrm{n}(k)$ and of $\mathbf{B}_\mathrm{s}(k)$ in the lower signal path of Fig. 5.4 and the cascade of $\mathbf{H}_\mathrm{n}(k)$ and of $\mathbf{W}_{\mathrm{c,s}}(k)$ in the upper path. Note that the upper signal path and the lower signal path, generally, do not have common zeroes despite presence of the matrix of room impulse responses $\mathbf{H}_\mathrm{n}(k)$ in both signal paths. In the upper signal path, the zeroes are changed by the summation over the M sensor channels by the quiescent weight vector $\mathbf{W}_{\mathrm{c,s}}(k)$. In the lower signal path, the zeroes are changed by the linear combination of the sensor signals by the blocking matrix $\mathbf{B}_\mathrm{s}(k)$. It follows that, generally, common zeroes in the upper signal path and in the lower signal path do not exist, and that the optimum weight vector $\mathbf{w}_{\mathrm{a,s}}(k)$ will model inverses of room impulse responses, which are non-minimum phase. However, as opposed to the SIMO blocking matrix filters $\mathbf{B}_\mathbf{b}(k)$, the interference canceller $\mathbf{w}_{\mathrm{a,s}}(k)$ is a MISO system. Hence, $\mathbf{w}_{\mathrm{a,s}}(k)$ will find an exact model with a limited number of filter coefficients if the number of interferers L is less than the number of sensors M according to the MINT (Sect. 3.2.2). The required number of filter taps N can be determined with (3.43) for $P' = L$. The number of filter taps $N_\mathbf{g}$ is equal to the number of filter taps of the cascade of $\mathbf{H}_\mathrm{n}(k)$ and $\mathbf{B}_\mathrm{s}(k)$. Since $\mathbf{B}_\mathrm{s}(k)$ is the cascade of $\mathbf{w}_\mathrm{c}(k)$ and $\mathbf{B}_\mathbf{b}(k)$, we

[2]Note that the mutual correlation of mutually correlated interferers can be captured in linear impulse responses (Sect. 2.2.1). In this case, $\mathbf{H}_\mathrm{n}(k)$ would represent the mutual correlation of the source signals and the room impulse responses, so that the assumption of mutually uncorrelated interference source signals is not a restriction.

find $N_{\mathbf{g}} = N_{\mathbf{h}_\mathrm{n}} + 2N - 2$. Next, condition (3.42) must be met. We identify the number of filter taps of the cascade $\mathbf{H}_\mathrm{n}(k)$ and $\mathbf{W}_{\mathrm{c,s}}(k)$ as $N_{\mathbf{g}'} = N_{\mathbf{h}_\mathrm{n}} + N - 1$. It is thus necessary to append zeroes to the impulse responses in the upper signal path of Fig. 5.4 for fulfilling (3.42). The length of the filters $\mathbf{w}_{\mathrm{a,s}}(k)$ directly depends on the number of interferers. With increasing difference between the number of interferers L and the number of sensors M, the necessary filter length N reduces hyperbolically. If the number of filter taps in $\mathbf{w}_{\mathrm{a,s}}(k)$ is greater than the value given by (3.43), the optimum weight vector $\mathbf{w}_{\mathrm{a,s,o}}(k)$ is not unique (assuming that the upper signal path in Fig. 5.4 has no common zeroes with the lower signal path). For less filter coefficients, the optimum weight vector $\mathbf{w}_{\mathrm{a,s,o}}(k)$ is unique. This discussion shows that it is possible to trade off spatial degrees of freedom against temporal degrees of freedom. By increasing the number of sensors, the length of the weight vector $\mathbf{w}_{\mathrm{a,s,o}}(k)$ can be reduced and vice versa. This is especially advantageous for adaptive realizations of beamformers where the length of the adaptive filters influences the rate of convergence.

An example of optimum filter coefficients $w_{\mathrm{a,o},n,0}(k)$ over n is depicted in Fig. 5.3b. The synchronization delay is set to $\kappa_{\mathbf{w}_\mathrm{a}} = \kappa_{\mathbf{B}_\mathrm{b}} + N/2$, which is motivated by the symmetry of $\mathbf{w}_{\mathrm{a,s,o,0}}(k)$. We notice the different decays of the magnitudes of $b_{\mathrm{o},n,0}(k)$ and of $w_{\mathrm{a,o},n,0}(k)$, which was predicted using the MINT.

Transient Conditions

We have seen that the weight vector of the interference canceller depends on the room impulse responses between the sources, on the quiescent weight vector, and on the blocking matrix. If these quantities are time-varying, it has to be considered that the interference canceller has to track this time-variance. Especially, for adaptive realizations using recursive adaptation algorithms, this time-variance may be a limiting factor due to insufficient convergence speed or insufficient tracking capability of the adaptive filters. Usage of shorter filters in combination with fast-converging adaptation algorithms for the interference canceller may resolve these problems, since the convergence speed is increased. One would typically limit the number of filter taps to $N = 64 \dots 512$ for a practical realization depending on the time-variance of the acoustic conditions.

5.5 Experimental Results for Stationary Acoustic Conditions

In this section, we study the interference rejection and the cancellation of the desired signal by the blocking matrix of the RGSC for stationary wideband signals and time-invariant environments experimentally. Only time-invariant acoustic scenarios are examined, since the performance of the RGSC is independent of the implementation for stationary conditions, if the blocking matrix and the interference canceller fulfill the LSE optimization criterion. For non-stationary signals, the performance of the RGSC depends on the implementation. Therefore, transient signals and time-varying acoustic environments are considered later in Chap. 7, when our realization of the RGSC is introduced. The experimental results that we describe below match well with the results of the theoretical considerations above.

In Sect. 5.5.1, we define the interference rejection and the cancellation of the desired signal for the context of the RGSC. Section 5.5.2 summarizes the setup of the experiments. In Sect. 5.5.3, the interference rejection of the RGSC is analyzed as a function of parameters which are important for practical applications. Section 5.5.4 illustrates the cancellation of the desired signal by the blocking matrix for various conditions. The cancellation of the desired signal by the RGSC is not considered here since the results would be less significant for practical scenarios due to the assumed time-invariance.

5.5.1 Performance Measures in the Context of the RGSC

In (4.75) and (4.81), the cancellation of the desired signal, $DC(k)$, and the interference rejection, $IR(k)$, were introduced as performance measures for optimum data-dependent beamformers. In this section, we introduce further performance measures which are relevant in the context of the RGSC.

It cannot be expected that the blocking matrix $\mathbf{B}_\mathrm{s}(k)$ cancels the desired signal completely in practical situations. Hence, for measuring the *cancellation of the desired signal by the blocking matrix*, we introduce $DC_{\mathbf{B}_\mathrm{s}}(k)$ (in dB), which is defined as the ratio of the variance of the desired signal at the sensors averaged over the M sensor signals and the variance of the desired signal at the outputs of the blocking matrix averaged over the M output channels of the blocking matrix:

$$DC_{\mathbf{B}_\mathrm{s}}(k) = 10\log_{10}\frac{M\mathrm{tr}\{\mathbf{R}_{\mathbf{dd},\mathrm{s}}(k)\}}{\mathrm{tr}\{\mathbf{B}_\mathrm{s}^T(k)\mathbf{R}_{\mathbf{dd},\mathrm{s}}(k)\mathbf{B}_\mathrm{s}(k)\}}, \tag{5.58}$$

where $\mathbf{R}_{\mathbf{dd},\mathrm{s}}(k) = \mathcal{E}\{\mathbf{d}_\mathrm{s}(k)\mathbf{d}_\mathrm{s}^T(k)\}$, and where $\mathbf{d}_\mathrm{s}(k)$ is defined analogously to (5.16) with $\mathbf{x}(k)$ replaced by $\mathbf{d}(k)$. Correspondingly, the blocking matrix may cancel interference-plus-noise components. Therefore, we define the *interference rejection of the blocking matrix* $IR_{\mathbf{B}_\mathrm{s}}(k)$ (in dB) as the ratio of the variance of interference-plus-noise at the sensors averaged across the M sensor signals and the variance of interference-plus-noise at the outputs of the blocking matrix averaged over the M output channels of the blocking matrix:

$$IR_{\mathbf{B}_\mathrm{s}}(k) = 10\log_{10}\frac{M\mathrm{tr}\{\mathbf{R}_{\mathbf{nn},\mathrm{s}}(k)\}}{\mathrm{tr}\{\mathbf{B}_\mathrm{s}^T(k)\mathbf{R}_{\mathbf{nn},\mathrm{s}}(k)\mathbf{B}_\mathrm{s}(k)\}}, \tag{5.59}$$

where $\mathbf{R}_{\mathbf{nn},\mathrm{s}}(k) = \mathcal{E}\{\mathbf{n}_\mathrm{s}(k)\mathbf{n}_\mathrm{s}^T(k)\}$, and where $\mathbf{n}_\mathrm{s}(k)$ is defined analogously to (5.16) with $\mathbf{x}(k)$ replaced by $\mathbf{n}(k)$. For measuring the cancellation of the desired signal by the RGSC, we define $DC_{\mathbf{w}_{\mathrm{RGSC}}}(k)$ analogously to (4.75) with $\mathbf{R}_{\mathbf{dd}}(k)$ and with $\mathbf{w}(k)$ replaced by $\mathbf{R}_{\mathbf{dd},\mathrm{s}}(k)$ and by $\mathbf{w}_{\mathrm{RGSC}}(k)$, respectively. Furthermore, for comparing the performance of the quiescent weight vector $\mathbf{w}_{\mathrm{c}}(k)$ in the reference path with the performance of the sidelobe cancelling path, we define the interference rejection of the quiescent weight vector, $IR_{\mathbf{w}_{\mathrm{c}}}(k)$, analogously to (4.81) with $\mathbf{w}(k)$ replaced by $\mathbf{w}_{\mathrm{c}}(k)$, and the interference rejection of the RGSC, $IR_{\mathbf{w}_{\mathrm{RGSC}}}(k)$, analogously to (4.81) with $\mathbf{R}_{\mathbf{nn}}(k)$ and with $\mathbf{w}(k)$ replaced by $\mathbf{R}_{\mathbf{nn},\mathrm{s}}(k)$ and by $\mathbf{w}_{\mathrm{RGSC}}(k)$, respectively. For WSS signal, we drop the dependency on k. For narrowband signals, the performance measures are transformed into the DTFT domain, yielding $DC_{\mathrm{f},\mathbf{B}_{\mathrm{s}}}(\omega)$, $DC_{\mathrm{f},\mathbf{w}_{\mathrm{RGSC}}}(\omega)$, $IR_{\mathrm{f},\mathbf{w}_{\mathrm{c}}}(\omega)$, and $IR_{\mathrm{f},\mathbf{w}_{\mathrm{RGSC}}}(\omega)$ for $DC_{\mathbf{B}_{\mathrm{s}}}(k)$, $DC_{\mathbf{w}_{\mathrm{RGSC}}}(k)$, $IR_{\mathbf{w}_{\mathrm{c}}}(k)$, and for $IR_{\mathbf{w}_{\mathrm{RGSC}}}(k)$, respectively. Estimates of $DC_{\mathbf{B}_{\mathrm{s}}}(k)$, $DC_{\mathbf{w}_{\mathrm{RGSC}}}(k)$, $IR_{\mathbf{w}_{\mathrm{c}}}(k)$, $IR_{\mathbf{w}_{\mathrm{RGSC}}}(k)$, $DC_{\mathrm{f},\mathbf{B}_{\mathrm{s}}}(\omega)$, $DC_{\mathrm{f},\mathbf{w}_{\mathrm{RGSC}}}(\omega)$, and $IR_{\mathrm{f},\mathbf{w}_{\mathrm{RGSC}}}(\omega)$ for the experiments are obtained as described in Sect. 4.2.3.

5.5.2 Experimental Setup

For illustrating the properties of the RGSC, the analytically optimum weight vectors of the RGSC according to (5.27) and (5.36) require prohibitively high computational complexity because of the inversion of matrices of very high dimension (e.g., matrices of size $MN^3 \times MN^3$ for the interference canceller (5.36)). Since recursive optimization of the weight vectors circumvents matrix inversions, we use the realization of Chap. 7 which is based on recursive adaptive filtering for computing the optimum weight vectors. We only consider stationary signal characteristics, for which the behavior of the RGSC is independent of the realization. Analogous results can thus also be obtained with the time-domain realization after [HSH99] or with all other implementations.

For all scenarios, the blocking matrix is adapted until the increase of the cancellation of the desired signal, $DC_{\mathbf{B}_{\mathrm{s}}}(k)$, over a data block of length 20000 samples is less than 0.02 dB with a stepsize of 0.1. Correspondingly, the interference canceller is adapted until the interference rejection $IR_{\mathbf{w}_{\mathrm{RGSC}}}(k)$ increases by less than 0.02 dB over a data block of the same length using the same stepsize. For estimating the performance measures described in Sect. 5.5.1, it is necessary to have access to the individual mixing components at the sensors and at the desired output signals. Since in a realistic system, the mixing components are not known, we recorded the sensor signals individually, and we used the simulation structure described in App. B. The experiments are conducted in the environment with $T_{60} = 250$ ms for presence of temporally white noise, if not mentioned otherwise. (See App. B for a detailed description of the experimental conditions.) Sensor signals are highpass-filtered with cut-off frequency 200 Hz. The desired source is located in broadside direction ($\theta_{\mathrm{d}} = \pi/2$) at a distance of 60 cm from the array center. The positions of the interferers depend on the scenario. The quiescent weight vector is realized as a uniformly weighted beamformer: $\mathbf{w}_{\mathrm{c}}(k) = 1/M \cdot \mathbf{1}_{M\times 1}$.

5.5.3 Interference Rejection of the RGSC

In this section, we study the interference rejection of the RGSC as a function of various parameters. First, we illustrate the interference rejection as a function of the number of interferers L, of the number of sensors M, and of the filter length N. We will see that the MINT explains the dependencies of the interference rejection of the RGSC on these parameters. Second, the dependency of frequency is examined. Third, we analyze the influence of the position of the interferer.

Dependency of the Number of Interferers, of the Number of Sensors, and of the Filter Length

In Sect. 5.4.2, we have seen that the length N of the optimum weight vector of the interference canceller strongly depends on the number of interferers L and on the number of sensors M. Generally, for temporally WSS signals, the interference rejection of the RGSC increases with increasing number of spatial and temporal degrees of freedom, i.e., with increasing number of sensors and with increasing filter lengths.

Figure 5.5 and Fig. 5.6a show the interference rejection of the RGSC, $IR_{\mathbf{w}_{\mathrm{RGSC}}}$, as a function of the number of interferers L, of the number of

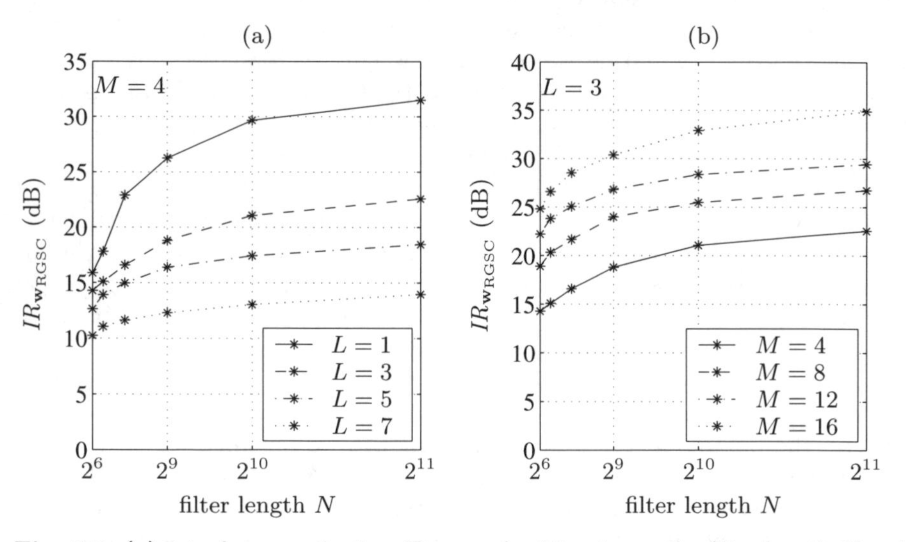

Fig. 5.5. (a) Interference rejection $IR_{\mathbf{w}_{\mathrm{RGSC}}}$ for $M = 4$ over the filter length N and as a function of the number of interferers L; **(b)** interference rejection $IR_{\mathbf{w}_{\mathrm{RGSC}}}$ for $L = 3$ over the filter length N as a function of the number of sensors M; multimedia room with $T_{60} = 250$ ms

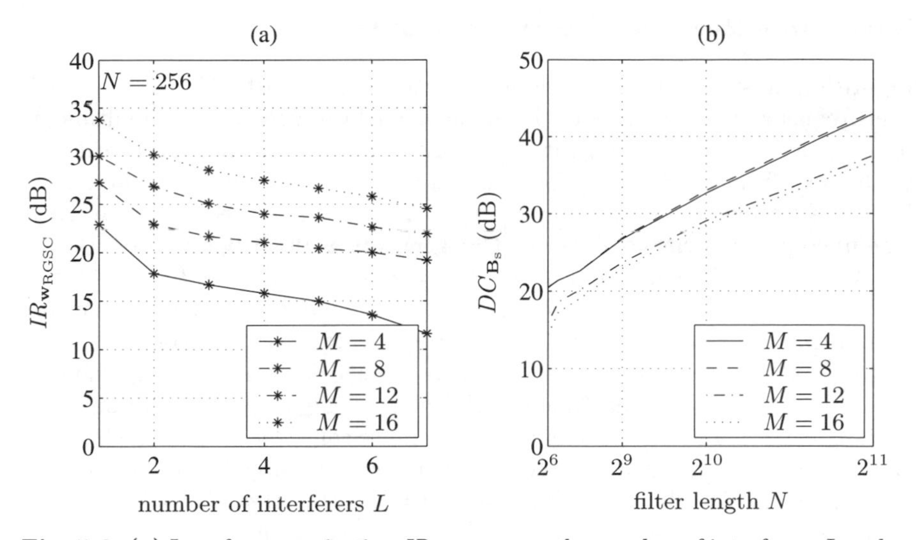

Fig. 5.6. (a) Interference rejection $IR_{\mathbf{w}_{\mathrm{RGSC}}}$ over the number of interferers L and as a function of the number of sensors M; **(b)** cancellation $DC_{\mathbf{B}_s}$ of the desired signal by the blocking matrix over the filter length N and as a function of the number of the number of sensors M; multimedia room with $T_{60} = 250\,\mathrm{ms}$

sensors M, and of the filter length N of the interference canceller. The interferers are located at a distance of 1.2 m from the array center at directions $\theta_n = 0, \pi/18, \ldots, \pi/3$ with the first interferer being in endfire position. The length of the filters of the blocking matrix is 256. The inter-dependencies which have been described earlier in Sect. 5.4.2 can clearly be noticed: Consider, for example, Fig. 5.5a, where $IR_{\mathbf{w}_{\mathrm{RGSC}}}$ is depicted over the filter length N of the interference canceller as a function of the number of interferers L for $M = 4$. It may be seen that for $L = 3$ and for $N = 2048$, approximately the same interference rejection of about $IR_{\mathbf{w}_{\mathrm{RGSC}}} = 22.6\,\mathrm{dB}$ is obtained as with $L = 1$, $N = 256$. The filters must be greater by a factor of 2^3 for $L = 3$ than for $L = 1$ in order to obtain the same interference rejection in our simulation. The theoretical relative filter lengths can be predicted using (3.43). We set $L := P'$, we determine N for both numbers of interferers $L = 1, 3$, and we calculate the ratio of the resulting filter lengths N. The common factor $(N_{\mathbf{g}} - 1)$ cancels out, and we obtain the theoretical relative factor of 9 which matches well with the experimental factor 2^3. The same procedure can be applied for other combinations of L, M, and N as well, yielding corresponding results.[3]

[3]The different asymptotes in Fig. 5.5a for large N ($N = 2048$) for different numbers of interferers L can be explained by the different positions of the interferers and by the different convergence behavior of the adaptive weight vector. Similarly, the inconsistent asymptotic behavior in Fig. 5.5b can be explained by the different array geometries. In Fig. 5.6a, the different asymptote for $M = 4$ for large L, results from

Interference Rejection as a Function of Frequency

Figure 5.7 illustrates the interference rejection of the RGSC, $IR_{\mathrm{f},\mathbf{w}_{\mathrm{RGSC}}}(\omega)$, over frequency as a function of the acoustic environment. For the multimedia room with $T_{60} = 250$ ms and with $T_{60} = 400$ ms, the interference is directional white noise from endfire direction at a distance of 1.2 m from the array center. For the passenger cabin of the car, realistic car noise is used as interference. The microphone array consists of $M = 4$ sensors as described in App. B.

For the multimedia room with $T_{60} = 250$ ms and with $T_{60} = 400$ ms, it can first be observed that $IR_{\mathrm{f},\mathbf{w}_{\mathrm{RGSC}}}(\omega)$ is minimum at low frequencies ($f < 2$ kHz). This can be explained by the fact that the reverberation time is longer at low frequencies than at high frequencies since high frequencies are better absorbed by the paneling of the room. Longer reverberation times require longer filter lengths N of the interference canceller so that the interference rejection of the RGSC reduces for a given fixed filter length N. Second, $IR_{\mathrm{f},\mathbf{w}_{\mathrm{RGSC}}}(\omega)$ is locally maximum around $f = 2.5$ kHz. Here, the beampattern of the quiescent weight vector (uniformly weighted beamformer) has a minimum for the DOA of the interferer so that interference is better suppressed by the quiescent weight vector. (See Fig. 4.7a.) For a limited number of filter taps of the interference canceller, this translates to a better interference rejection of the RGSC. For $f > 3$ kHz the interference rejection of the RGSC is reduced. Here, the beampattern of the quiescent weight vector has maxima for the DOA of the interferer due to spatial aliasing.

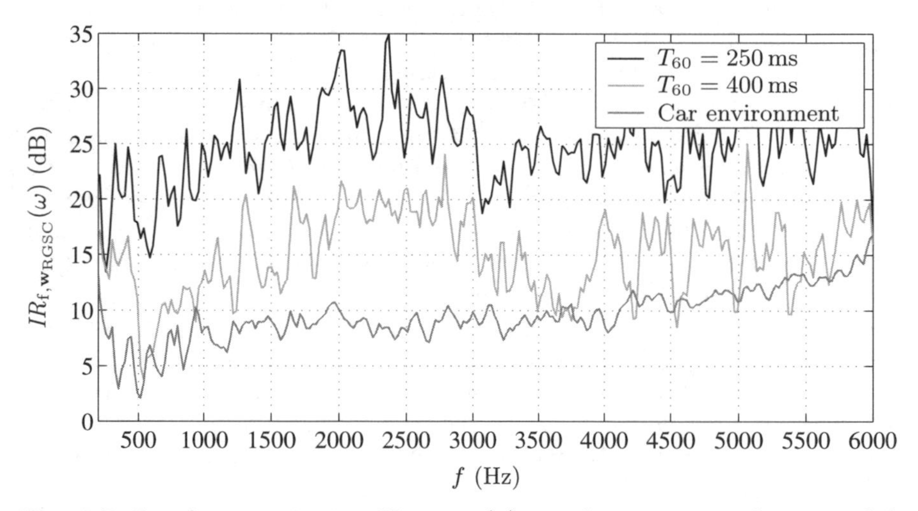

Fig. 5.7. Interference rejection $IR_{\mathrm{f},\mathbf{w}_{\mathrm{RGSC}}}(\omega)$ over frequency as a function of the acoustic environment; $M = 4$, $N = 256$, $L = 1$

---

the insufficient number of spatial degrees of freedom of the beamformer for $L \geq M$, where the MINT cannot be applied.

For the passenger cabin of the car, the interference rejection is lower than for the multimedia room due to the diffuse character of the car noise. The minimum at low frequencies results from the low directivity of the beamformer at low frequencies. The increase of the interference rejection with increasing frequencies results from directional signal components of the car noise, which is used for this experiment.

Dependency of the Position of the Interferer

In this experiment, we study the influence of the DOA of the interference on the interference rejection of the RGSC compared to a simple uniformly weighted beamformer. We consider presence of temporally white noise, a male speaker, and a female speaker. The distance between the array center and the position of the interferers is 1.2 m. The sensor array with $M = 4$ is used. The speech signals have stronger low-frequency components than the white noise signal, as it is typical for speech signals. The female voice has less low-frequency components than the male voice [RS78].

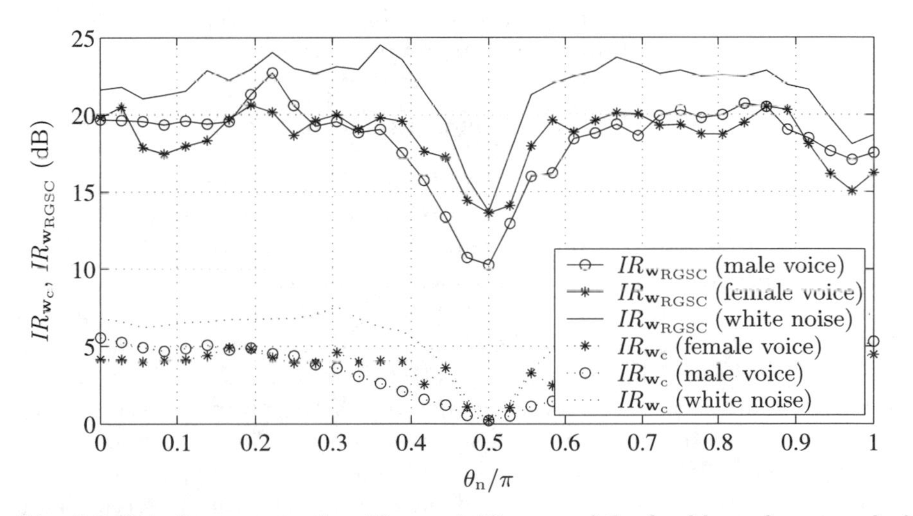

Fig. 5.8. Interference rejection $IR_{\mathbf{w}_c}$ and $IR_{\mathbf{w}_{RGSC}}$ of the fixed beamformer and of the RGSC, respectively, as a function of the DOA of the interferer for fixed steering direction; multimedia room with $T_{60} = 250$ ms, $M = 4$, $N = 256$

The results are depicted in Fig. 5.8. As we would expect, $IR_{\mathbf{w}_c}$ is much less than $IR_{\mathbf{w}_{RGSC}}$. Then, $IR_{\mathbf{w}_{RGSC}}$ is lower for speech than for temporally white noise due to the combination of stronger low-frequency components of the source signals and reduced interference rejection of the beamformer at low frequencies. It can further be noticed that $IR_{\mathbf{w}_{RGSC}} \neq 0$ for $\theta_n = \pi/2$, despite presence of the 'broadside' constraint for the desired signal. This can be explained by the different distances from the sources to the array center, which is 1.2 m for the interferer and 0.6 m for the desired source. Since both sources are located in the near field of the array, the array is spatially selective for range. The minimum of $IR_{\mathbf{w}_{\mathrm{RGSC}}}$ for broadside direction results from the very small differences of the inter-sensor propagation delays for the two distances from the sources to the array center. The male voice is less suppressed than the female voice around the steering direction of the array ($\theta_{\mathrm{n}} = \pi/2$), since the male voice has stronger components at low frequencies than the female voice and since the spatial selectivity of the array is minimum at low frequencies.

5.5.4 Cancellation of the Desired Signal by the Blocking Matrix

In this section, the cancellation of the desired signal by the blocking matrix is analyzed. First, the influence of the filter length N and of the number of sensors M on the cancellation of the desired signal by the blocking matrix is studied. Second, we illustrate the cancellation of the desired signal as a function of frequency. Third, the dependency of the position of the desired source is examined. Finally, the performance of the blocking matrix is illustrated for simultaneous presence of desired signal and interference.

Dependency of the Number of Sensors and of the Filter Length

Figure 5.6b shows the cancellation of the desired signal by the blocking matrix over the filter length N of the filters of the blocking matrix as a function of the number of sensors M. Interestingly, the cancellation of the desired signal $DC_{\mathbf{B}_{\mathrm{s}}}$ does not necessarily increase with an increasing number of sensors. This can be explained by considering that the blocking matrix with the desired source signal $s_{\mathbf{p}_{\mathrm{d}}}(kT_{\mathrm{s}})$ as input signal can be modeled as the SIMO system after Fig. 5.2. The optimum blocking matrix filters will model the combined system of room impulse responses $\mathbf{H}_{\mathrm{d}}(k)$ and of the inverse of the cascade of $\mathbf{H}_{\mathrm{d}}(k)$ and of the quiescent weight vector $\mathbf{w}_{\mathrm{c}}(k)$. The cascade of $\mathbf{H}_{\mathrm{d}}(k)$ and $\mathbf{w}_{\mathrm{c}}(k)$ is a SISO system $\mathbf{g}(k)$ with the room impulse responses $\mathbf{h}_{\mathrm{d},m}(k)$ convolved with the filters $\mathbf{w}_{\mathrm{c},m}(k)$ and summed up over the M sensor channels. The summation does not affect the length of the impulse response of the cascade of $\mathbf{H}_{\mathrm{d}}(k)$ and $\mathbf{w}_{\mathrm{c}}(k)$. However, the distribution of the zeroes of the corresponding room transfer function is changed. This statement is visualized in Fig. 5.9a, which shows the cumulative relative frequency of the absolute value of the zeroes of the transfer function of $\mathbf{g}(k)$ for the interval [0.9975; 0.999].

It can be seen that the distance between the unit circle and the zeroes of $\mathbf{g}(k)$ is largest for $M = 8$ and that the distance decreases for $M = 4$, 12, 16.[4]

[4] A trend of the dependency of $DC_{\mathbf{B}_{\mathrm{s}}}$ on the number of sensors could not be found by experiments in different acoustic environments with different geometrical setups.

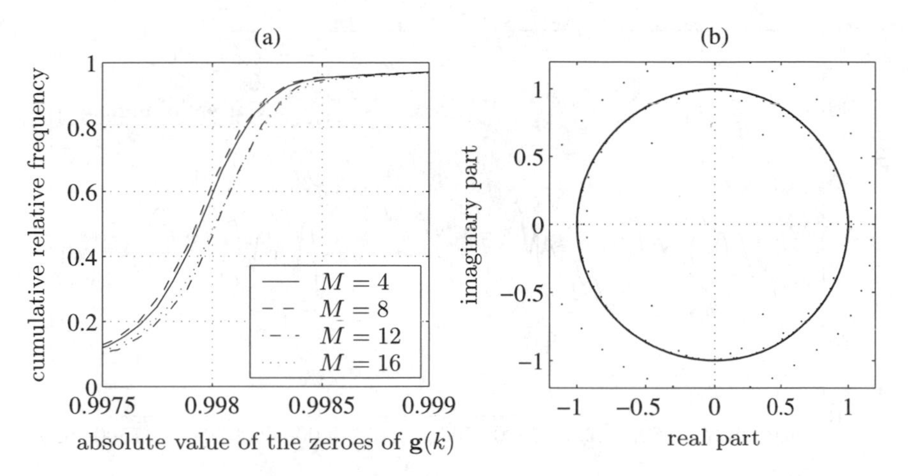

Fig. 5.9. (a) Cumulative relative frequency of the absolute value of the zeroes of the transfer function of the SISO system $\mathbf{g}(k)$ as a function of the number of sensors M, **(b)** distribution of the zeroes of the transfer function of $\mathbf{g}(k)$ around the unit circle for $M = 4$; multimedia room with $T_{60} = 250\,\mathrm{ms}$

This means that the number of filter coefficients for modeling the inverse of $\mathbf{g}(k)$ increases for $M = 4, 12, 16$. This yields in our example the behavior that is illustrated in Fig. 5.6b: The cancellation of the desired signal by the blocking matrix, $DC_{\mathbf{B}_s}$, is highest for $M = 8$ and decreases for $M = 4, 12, 16$. Figure 5.9b shows the distribution of the zeroes of the transfer function of $\mathbf{g}(k)$ around the unit circle for $M = 4$ in order to show the distribution over a wider range of absolute values of the zeroes than in Fig. 5.9a.

Cancellation of the Desired Signal as a Function of Frequency

Figure 5.10 shows the cancellation of the desired signal by the blocking matrix, $DC_{\mathrm{f},\mathbf{B}_s}(\omega)$, over frequency as a function of the acoustic environment. We use the array with $M = 4$ sensors and the RGSC with $N = 256$. As one would expect, $DC_{\mathrm{f},\mathbf{B}_s}(\omega)$ decreases with increasing reverberation time of the environment (which implies stronger reflections from other DOAs). The cancellation of the desired signal $DC_{\mathrm{f},\mathbf{B}_s}(\omega)$ is highest for the passenger cabin of the car with $T_{60} = 50\,\mathrm{ms}$, and it is lowest for the multimedia room with $T_{60} = 400\,\mathrm{ms}$.

Dependency of the Position of the Desired Source

Next, we consider the cancellation of the desired signal by the blocking matrix as a function of the DOA for white noise, for male speech, and for female speech. The steering direction of the RGSC is always broadside direction.

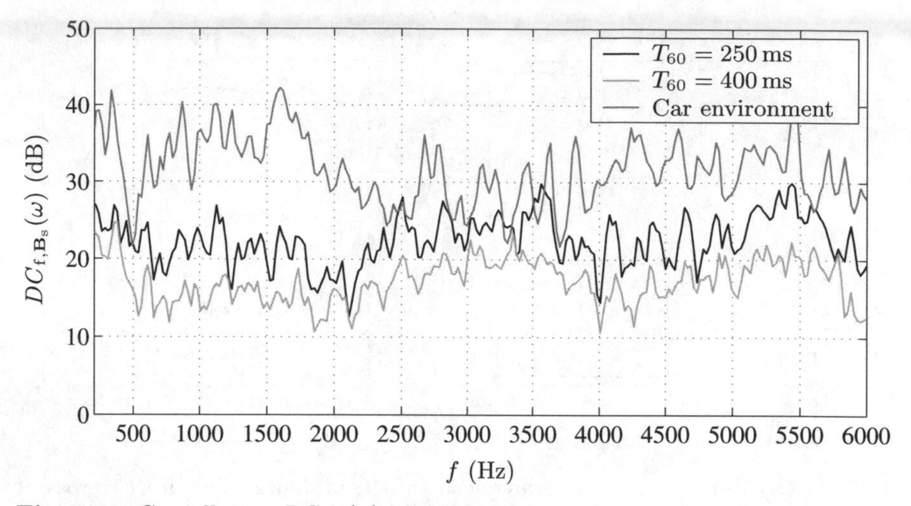

Fig. 5.10. Cancellation $DC_{\mathbf{B}_s}(\omega)$ of the desired signal by the blocking matrix over frequency as a function of the acoustic environment; $M = 4$, $N = 256$

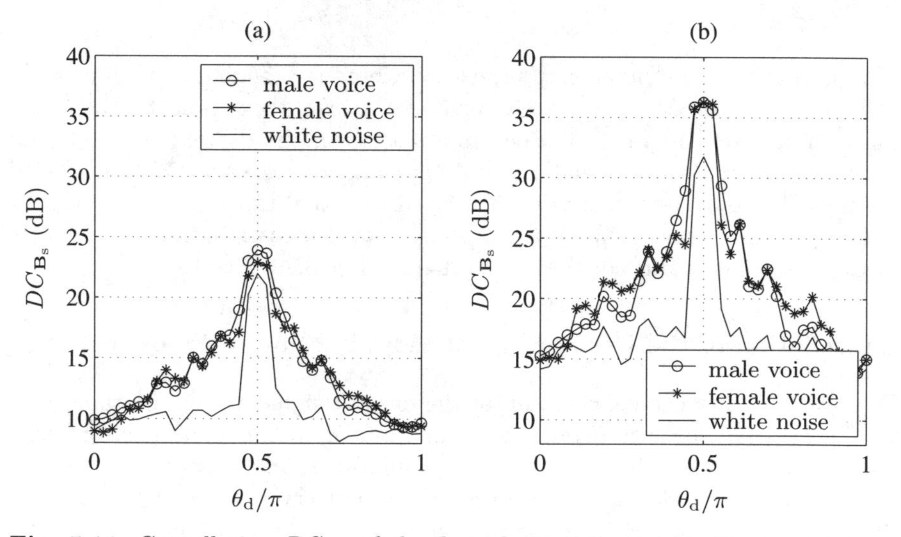

Fig. 5.11. Cancellation $DC_{\mathbf{B}_s}$ of the desired signal by the blocking matrix as a function of the DOA of the desired signal: **(a)** $N = 256$, **(b)** $N = 1024$; multimedia room with $T_{60} = 250$ ms, $M = 4$

The filters of the blocking matrix are adapted for each DOA of the desired signal until convergence.

Figure 5.11a and Fig. 5.11b show $DC_{\mathbf{B}_s}$ over θ_d for $N = 256$ and for $N = 1024$, respectively. We see that the filters of the blocking matrix have a high spatial selectivity. Signals from the steering direction are much better suppressed than signals from other directions. Obviously, if the direct path of the desired signal is summed up incoherently by the quiescent weight vector, more filter coefficients are required for the same $DC_{\mathbf{B}_s}$ than for coherent summation. This observation is reflected by the analytical solution of the optimum filters of the blocking matrix (5.47). For coherent superposition, the denominator of (5.47) is greater than for incoherent superposition. This yields optimum weight vectors of the blocking matrix that decay faster than for incoherent superposition, which results in the observed behavior of $DC_{\mathbf{B}_s}$ for coefficient vectors of limited length.

Comparing Fig. 5.11a with $N = 256$ and Fig. 5.11b with $N = 1024$, we see that $DC_{\mathbf{B}_s}$ for $N = 1024$ is greater for all θ_d than for $N = 256$. However, the spatial selectivity of the blocking matrix is preserved. This selectivity only vanishes for infinitely long weight vectors since the optimum filters for total suppression of the desired signal are infinitely long. (See Sect. 5.4.1.) For presence of male speech and female speech, the spatial selectivity of the blocking matrix is reduced due to stronger low-frequency components and due to the consequentially reduced spatial selectivity of the quiescent weight vector.

This spatial selectivity has two consequences on the design of the RGSC for practical applications: First, for maximum robustness against cancellation of the desired signal by the RGSC, the spatio-temporal blocking matrix only allows for variations in a limited interval of DOAs around the steering direction of the array. This interval can be adjusted in a certain range by choosing a design for the quiescent weight vector which is more robust against steering errors than the simple uniformly weighted beamformer. (See Sect. 4.3.) Second, consider that the desired source is located within the interval of directions with high desired signal suppression and that interferers are located in direction with low signal suppression. Then, robustness against interference rejection by the blocking matrix, $IR_{\mathbf{B}_s}$, is partly included in the spatio-temporal blocking matrix. This may yield greater interference rejection of the RGSC, $IR_{\mathbf{w}_{\mathrm{RGSC}}}$, in situations where adaptation of the sidelobe cancelling path is difficult.

Determination of the Blocking Matrix During Double-Talk

In Sect. 5.2.2, we have seen that the blocking matrix is determined for presence of only the desired signal in order to suppress only desired signal and not interference or noise. However, in practical situations, interference may be active during adaptation of the blocking matrix, which may lead to cancellation of the interference by the blocking matrix. This reduces the interference rejection of the RGSC since interference components which are cancelled by the blocking matrix cannot be suppressed by the interference canceller at the output of the RGSC. Therefore, we study in this section the dependency of the cancellation of the desired signal, $DC_{\mathbf{B}_s}$, and of the interference rejection by the blocking matrix, $IR_{\mathbf{B}_s}$, during adaptation of the blocking matrix on the SINR at the sensors, $SINR_{\mathrm{in}}$.

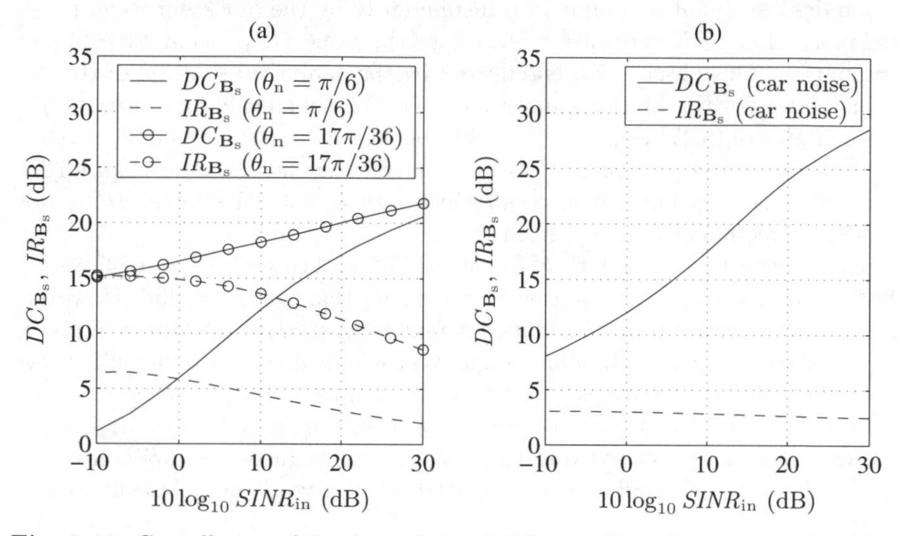

Fig. 5.12. Cancellation of the desired signal $DC_{\mathbf{B}_s}$ and interference rejection $IR_{\mathbf{B}_s}$ by the blocking matrix as a function of $10 \log_{10} SINR_{\mathrm{in}}$ for **(a)** the multimedia room with $T_{60} = 250$ ms with directional temporally white interference and for **(b)** the passenger cabin of the car for presence of car noise; $M = 4$, $N = 256$

The blocking matrix is, thus, adapted for variable $SINR_{\mathrm{in}}$. In Fig. 5.12a, the multimedia room with $T_{60} = 250$ ms is considered for presence of temporally white noise that arrives from a direction that is far from the steering direction of the beamformer ($\theta_n = \pi/6$) and for a direction that is close to the steering direction of the beamformer ($\theta_n = 17\pi/36$). For $\theta_n = \pi/6$, it may be seen that $DC_{\mathbf{B}_s}$ and $IR_{\mathbf{B}_s}$ are small for low $SINR_{\mathrm{in}}$. The blocking matrix with its single spatial degree of freedom concentrates on the suppression of the stronger interference, and $DC_{\mathbf{B}_s}$ is low in turn. However, due to the high spatial selectivity of the blocking matrix, the interference is not efficiently suppressed either, and $IR_{\mathbf{B}_s}$ is low, too. With increasing $SINR_{\mathrm{in}}$, $DC_{\mathbf{B}_s}$ increases, and $IR_{\mathbf{B}_s}$ reduces. The blocking matrix concentrates on the stronger desired signal. The suppression of the desired signal is not disturbed by the spatial selectivity of the blocking matrix. For $\theta_n = 17\pi/36$, the interferer is close to the desired source in broadside direction, where the blocking matrix provides maximum cancellation of the sensor signals in our setup. For $10 \log_{10} SINR_{\mathrm{in}} = -10$ dB, desired signal and interferer are equally well suppressed, i.e., $DC_{\mathbf{B}_s} \approx IR_{\mathbf{B}_s} \approx 15$ dB. The blocking matrix places a spatial zero between $\theta_{\mathrm{n}} = 17\pi/36$ and $\theta_{\mathrm{d}} = \pi/2$ such that $DC_{\mathbf{B}_{\mathrm{s}}}$ and $IR_{\mathbf{B}_{\mathrm{s}}}$ are maximized. With increasing $SINR_{\mathrm{in}}$, $DC_{\mathbf{B}_{\mathrm{s}}}$ increases, and $IR_{\mathbf{B}_{\mathrm{s}}}$ reduces, since the blocking matrix concentrates more and more on the cancellation of the stronger desired signal. However, since interferer and desired source are so close to each other, $IR_{\mathbf{B}_{\mathrm{s}}}$ is larger than for $\theta_{\mathrm{n}} = \pi/6$ even for high values of $SINR_{\mathrm{in}}$.

In Fig. 5.12b, the car environment with presence of car noise is considered. In contrast to Fig. 5.12a, $DC_{\mathbf{B}_{\mathrm{s}}}$ is greater for all $SINR_{\mathrm{in}}$, while $IR_{\mathbf{B}_{\mathrm{s}}}$ is lower for all $SINR_{\mathrm{in}}$. This result reflects that diffuse car noise disturbs the function of the blocking matrix less than a second directional sound source. Diffuse interference can be efficiently suppressed by the interference canceller (as indicated in previous experiments) though the blocking matrix may be adapted for presence of interference and noise. However, $DC_{\mathbf{B}_{\mathrm{s}}}$ is still low for small values of $SINR_{\mathrm{in}}$, so that robustness against cancellation of the desired signal by the RGSC is decreased. Therefore, robustness improvement of the RGSC is particularly important for such noise fields for real-world implementations.

In conclusion, this experiment shows that directional interference has to be excluded from the adaptation of the blocking matrix in order to assure high interference rejection and high robustness against cancellation of the desired signal of the RGSC. Diffuse interference, as, e.g., car noise, disturbs the function of the blocking matrix less so that the exclusion of diffuse noise is less important. This experimental result will be used in the following section about the adaptation control strategy of the RGSC.

5.6 Strategy for Determining Optimum RGSC Filters

In this section, an adaptation strategy in the eigenspace of the desired signal and of interference-plus-noise is described for determining the optimum blocking matrix and the optimum interference canceller. The adaptation strategy is based on non-stationarity of the sensor signals. This method will be applied in Chap. 7 in the DFT domain for realizing the RGSC for real-time applications. The determination of the blocking matrix is explained in Sect. 5.6.1, the calculation of the interference canceller is described in Sect. 5.6.2. The influence of stationarity of the desired signal and of interference-plus-noise is shown.

5.6.1 Determination of the Optimum Filters for the Blocking Matrix

Due to the dependency of the optimum coefficient vectors $\mathbf{B}_{\mathbf{b},\mathrm{o}}(k)$ of the blocking matrix on the spatio-temporal correlation matrix $\mathbf{\Phi}_{\mathbf{dd},\mathrm{s}}(k)$ w.r.t. the desired signal, the spatio-temporal correlation matrix $\mathbf{\Phi}_{\mathbf{dd},\mathrm{s}}(k)$ must be known for calculation of $\mathbf{B}_{\mathbf{b},\mathrm{o}}(k)$. For realistic situations, only $\mathbf{\Phi}_{\mathbf{xx},\mathrm{s}}(k) = \mathbf{\Phi}_{\mathbf{dd},\mathrm{s}}(k) + \mathbf{\Phi}_{\mathbf{nn},\mathrm{s}}(k)$ can be determined, which contains both desired signal and interference. The sample spatio-temporal correlation matrix $\mathbf{\Phi}_{\mathbf{nn},\mathrm{s}}(k)$ is defined according to (5.24) with $\mathbf{D}_{\mathrm{s}}(k)$ replaced by $\mathbf{N}_{\mathrm{s}}(k)$. If $\mathbf{B}_{\mathbf{b},\mathrm{o}}(k)$ is determined for both desired signal presence and interference presence, desired signal and interference are suppressed by the blocking matrix, which reduces the array gain of the RGSC.[5] Therefore, $\mathbf{B}_{\mathbf{b},\mathrm{o}}(k)$ should only be determined if only the desired signal is active. However, this yields a very low tracking capability, so that non-stationary signal characteristics cannot be tracked during double-talk. For enabling the calculation of $\mathbf{B}_{\mathbf{b},\mathrm{o}}(k)$ during double-talk, and for improving the tracking capability in turn, we propose to detect eigenvectors of $\mathbf{\Phi}_{\mathbf{xx},\mathrm{s}}(k)$, which are orthogonal to $\mathbf{\Phi}_{\mathbf{nn},\mathrm{s}}(k)$, and to determine $\mathbf{B}_{\mathbf{b},\mathrm{o}}(k)$ only in the space which is spanned by these eigenvectors. The eigendecomposition of $\mathbf{\Phi}_{\mathbf{xx},\mathrm{s}}(k)$ is introduced as

$$\mathbf{\Phi}_{\mathbf{xx},\mathrm{s}}(k) = \mathbf{U}_{\mathbf{X}_{\mathrm{s}}}(k)\mathbf{\Sigma}^2_{\mathbf{X}_{\mathrm{s}}}(k)\mathbf{U}^T_{\mathbf{X}_{\mathrm{s}}}(k)\,. \tag{5.60}$$

The eigenvectors are arranged such that $\mathbf{U}_{\mathbf{X}_{\mathrm{s}}}(k)$ and $\mathbf{\Sigma}_{\mathbf{X}_{\mathrm{s}}}(k)$ can be described by

$$\mathbf{U}_{\mathbf{X}_{\mathrm{s}}}(k) = \left(\mathbf{U}_{\mathbf{D}_{\mathrm{s}}}(k),\, \mathbf{U}_{\mathbf{N}_{\mathrm{s}}}(k),\, \mathbf{U}_{\mathbf{D}_{\mathrm{s}}+\mathbf{N}_{\mathrm{s}}}(k)\right)\,, \tag{5.61}$$

$$\mathbf{\Sigma}_{\mathbf{X}_{\mathrm{s}}}(k) = \mathrm{diag}\left\{\left(\mathbf{\Sigma}_{\mathbf{D}_{\mathrm{s}}}(k),\, \mathbf{\Sigma}_{\mathbf{N}_{\mathrm{s}}}(k),\, \mathbf{\Sigma}_{\mathbf{D}_{\mathrm{s}}+\mathbf{N}_{\mathrm{s}}}(k)\right)\right\}\,, \tag{5.62}$$

respectively. The matrices $\mathbf{U}_{\mathbf{D}_{\mathrm{s}}}(k)$ and $\mathbf{U}_{\mathbf{N}_{\mathrm{s}}}(k)$ capture the eigenvectors which span subspaces of the eigenspace of $\mathbf{\Phi}_{\mathbf{dd},\mathrm{s}}(k)$ and $\mathbf{\Phi}_{\mathbf{nn},\mathrm{s}}(k)$, respectively, such that $\mathbf{U}_{\mathbf{D}_{\mathrm{s}}}(k)$ is orthogonal to $\mathbf{U}_{\mathbf{N}_{\mathrm{s}}}(k)$, i.e.,

$$\mathbf{U}^T_{\mathbf{D}_{\mathrm{s}}}(k)\mathbf{U}_{\mathbf{N}_{\mathrm{s}}}(k) = \mathbf{0}_{MN^2 \times MN^2}\,. \tag{5.63}$$

The columns of $\mathbf{U}_{\mathbf{D}_{\mathrm{s}}+\mathbf{N}_{\mathrm{s}}}(k)$ then span the space

$$\begin{aligned} &\mathrm{span}\left\{\mathbf{U}_{\mathbf{D}_{\mathrm{s}}+\mathbf{N}_{\mathrm{s}}}(k)\right\} = \\ &= \mathrm{span}\left\{\mathbf{\Phi}_{\mathbf{xx},\mathrm{s}}(k)\right\} \cap \left(\mathrm{span}\left\{\mathbf{U}_{\mathbf{D}_{\mathrm{s}}}(k)\right\} \cup \mathrm{span}\left\{\mathbf{U}_{\mathbf{N}_{\mathrm{s}}}(k)\right\}\right)^{\perp}\,. \end{aligned} \tag{5.64}$$

Span$\{\mathbf{U}_{\mathbf{D}_{\mathrm{s}}}(k)\}$ and span$\{\mathbf{U}_{\mathbf{N}_{\mathrm{c,s}}}(k)\}$ thus correspond to the space which contain only desired signal and only interference, respectively. The space span$\{\mathbf{U}_{\mathbf{D}_{\mathrm{s}}+\mathbf{N}_{\mathrm{s}}}(k)\}$ is the space, which contains both desired signal and interference components. The diagonal matrices $\mathbf{\Sigma}_{\mathbf{D}_{\mathrm{s}}}(k)$, $\mathbf{\Sigma}_{\mathbf{N}_{\mathrm{s}}}(k)$, and $\mathbf{\Sigma}_{\mathbf{D}_{\mathrm{s}}+\mathbf{N}_{\mathrm{s}}}(k)$ capture the singular values, which correspond to the left singular vectors $\mathbf{U}_{\mathbf{D}_{\mathrm{s}}}(k)$, $\mathbf{U}_{\mathbf{N}_{\mathrm{s}}}(k)$, and $\mathbf{U}_{\mathbf{D}_{\mathrm{s}}+\mathbf{N}_{\mathrm{s}}}(k)$, respectively.

For calculation of $\mathbf{B}_{\mathbf{b},\mathrm{o}}(k)$ only for the desired signal without knowledge of $\mathbf{\Phi}_{\mathbf{dd},\mathrm{s}}(k)$, we thus replace in (5.25) the matrix $\mathbf{\Phi}_{\mathbf{dd},\mathrm{s}}(k)$ by the spatio-temporal correlation matrix

$$\tilde{\mathbf{\Phi}}_{\mathbf{dd},\mathrm{s}}(k) = \mathbf{U}_{\mathbf{D}_{\mathrm{s}}}(k)\mathbf{\Sigma}^2_{\mathbf{D}_{\mathrm{s}}}(k)\mathbf{U}^T_{\mathbf{D}_{\mathrm{s}}}(k)\,. \tag{5.65}$$

Generally, span$\{\mathbf{U}_{\mathbf{D}_{\mathrm{s}}+\mathbf{N}_{\mathrm{s}}}(k)\} \neq \{0\}$, which means that desired signal components which are contained in span$\{\mathbf{U}_{\mathbf{D}_{\mathrm{s}}+\mathbf{N}_{\mathrm{s}}}(k)\}$ cannot be suppressed by the

[5]Sensor noise can generally be neglected within this context, since in typical applications the SNR at the sensors is high.

blocking matrix. These desired signal components leak through the blocking matrix and cause distortion of the desired signal at the output of the RGSC.

In order to reduce these distortions, $\tilde{\mathbf{\Phi}}_{\mathbf{dd},\mathrm{s}}(k)$ can be augmented by exponentially weighted past spatio-temporal correlation matrices $\tilde{\mathbf{\Phi}}_{\mathbf{dd},\mathrm{s}}(k-i)$, $i = 1, 2, \ldots, k$. That is,

$$\hat{\mathbf{\Phi}}_{\mathbf{dd},\mathrm{s}}(k) = \sum_{i=1}^{k} \beta^{k-i} \tilde{\mathbf{\Phi}}_{\mathbf{dd},\mathrm{s}}(i) , \tag{5.66}$$

where $0 < \beta \leq 1$ is the exponential forgetting factor. This incorporates more information about $\mathbf{\Phi}_{\mathbf{dd},\mathrm{s}}(k)$ and extends $\mathrm{span}\{\tilde{\mathbf{\Phi}}_{\mathbf{dd},\mathrm{s}}(k)\}$. This exponential averaging over past data blocks has a spatial and a temporal whitening effect if $\mathbf{\Phi}_{\mathbf{dd},\mathrm{s}}(k)$ is temporally non-stationary, i.e., if the spatial and temporal characteristics of the desired signal are time-varying. Some of the flexibility of the constraints (5.9) and (5.10) (see Sect. 5.2.1), which depend on the characteristics of the desired signal, are traded off against improved output signal quality of the RGSC. Obviously, this averaging fails if the desired signal components contained in $\mathrm{span}\{\mathbf{U}_{\mathbf{D}_\mathrm{s}+\mathbf{N}_\mathrm{s}}(k)\}$ change abruptly and are not contained in $\mathrm{span}\{\hat{\mathbf{\Phi}}_{\mathbf{dd},\mathrm{s}}(k)\}$.

If the interference contains stationary signal components which are not orthogonal to the desired signal, the corresponding components of the desired signal cannot be estimated at all. The blocking matrix cannot cancel these components, which leads to distortion of the desired signal. As a solution, interference components may be suppressed by an efficient single-channel noise-reduction technique that is integrated into the input signals of the blocking matrix filters $\mathbf{B}_\mathbf{b}(k)$ so that these interference components are not seen by the blocking matrix. Stationarity of the desired signal has no influence here. However, the time-variance of the noise-reduction may affect the convergence of the blocking matrix.[6]

5.6.2 Determination of the Optimum Filters for the Interference Canceller

As we have seen, it can not be assured that the desired signal is always cancelled by the blocking matrix. Therefore, the weight vector $\mathbf{w}_{\mathrm{a,s,o}}(k)$ should only be determined in the subspace which is orthogonal to $\mathrm{span}\{\mathbf{\Phi}_{\mathbf{dd},\mathrm{ss}}(k)\}$, where $\mathbf{\Phi}_{\mathbf{dd},\mathrm{ss}}(k)$ is defined according to (5.35) with $\mathbf{X}_\mathrm{s}(k)$ replaced by $\mathbf{D}_\mathrm{s}(k)$. Since the procedure for calculating $\mathbf{w}_{\mathrm{a,s,o}}(k)$ in the subspace of interference-plus-noise is identical to the method for calculating $\mathbf{B}_{\mathbf{b},\mathrm{o}}(k)$ in the subspace of $\tilde{\mathbf{\Phi}}_{\mathbf{dd},\mathrm{s}}(k)$, we do not give detailed equations in this case. For improving the

[6]Note that single-channel noise-reduction techniques depend on the spectral characteristics of the interference and of the desired signal. Since the PSD of the desired signal is generally strongly time, single-channel noise-reduction is generally strongly time-varying, too.

suppression of transient interference, exponential weighting can be introduced according to (5.66), which yields $\hat{\mathbf{\Phi}}_{\mathbf{nn},\mathrm{ss}}(k)$ as the sample spatio-temporal correlation matrix w.r.t. interference-plus-noise for the calculation of $\mathbf{w}_{\mathrm{a,s,o}}(k)$.

If the desired signal contains stationary components, the interference canceller cannot be calculated for only interference-plus-noise, which means that robustness against cancellation of the desired signal is reduced. Stationarity of the interference is advantageous, since changing signal statistics need not to be tracked. For better robustness against cancellation of the desired signal, diagonal loading may be used as proposed in [HSH99]. As we have seen in Sect. 4.4.2, diagonal loading limits the norm of the weight vector of the beamformer, which translates to the norm of the weight vector of the interference canceller [Tre02] for the GSC. Assuming that components of the desired signal leak through the blocking matrix, but that the level of interference at the output of the blocking matrix is higher than for the desired signal. The quiescent weight vector attenuates interference relative to the desired signal. Then, for cancelling the desired signal at the output of the RGSC, the L_2-norm of the weight vector of the interference canceller must be greater than for cancelling interference. Therefore, limitation of the L_2-norm (diagonal loading) reduces cancellation of the desired signal at the output of the RGSC without impairing suppression of the interference.

As discussed in Sect. 5.4.2, the time-variance of the blocking matrix has to be tracked. Assuming time-variance, the blocking matrix changes when updating $\mathbf{B}_{\mathbf{b}}(k)$ in $\mathrm{span}\{\hat{\mathbf{\Phi}}_{\mathbf{dd},\mathrm{s}}(k)\}$. The interference canceller, which depends on $\mathbf{B}_{\mathbf{b}}(k)$ can generally not follow immediately, since $\mathbf{w}_{\mathrm{a,s,o}}(k)$ is only adapted in $\mathrm{span}\{\hat{\mathbf{\Phi}}_{\mathbf{nn},\mathrm{ss}}(k)\}$, which generally does not span the whole subspace of interference-plus-noise due to double-talk. As a result, the suppression of interference-plus-noise may be reduced during double-talk relative to periods where only interference-plus-noise is present.

5.7 Relation to Alternative GSC Realizations

A spatio-temporal blocking matrix for cancelling the desired signal at the input of an interference canceller was also studied in other works: In [FMB89], a two-channel system was proposed for enhancing speech corrupted by additive noise. The structure is equivalent to Fig. 5.1 for $M = 2$, except that the reference path consists of a single sensor signal and not of the output signal of a quiescent weight vector. It is suggested to learn the filter of the blocking matrix during an initialization phase without noise presence and to keep the filters fixed when the system is running. This, of course, limits the usage to situations with time-invariant propagation characteristics w.r.t. the desired signal.

In [Com90], the system of [FMB89] is extended to an arbitrary number of sensors. Instead of choosing the filters of the quiescent weight vector independently of the filters for the blocking matrix, they are the same. The steering of the array to the position of the desired source is implicitly included in the quiescent weight vector. The problem of adapting the blocking matrix and the interference canceller is addressed by switching between the adaptation of the blocking matrix and the adaptation of the interference canceller depending on presence of the desired signal and presence of interference, respectively. The decision is performed by a full-band energy-based voice activity detector.

The way to compute the filters for the blocking matrix and for the quiescent weight vector is altered in [GBW01]. As discussed in Sect. 5.4.1, the determination of optimum filters for the blocking matrix can be seen from a system identification point of view. Using this interpretation, it is suggested in [GBW01] to estimate the blocking matrix by a system identification technique which exploits the non-stationarity of the desired signal. As in [Com90], speech pauses are required to determine the blocking matrix filters. The tracking problem for the sidelobe cancelling path is thus not resolved.

5.8 Discussion

In this chapter, the RGSC after [HSH99] was introduced as an optimum LCLSE beamformer with time-varying spatio-temporal constraints. In contrast to purely spatial constraints, spatio-temporal constraints allow to design blocking matrices which cancel not only the direct signal path but also reverberation w.r.t. the desired signal (as long as the reverberation can be modeled by room impulse responses). Since non-ideal characteristics of sensors, as, e.g., tolerances of the positions or of the complex gain, can be captured in the transmission of the desired source to the sensor signals, the spatio-temporal constraints can take such array perturbations into account. Generally, the position of the desired source, the acoustic environment, or the transducer characteristics are time-varying, so that the spatio-temporal constraints have to be time-varying, too. The tracking of these variations by the spatio-temporal blocking matrix is obtained by continuously optimizing the filters of the blocking matrix for the present conditions. For minimum computational complexity, the optimization can be performed with recursive adaptation algorithms which minimize the energy of the output signals of the blocking matrix w.r.t. the desired signal. The spatio-temporal blocking matrix can thus be interpreted as an optimum SIMO system, where the filter coefficients depend on the characteristics of the acoustic conditions. The filters of the blocking matrix for complete suppression of the desired signal are infinitely long. Therefore, maximum cancellation of the desired signal by the blocking matrix, or, equivalently, filter length, has to be traded off against convergence speed, or suppression of the desired signal in transient conditions.

For the interference canceller of the RGSC, the interpretation as an optimum MISO filter showed the relationship between the number of sensors, the number of interferers, and the number of filter coefficients of the interference canceller.

It was shown that the blocking matrix should be adapted for presence of the desired signal only in order to suppress only the desired signal. For presence of stationary or slowly time-varying wideband interference, as, e.g., car noise in the passenger cabin of a car, adaptation of the blocking matrix for desired signal only is not possible. However, the filters of the blocking matrix have an implicit spatial selectivity for directional interferers because of the limited number of filter coefficients. Diffuse interference, as, e.g., car noise in the passenger cabin of a car, cannot be suppressed by the blocking matrix since only a single spatial degree of freedom is available. These effects lead to an implicit robustness against interference rejection by the blocking matrix for presence of slowly time-varying wideband interference. However, adaptation of the blocking matrix for double-talk reduces the cancellation of the desired signal by the blocking matrix, and, thus, reduces robustness against cancellation of the desired signal at the output of the RGSC. Therefore, the interference canceller should only be adapted for presence of interference only. For maximizing the number of adaptations per time unit of the blocking matrix and of the interference canceller, an adaptation in the eigenspace of desired signal and of interference, respectively, yields maximum tracking capability for time-varying signals and for time-varying environments. This adaptation strategy requires a classifier, which detects presence of the desired signal only, presence of interference only, and double-talk in the eigenspace of the sensor signals (Chap. 7).

6

Beamforming Combined with Multi-channel Acoustic Echo Cancellation

For audio signal acquisition, beamforming microphone arrays can be efficiently used for enhancing a desired signal while suppressing interference-plus-noise. For full-duplex communication systems, not only local interferers and noise corrupt the desired signal, but also acoustic echoes of loudspeaker signals. So far, we did not distinguish between local interferers and acoustic echoes. However, for suppressing acoustic echoes, more efficient techniques exist, which exploit the available loudspeaker signals as reference information. These methods are called *acoustic echo cancellers* (AECs) [SK91, BDH+99, GB00, BH03]. To cancel the acoustic echoes in the sensor channels, replicas of the echo signals are estimated and subtracted from the sensor signals. Acoustic echo cancellation is an application of system identification (Sect. 3.2.1). While the problem of monophonic acoustic echo cancellation has been studied for many years now, acoustic echo cancellation was only recently extended to more than one reproduction channel [SMH95, BBK03].

For optimally suppressing local interferers and acoustic echoes, it is thus desirable to combine acoustic echo cancellation with beamforming in the acoustic human/machine interface [MV93, Mar95, Kel97, HK00, Kel01]. For optimum performance, synergies between the AECs and the beamforming microphone array should be maximally exploited while the computational complexity should be kept moderate. Therefore, we study in this chapter various combinations of multi-channel acoustic echo cancellation and beamforming. We will see that each of them has advantages for different acoustic conditions. Compared to [MV93, Mar95, Kel97, HK00, Kel01], we discuss the different options in more details. Especially, a new structure is presented, which uses the RGSC as a basis, and which outperforms the other combinations for highly transient echo paths due to a better tracking capability of the AEC while keeping the computational complexity moderate.

In Sect. 6.1, multi-channel acoustic echo cancellation is introduced. Section 6.2 presents two generic combinations of AEC and beamforming, with the echo canceller in front of the beamformer and with the echo canceller after the beamformer. For beamformers in the GSC structure, two methods for integration of the AEC into the beamformer are possible (Sect. 6.3): First, the AEC can be placed in the reference path of the GSC after the quiescent weight vector [HK00, HK01d, Kel01]. Second, the AEC can be integrated into the interference canceller. This second possibility has not been studied so far.

6.1 Multi-channel Acoustic Echo Cancellation

In this section, the problem of and the solution to multi-channel acoustic echo cancellation is summarized. In Sect. 6.1.1, the structure of an acoustic echo canceller is presented. Sect. 6.1.2 illustrates the specific problems of multi-channel acoustic echo cancellation and shows possible solutions.

6.1.1 Problem Statement

The principle of multi-channel acoustic echo cancellation is illustrated in Fig. 6.1 for one microphone (the m-th sensor of a given microphone array).

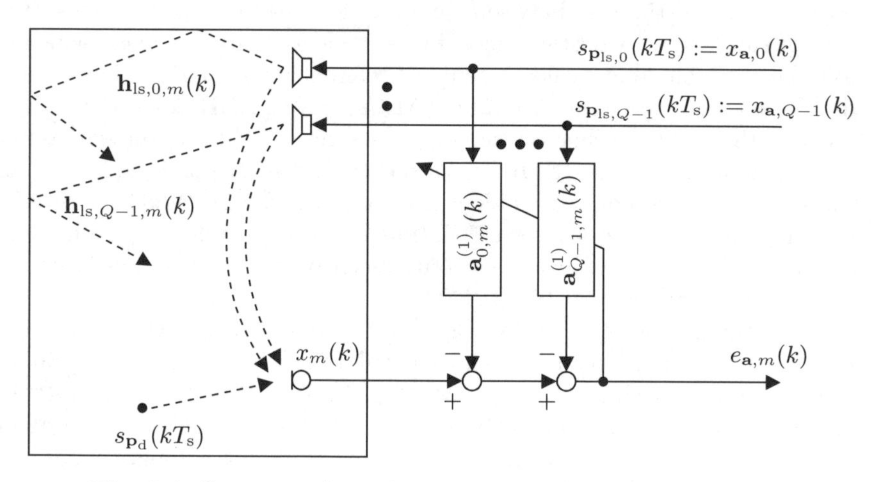

Fig. 6.1. Structure of a multi-channel acoustic echo canceller

For now, we assume that local interferers are not present. Loudspeakers should be 'ideal' transducers such that the signals of the loudspeakers at positions $\mathbf{p}_{\mathrm{ls},q}$ can be modeled as point sources and such that the signals $s_{\mathbf{p}_{\mathrm{ls},q}}(t)$ emitted by the loudspeakers are equivalent to the signals in the loudspeaker channels $x_{\mathbf{a},q}(k) = s_{\mathbf{p}_{\mathrm{ls},q}}(kT_\mathrm{s})$, $q = 0, 1, \ldots, Q-1$. The loudspeaker signals $s_{\mathbf{p}_{\mathrm{ls},q}}(t)$ are fed back to the microphone(s) and appear as acoustic echoes in the sensor channel(s). As long as nonlinearities can be neglected, the acoustic echo path between the q-th loudspeaker and the m-th microphone can be modeled as a linear room impulse response $\mathbf{h}_{\mathrm{ls},q,m}(k)$. (See, e.g.,

[Ste00, KK02b, KK03] for the nonlinear case.) The room impulse responses $\mathbf{h}_{\mathrm{ls},q,m}(k)$, $q = 0, 1, \ldots, Q-1$, $m = 0, 1, \ldots, M-1$, of length $N_{\mathbf{h}_{\mathrm{ls}}}$ are defined according to the vectors $\mathbf{h}_{\mathrm{n},l,m}(k)$ in (5.55). For cancelling the acoustic echoes $n_{\mathrm{ls},q,m}(k)$ in the sensor channel, adaptive filters $\mathbf{a}^{(1)}_{q,m}(k)$ of length $N_{\mathbf{a}}$ are placed in parallel to the echo paths between the loudspeakers and the microphones with the loudspeaker signals $x_{\mathbf{a},q}(k)$ as reference. The output signals of the adaptive filters are summed up and subtracted from the sensor signal $x_m(k)$. The adaptive filters $\mathbf{a}^{(1)}_{q,m}(k)$ form replica of the echo paths such that the output signals of the adaptive filters are replica of the acoustic echoes. Subtraction from the sensor signal(s) suppresses the echo signals, which yields the output signals $e_{\mathbf{a},m}(k)$ of the AEC. Acoustic echo cancellation is thus a system identification problem (Sect. 3.2.1), where the echo paths $\mathbf{h}_{\mathrm{ls},q,m}(k)$ must be identified. For specifying the mismatch between the true echo paths and the replicas, the system error $\Delta W_{\mathrm{rel}}(k)$ after (3.37) is generally used.

A common performance measure for acoustic echo cancellation is the *echo return loss enhancement* (ERLE) (in dB). The ERLE is defined as the ratio of the variance of the echo signals before and after the AEC on a logarithmic scale [BDH+99, HS03]. Without presence of the desired signal, interference, or noise, the ERLE is given by

$$ERLE_{\mathrm{AEC}}(k) = 10 \log_{10} \frac{\sum_{m=0}^{M-1} \mathcal{E}\{x_m^2(k)\}}{\sum_{m=0}^{M-1} \mathcal{E}\{e_{\mathbf{a},m}^2(k)\}} . \tag{6.1}$$

The ERLE is thus the analogon to the interference rejection after (4.81).

6.1.2 Challenges

In this section, the problems are summarized which come along with acoustic echo cancellers realized by adaptive filters. First, the challenges for monophonic AECs are presented. Second, we extend the discussion to the multi-channel case.

Monophonic Acoustic Echo Cancellers (AECs)

Acoustic echo cancellation, in general, involves several problems that have to be considered when designing adaptive algorithms for identifying the echo paths [BDH+99, HS03]. First, a good tracking performance is required, since $\mathbf{h}_{\mathrm{ls},q,m}(k)$ may vary strongly over time.

Then, the rate of convergence has to be taken into account. The rate of convergence is limited, since the number of taps of the impulse responses $\mathbf{h}_{\mathrm{ls},q,m}(k)$ is typically large. The length $N_{\mathbf{a}}$ of the modeling filters $\mathbf{a}^{(1)}_{q,m}(k)$ for a desired ERLE (in dB) can be approximated by [BDH+99]

$$N_{\mathbf{a}} \approx \frac{ERLE_{\mathrm{AEC}}}{60} f_{\mathrm{s}} T_{60} . \tag{6.2}$$

E.g., for $ERLE_{\mathrm{AEC}} = 20\,\mathrm{dB}$, $N_{\mathbf{a}} = 1600$ filter taps are required in an environment with $T_{60} = 400$ ms and $f_{\mathrm{s}} = 12\,\mathrm{kHz}$. The rate of convergence is also influenced by the high auto-correlation of the loudspeaker signals. With increasing auto-correlation, the condition number of the correlation matrix of the loudspeaker signal increases, which reduces the rate of convergence of many adaptation algorithms in turn [Hay96].

Activity of the desired signal $s_{\mathbf{p}_{\mathrm{d}}}(t)$ (or of local interference) disturbs the system identification and may lead to instabilities of the adaptive filters. In order to prevent these instabilities, adaptation control mechanisms are required, which adjust the adaptation speed to the present acoustic conditions [BDH+99, MPS00].

AECs for More Reproduction Channels

Acoustic echo cancellation for more than one reproduction channel, additionally, has to deal with the problem of highly cross-correlated loudspeaker signals, since multi-channel loudspeaker signals are often filtered signals of a few mutually uncorrelated sources [SMH95, BMS98, GB02]. The filtering is a consequence of the fact that the multitude of loudspeaker signals often carry only spatial information about the sources. Generally, with increasing number of reproduction channels, the condition number of the correlation matrix of the loudspeaker signals increases [BK01, BBK03]. As already stated for the monophonic case, an increasing condition number of the correlation matrix of the loudspeaker signals reduces the convergence speed of many adaptive algorithms. With increasing number Q of loudspeaker channels, the choice of the adaptation algorithm is more and more crucial.

For increasing the convergence speed of the adaptive algorithm and for facilitating the identification of the echo paths in turn, two measures are known, which are generally used in combination:

1. The loudspeaker signals can be de-crosscorrelated before the AEC filters for reducing the condition number of the sample correlation matrix of the loudspeaker signals. So far, complete de-crosscorrelation is not possible without noticeable modification or distortion of the loudspeaker signals, which is, of course, not desired. Partial de-crosscorrelation of the loudspeaker signals can be obtained by using inaudible nonlinearities [SMH95, BMS98, GB02], by adding inaudible uncorrelated noise to the loudspeaker signals [GT98, GE98], or by inaudible time-varying filtering [Ali98, JS98, TCH99].
2. For identification of the echo paths, adaptive algorithms can be used, which maximally exploit the characteristics of the correlation matrix of the loudspeaker signals. For the RLS algorithm, the mean-squared error convergence is independent of the condition number of the sample correlation matrix of the loudspeaker signals [Hay96], which makes the RLS algorithm the optimum choice for multi-channel AECs. However, its complexity is

prohibitively high for real-time consumer applications for the near future. Optimum results in terms of maximum echo suppression with minimum computational complexity are obtained with multi-channel adaptive filters in the DFT domain with RLS-like properties [BM00, BBK03] (Chap. 7).

6.2 Combination of Beamforming and Acoustic Echo Cancellation

Combinations of beamforming and acoustic echo cancellation can be devised in two obvious ways as depicted in Fig. 6.2: The AEC can be placed in the sensor channels in front of the beamformer ('AEC first', Sect. 6.2.1) or after the beamformer ('beamformer first', Sect. 6.2.2). Depending on the succession of acoustic echo cancellation and beamforming, interactions have to be considered in order to optimally exploit synergies. As an example for the beamformer, we will use the RGSC after Chap. 5. However, the discussion is equally valid for other adaptive beamformer realizations.

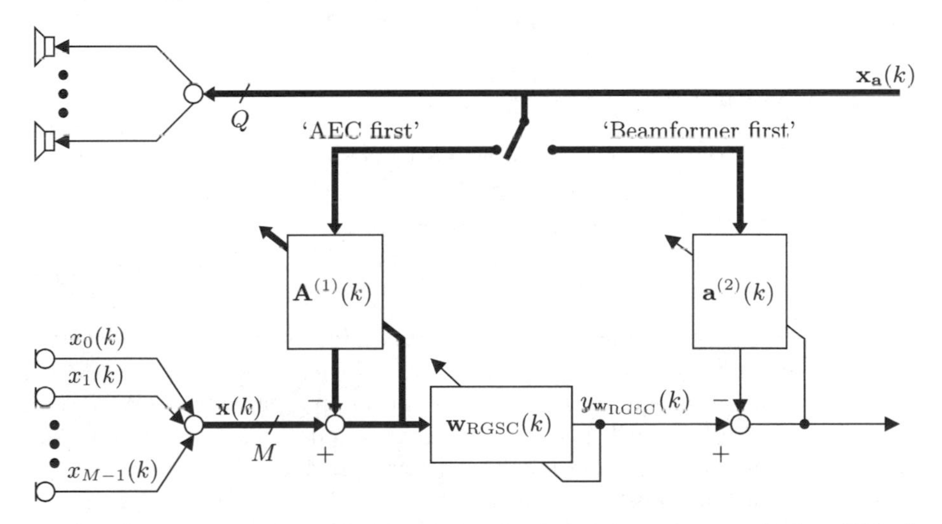

Fig. 6.2. Generic structures for combining AEC with beamforming [MV93, Kel01]

We now assume that both local interference and acoustic echoes are present, so that we distinguish between them. $n_{c,l,m}(k)$, $l = 0, 1, \ldots, L-1$, $m = 0, 1, \ldots, M-1$, should denote the contribution of the l-th local interferer to the m-th sensor signal. $n_{\mathrm{ls},q,m}(k)$, $q = 0, 1, \ldots, Q-1$, $m = 0, 1, \ldots, M-1$, denotes the contribution of the q-th loudspeaker signal to the m-th sensor signal as introduced above. Sensor noise is not present, i.e., $n_{\mathrm{w},m}(k) = 0$. As a consequence, the interference rejection $IR_{\mathbf{w}_{\mathrm{RGSC}}}(k)$ only measures the suppression of local interference. For measuring the echo suppression of the AEC, we use $ERLE_{\text{AEC}}(k)$ after (6.1). $ERLE_{\text{AEC}}(k)$ measures the ERLE with the sensor signals as reference. However, for 'beamformer first', the AEC does not operate on the sensor signals but on the beamformer output so that $ERLE_{\text{AEC}}(k)$ after (6.1) would measure the ERLE of the beamformer and of the AEC. To avoid conflicts, we always measure $ERLE_{\text{AEC}}(k)$ with the input signals of the AEC as reference, as, e.g., $y_{\mathbf{w}_{\text{RGSC}}}(k)$ for 'beamformer first'. For measuring the ERLE of the beamformer, we introduce $ERLE_{\mathbf{w}_{\text{RGSC}}}(k)$ (in dB) as the ratio of the average variance of the acoustic echoes at the input signals of the RGSC and the variance of the acoustic echoes at the output of the RGSC, analogously to $IR_{\mathbf{w}_{\text{RGSC}}}(k)$. The echo suppression of the joint system of AEC and beamformer is measured by $ERLE_{\text{tot}}(k) = ERLE_{\text{AEC}}(k) + ERLE_{\mathbf{w}_{\text{RGSC}}}(k)$ (in dB). For wide-sense stationarity of the sensor signals, the dependency of $ERLE_{\text{AEC}}(k)$, $ERLE_{\mathbf{w}_{\text{RGSC}}}(k)$, and $ERLE_{\text{tot}}(k)$ on the discrete time k is dropped. For narrowband signals and time-invariance of the adaptive filters of the AEC and of the RGSC, $ERLE_{\text{AEC}} := ERLE_{\text{AEC}}(k)$ and $ERLE_{\text{tot}} := ERLE_{\text{tot}}(k)$ are transformed into the DTFT domain, yielding $ERLE_{\text{f,AEC}}(\omega)$ and $ERLE_{\text{f,tot}}(\omega)$, respectively, analogously to $IR_{\text{f}}(\omega)$ in Sect. 4.2.3. For the experiments, estimates of $ERLE_{\text{AEC}}(k)$, $ERLE_{\mathbf{w}_{\text{RGSC}}}(k)$, $ERLE_{\text{tot}}(k)$, $ERLE_{\text{AEC}}$, and $ERLE_{\text{tot}}$ are obtained as described in Sect. 4.2.3.

6.2.1 'AEC First'

For 'AEC first', the AEC is realized as a MIMO system $\mathbf{A}^{(1)}(k)$ with Q input signals and M output signals in the sensor channels. The matrix $\mathbf{A}^{(1)}(k)$ of size $QN_{\mathbf{a}} \times M$ of AEC filters $\mathbf{a}_{q,m}^{(1)}(k)$ is defined as

$$[\mathbf{A}^{(1)}(k)]_{q,m} = \mathbf{a}_{q,m}^{(1)}(k) \,. \tag{6.3}$$

The computational complexity increases with an increasing number of sensors M, since at least the filtering and the filter coefficient update of the AEC in the sensor channels require M-fold computations compared to an AEC with a single recording channel. For real-time realizations, this may be a limiting factor especially with a large number of microphones ($M \geq 4$) even in the near future for many cost-sensitive systems.

Implications for the AEC and for Fixed Beamformers

With the AEC in the sensor channels, the AEC is independent of the beamformer, so that the characteristics known from multi-channel acoustic echo cancellation translate to the combination with beamforming.

The beamformer – independently of an adaptive or a fixed realization – is not adversely affected by the AEC either. For a fixed beamformer, it is obvious: The echo suppression of the beamformer is simply added to the echo suppression of the AEC. Adaptive beamformers are discussed next.

Implications for Adaptive Beamformers

For adaptive data-dependent beamformers, positive synergies between acoustic echo cancellation and the beamforming occur. For $ERLE_{\mathrm{AEC}} = 20\,\mathrm{dB}$ in an environment with $T_{60} = 400\,\mathrm{ms}$, the length of the adaptive filters of the AEC is often as large as $N_{\mathbf{a}} = 1600$ at a sampling rate of $T_{\mathrm{s}} = 12\,\mathrm{kHz}$ according to (6.2). This leads to a slow convergence of the adaptive filters. For data-dependent beamformers, however, a typical filter length for $IR \approx 15\,\mathrm{dB}$ in an environment with $T_{60} = 400\,\mathrm{ms}$ is $N = 256$ (Chap. 7). Adaptive data-dependent beamformers thus converge considerably faster than AECs if comparable adaptation algorithms are used for the AEC and for the beamformer. This effect can be exploited if the beamformer is placed after the AEC. The beamformer does not feel any adverse effects of the time-variance of the AEC, since the beamformer can track the time-variance of the AEC due to the faster convergence. Whenever acoustic echoes leak through the AEC because of, e.g., changes of the echo paths, the beamformer suppresses both acoustic echoes and local interference (as long as the beamformer adaptation is not impaired due to presence of the desired signal). The beamformer thus uses the spatial degrees of freedom of the sensor array for both echo suppression and interference suppression. The number of spatial degrees of freedom for suppressing local interferers reduces. After convergence of the AEC, the AEC efficiently suppresses acoustic echoes, and the beamformer only sees residual echoes with low energy. The beamformer can concentrate on the suppression of local interference with all spatial degrees of freedom. The suppression of local interference increases.

Example Behavior

For illustrating the behavior of 'AEC first', we consider the combination of a stereophonic AEC with the RGSC. We assume that the beamformer can track the time-variance of the AEC and that mutual correlation of the loudspeaker signals can be neglected. The behavior of the beamformer is thus idealized in order to be independent of a given realization.[1]

Figure 6.3 shows an example behavior of 'AEC first' for $M = 4$, $N = 256$, $N_{\mathbf{a}} = 1024$ in the multimedia room with $T_{60} = 250\,\mathrm{ms}$. See App. B for a description of the sensor array with $M = 4$. The RGSC and the AEC are realized as described in Chap. 7. The desired source is located in broadside direction at a distance $d = 60\,\mathrm{cm}$. The two sources for the echo signals are placed at the two endfire positions at a distance $d = 60\,\mathrm{cm}$. A local interferer signal arrives from $\theta = 5\pi/6$, $d = 120\,\mathrm{cm}$. All source signals are mutually uncorrelated Gaussian white noise signals which are high-pass filtered with cut-off frequency 200 Hz. All signal levels at the sensors are equal.

---

[1]In Sect. 6.3.2, experimental results are provided for time-varying acoustic conditions for our realization of the system.

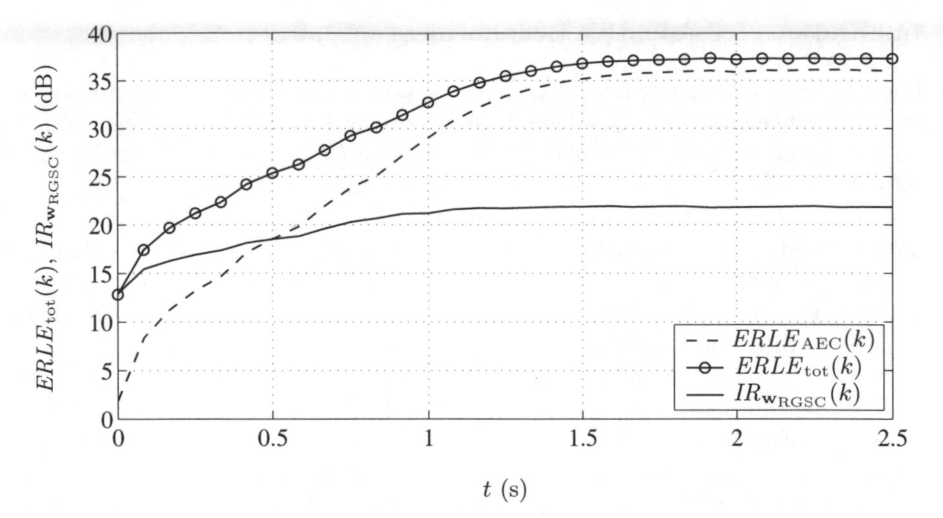

Fig. 6.3. Transient behavior of 'AEC first' with presence of acoustic echoes and local interference; multimedia room with $T_{60} = 250\,\mathrm{ms}$, $M = 4$

The blocking matrix is fixed after adaptation for presence of only the desired signal. After initialization with zeroes, the adaptive filters of the stereophonic AEC and of the interference canceller are updated in an alternating way: In each cycle, the AEC is first adapted for 1000 samples with presence of only acoustic echoes. After that, the interference canceller is adapted with presence of acoustic echoes and of local interference for a data block of 50000 samples, which leads to nearly complete convergence of the RGSC. Since the RGSC is nearly completely converged after each change of the AEC, it can be assumed that the behavior of 'AEC first' is independent of the realization of the adaptive beamformer. $ERLE_{\mathrm{AEC}}(k)$, $ERLE_{\mathrm{tot}}(k)$, and $IR_{\mathbf{w}_{\mathrm{RGSC}}}(k)$ are calculated after each iteration. The time axis in Fig. 6.3 corresponds to the adaptation time of the AEC. After initialization of the adaptive filters ($t = 0$), the AEC does not immediately suppress the acoustic echoes (dashed line). The beamformer suppresses the acoustic echoes and the local interference equally. With convergence of the adaptive filters of the AEC, $ERLE_{\mathrm{AEC}}(k)$ increases. The power ratio of acoustic echoes to local interference (EIR),

$$EIR_{\mathrm{in}}(k) = \frac{1/L \sum_{m=0}^{M-1} \sum_{l=0}^{L-1} \mathcal{E}\{n_{\mathrm{c},l,m}^2(k)\}}{1/Q \sum_{m=0}^{M-1} \sum_{q=0}^{Q-1} \mathcal{E}\{n_{\mathrm{ls},q,m}^2(k)\}}, \tag{6.4}$$

where we drop for stationarity of $n_{\mathrm{c},l,m}(k)$ and of $n_{\mathrm{ls},q,m}(k)$ the dependency on k, at the beamformer input decreases. As the number of degrees of freedom of the interference canceller is limited, $IR_{\mathbf{w}_{\mathrm{RGSC}}}(k)$ increases, and $ERLE_{\mathbf{w}_{\mathrm{RGSC}}}(k) = ERLE_{\mathrm{tot}}(k) - ERLE_{\mathrm{AEC}}(k)$ decreases. However, because of the increasing $ERLE_{\mathrm{AEC}}(k)$, $ERLE_{\mathrm{tot}}(k)$ increases. After convergence of the AEC, we see that the beamformer contributes little to the echo suppression, i.e., $ERLE_{\mathbf{w}_{\mathrm{RGSC}}}(k) = ERLE_{\mathrm{tot}}(k) - ERLE_{\mathrm{AEC}}(k) \approx 1.5\,\mathrm{dB}$. Compared to the RGSC alone, $ERLE_{\mathrm{tot}}(k)$ and $IR_{\mathbf{w}_{\mathrm{RGSC}}}(k)$ are 25 dB and 9 dB greater, respectively, when 'AEC first' is used. By using more microphones[2], the relative improvement decreases, since local interferers and acoustic echoes are better suppressed by the beamformer.

6.2.2 'Beamformer First'

For 'beamformer first' (Fig. 6.2), the AEC is realized as a MISO system $\mathbf{a}^{(2)}(k)$ with Q input channels. Relative to 'AEC first', computational complexity for acoustic echo cancellation is reduced to that of a single multi-channel AEC.

Implications for the Beamformer

The beamformer operates without any repercussions from the AEC. The performance of the beamformer thus corresponds to that discussed in Chap. 4 and in Chap. 5. For adaptive signal-dependent beamformers, it is possible to stop the adaptation of the beamformer if acoustic echoes are present. The beamformer is thus not optimized for the spatio-temporal characteristics of the acoustic echoes. On the one hand, this limits the echo suppression of the beamformer and reduces the tracking capability of the beamformer since the beamformer is adapted less frequently. On the other hand, however, more spatial degrees of freedom are available for interference suppression. Contrarily, the beamformer can be designed with fixed spatial nulls for the positions of the loudspeakers in order to provide optimum suppression of the direct signal path and of early reflections of the acoustic echoes even for strongly non-stationary environments. This is especially interesting for loudspeakers with known positions close to the array where the direct-path-to-reverberation ratio is high [DG01].

Implications for the AEC

The presence of the beamformer in the signal path of the echo canceller can be viewed from a simple noise-reduction perspective and from a system identification perspective.

Influence of the Beamformer from a Noise-Reduction Point of View

Due to the spatial filtering of the beamformer, the power ratio of acoustic echoes to the desired signal and to local interference is generally changed at the input of the AEC. A reduction of this power ratio makes the identification

[2]The length of the adaptive filters of the beamformer can be increased for better suppression of the interferers, too, but note that longer adaptive filters are problematic due to a slower convergence.

of the echo paths for the echo canceller more difficult, which generally reduces the convergence speed of the AEC and the echo suppression in time-varying environments [MPS00]. This aspect has to be especially considered for adaptive beamformers which provide high suppression of interfering signals. Consider, for example, acoustic echoes from a loudspeaker in an environment with spatially diffuse noise components, where the echo path is time-varying. An adaptive beamformer tracks the time-variance of the echo path and suppresses the spatially coherent acoustic echoes more efficiently than non-coherent noise. The power ratio of acoustic echoes to the diffuse noise decreases. The convergence speed of the echo canceller decreases, and the echo suppression reduces in turn. In contrast, if the beamformer is not optimized for suppression of acoustic echoes, the background noise is generally more efficiently suppressed than the acoustic echoes. The power ratio of acoustic echoes to diffuse noise increases, yielding faster convergence and a higher echo suppression of the AEC.

Influence of the Beamformer from a System-Identification Point of View [Kel01]

The AEC has to model the echo paths and the beamformer weight vector $\mathbf{w}_{\mathrm{RGSC}}(k)$. This generally increases the length of the AEC filters for a given echo suppression compared to 'AEC first' and shows all problems of longer adaptive filters. For adaptive beamformers, the necessity to model the beamformer weight vector raises severe difficulties, since the AEC cannot track the beamformer satisfactorily. Generally, adaptive beamformers are designed for fast convergence and fast tracking in order to optimally suppress time-varying interference-plus-noise and in order to track moving desired sources with time-varying second-order statistics. Essentially, this means that the length of the adaptive filters is moderate ($N = 64 \ldots 512$) and that fast-converging adaptation algorithms are used, as e.g., [BM00, BBK03]. (See Chap. 7.) The length of the AEC filters should be much longer because of the long reverberation time of acoustic environments and because the beamformer has to be modeled in addition to the room impulse responses for 'beamformer first'. Typically, the AEC filters are by a factor of $4 \ldots 8$ longer than the beamformer filters for the same environment, which yields the same reduction of the convergence speed of the AEC. Therefore, the AEC converges slowlier than the beamformer. For the AEC being able to converge, the beamformer has to be time-invariant for a sufficiently long period of time. This, of course, is not acceptable for time-varying acoustic conditions and for non-stationary signals, where the beamformer should be adapted continuously. As a result, the AEC will be almost ineffective for 'beamformer first'.

Example Behavior

Figure 6.4 illustrates this adaptation behavior for the realization of the RGSC after Chap. 7 with a stereophonic AEC placed after the RGSC, $N = 256$, $N_{\mathrm{a}} =$

1024. We assume the same experimental setup as in Sect. 6.2.1, except that the desired signal is male speech and that the echo signals are stereophonic pieces of music. The interferer signal is Gaussian white noise. All source signals are high-pass filtered with cut-off frequency 200 Hz. The temporal signals are shown in Figs. 6.4a, b. The adaptive modules – blocking matrix, interference canceller, and acoustic echo canceller – are adapted as illustrated in Fig. 6.4c. In Fig. 6.4d, the system misalignment $\Delta W_{\mathrm{rel}}(k)$ according to (3.37) of the AEC is depicted. Figure 6.4e shows the suppression of local interference and of acoustic echoes as a function of the discrete time k. Figure 6.4e illustrates the corresponding residual echo signals and local interference signals at the output of 'beamformer first'.

After the adaptation of the blocking matrix in Phase I (Fig. 6.4c), the suppression of local interference increases rapidly to $IR_{\mathbf{w}_{\mathrm{RGSC}}}(k) \approx 22\,\mathrm{dB}$ when local interference is active in Phase II. When the loudspeaker signals set in (Phase III), $IR_{\mathbf{w}_{\mathrm{RGSC}}}(k)$ decreases due to the limited number of degrees of freedom of the beamformer. Since acoustic echoes are stronger than the local interference, the suppression of acoustic echoes is greater than the interference rejection. The AEC cannot track the time-variance of the beamformer, and the system misalignment $\Delta W_{\mathrm{rel}}(k)$ remains small. After convergence of the beamformer in Phase IV, the misalignment $\Delta W_{\mathrm{rel}}(k)$ decreases. However, due to the low EIR, the additional gain of the AEC is only about 5 dB. When the loudspeaker signals are switched off in Phase V, the beamformer reconverges for optimally suppressing the local interferer. The misalignment $\Delta W_{\mathrm{rel}}(k)$ increases, because of the change of the echo path. During double-talk of desired signal, local interference, and acoustic echoes, the adaptation of beamformer and AEC is stopped (Phase VII). The AEC cannot adapt until the desired signal is inactive in Phase VIII. Since the beamformer suppresses local interference efficiently, but lets through the acoustic echoes, the EIR is now higher compared to Phases III, IV. The system misalignment is less than during Phases III, IV, and the additional gain of the AEC is higher. The beamformer concentrates on the suppression of local interference, the echo canceller suppresses the acoustic echoes. In conclusion, this example shows that positive synergies may be obtained with adaptive realizations of 'beamformer first' in idealized conditions, but that positive synergies cannot be expected in non-stationary realistic environments.

6.3 Integration of Acoustic Echo Cancellation into the GSC

The potentially high computational complexity of 'AEC first' and the tracking problematic for 'beamformer first' suggest to integrate the echo canceller into the time-varying beamformer such that optimum synergies between beamforming and acoustic echo cancellation are obtained with minimum computational complexity. For optimum synergies, we demand that the beamformer

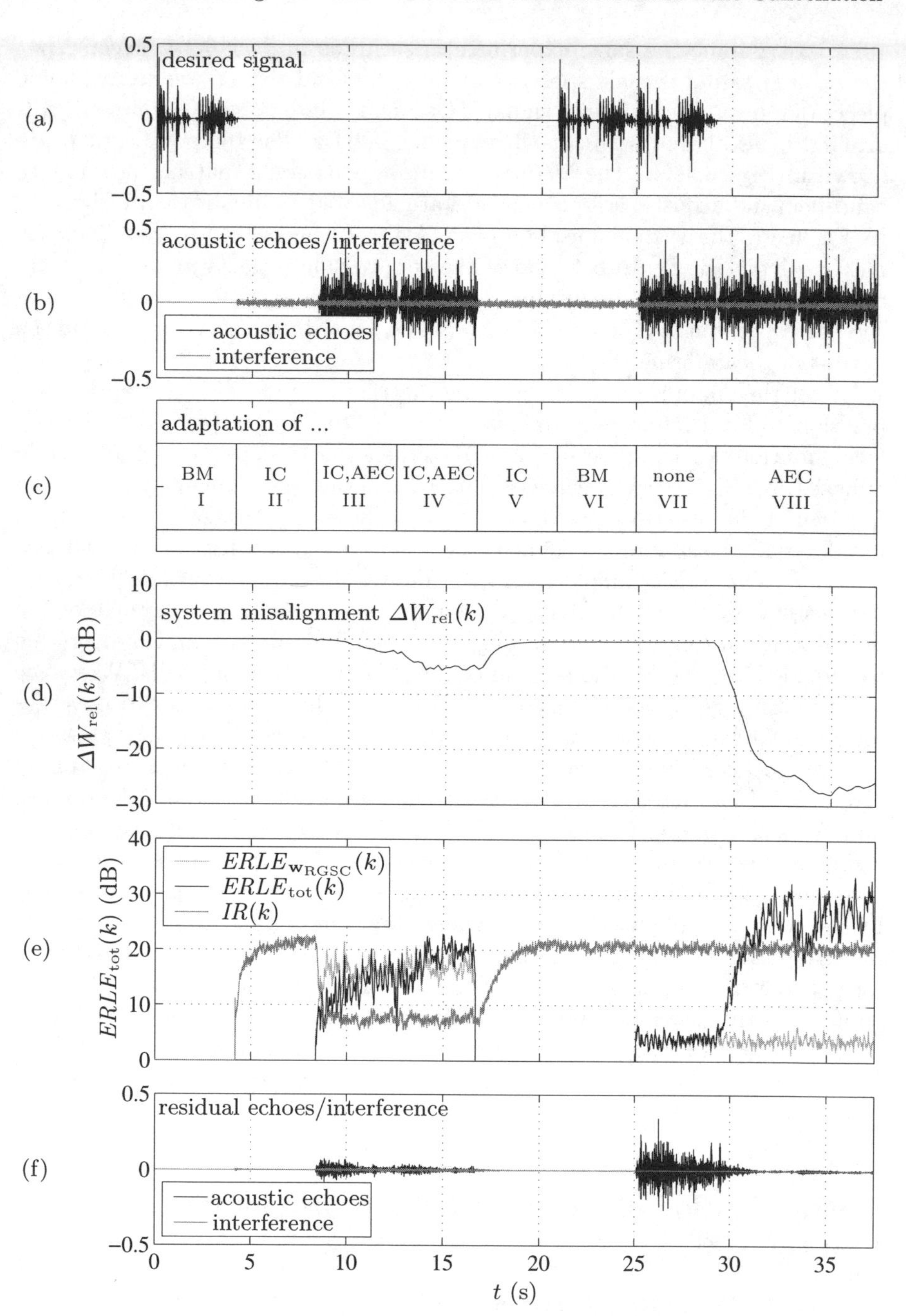

Fig. 6.4. Transient behavior of 'beamformer first' with presence of acoustic echoes, local interference, and desired signal; multimedia room with $T_{60} = 250\,\mathrm{ms}$, $M = 4$

does not 'see' acoustic echo signals after convergence of the AEC, so that the number of degrees of freedom for suppressing local interference is maximum.

The decomposition of the beamformer into a time-invariant part parallel to a time-varying part – according to the GSC structure with a time-invariant quiescent weight vector $\mathbf{w}_c$ – yields two options for integrating the AEC: First, the AEC can be cascaded with the fixed beamformer, yielding the GSAEC structure (Sect. 6.3.1). Second, the AEC can be combined with the interference canceller as additional channels of the sidelobe cancelling path (GEIC, Sect. 6.3.2).

6.3.1 AEC After the Quiescent Weight Vector (GSAEC)

With the AEC placed after the quiescent weight vector $\mathbf{w}_c$ (Fig. 6.5), we obviously obtain for the AEC the characteristics of 'beamformer first' with a time-invariant realization of the beamformer. Especially, the computational complexity is reduced to that of a single AEC for an arbitrary number of sensors. However, from Fig. 6.5, we see that acoustic echoes are still present in the adaptive sidelobe cancelling path while they may be efficiently cancelled by the AEC. Therefore, the interference canceller $\mathbf{w}_{a,s}(k)$ has to prevent leakage of acoustic echoes through the sidelobe cancelling path in addition to cancel local interference. The interference canceller has to cancel acoustic echoes before the subtraction of the output of the interference canceller from the output of the AEC. Interpreting the weight vector $\mathbf{w}_{a,s}(k)$ as a beamformer with the output signals of the blocking matrix $\mathbf{B}_s(k)$ as input signals, this means that $\mathbf{w}_{a,s}(k)$ has to place spatial nulls for the acoustic echo signals at the output of $\mathbf{B}_s(k)$. As a result, the number of spatial degrees of freedom for suppression of local interference is generally not increased relative to beamforming alone. However, positive synergies can be obtained, as will be discussed in the following.

For studying the echo suppression and the interference rejection of GSAEC, we distinguish between different propagation characteristics of acoustic echoes and interference and equal propagation characteristics.

Different Propagation Properties

For the case of different propagation properties, we use the room impulse response model. 'Different propagation characteristics' means that the room impulse responses w.r.t. acoustic echoes and w.r.t. interference do not have common zeroes. This is the case that will be encountered in practical situations. We distinguish between stationary conditions and transient conditions.

Stationary Conditions

For illustrating the behavior of GSAEC for stationary conditions, we measure $IR_{\mathbf{w}_{\mathrm{RGSC}}}$ and $ERLE_{\mathrm{tot}}$ of GSAEC after convergence of the adaptive filters as

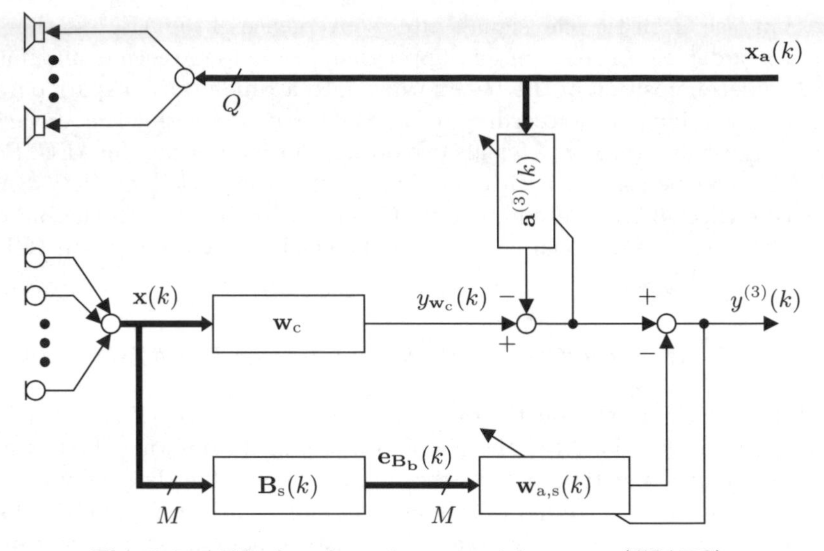

Fig. 6.5. AEC after the quiescent weight vector (GSAEC)

a function of the EIR at the sensors, EIR_{in} (Fig. 6.6). The experimental setup corresponds to Sect. 6.2.1. The AEC is adapted until $ERLE_{\text{AEC}} = 20\,\text{dB}$. In Fig. 6.6a, $IR_{\mathbf{w}_{\text{RGSC}}}$ as a function of EIR_{in} is depicted for GSAEC, for 'AEC first', and for the RGSC. $IR_{\mathbf{w}_{\text{RGSC}}}$ decreases with increasing EIR_{in}, since the beamformer concentrates on the suppression of the strongest interferer due to the limited number of degrees of freedom. Maximum interference rejection is obtained with 'AEC first' (solid line). $IR_{\mathbf{w}_{\text{RGSC}}}$ of GSAEC is up to 5 dB greater than for the RGSC but also up to 5 dB lower than for 'AEC first'. The improvement relative to the RGSC increases with increasing EIR_{in}.

Figure 6.6b illustrates the echo suppression, $ERLE_{\text{tot}}$ for the three structures. For low EIR_{in}, $ERLE_{\text{tot}}$ of 'AEC first' is considerably greater than for GSAEC and for RGSC. For 'AEC first', the acoustic echoes are suppressed by $ERLE_{\text{AEC}} = 20\,\text{dB}$ before they are seen by the beamformer so that acoustic echoes are efficiently suppressed though the beamformer concentrates on the suppression of the stronger local interference. For high EIR_{in}, $ERLE_{\text{tot}}$ of GSAEC is up to 10 dB greater than for the RGSC and only 3 dB less than for 'AEC first'.

Transient Conditions

The transient behavior of GSAEC is depicted in Fig. 6.7. The simulation setup is the same as in Sect. 6.2.1. The signal levels of the two acoustic echo signals and of the local interferer at the sensors are equal, so that $EIR_{\text{in}} = 6\,\text{dB}$. For $ERLE_{\text{AEC}}(k) = 20\,\text{dB}$ ($t = 0.75\,\text{s}$), we thus have $IR_{\mathbf{w}_{\text{RGSC}}}(k)$ and $ERLE_{\text{tot}}(k)$ according to Fig. 6.6a and (b) for $10\log_{10} EIR_{\text{in}} = 6\,\text{dB}$, respectively. The transient behavior of GSAEC is similar to the transient behavior of 'AEC

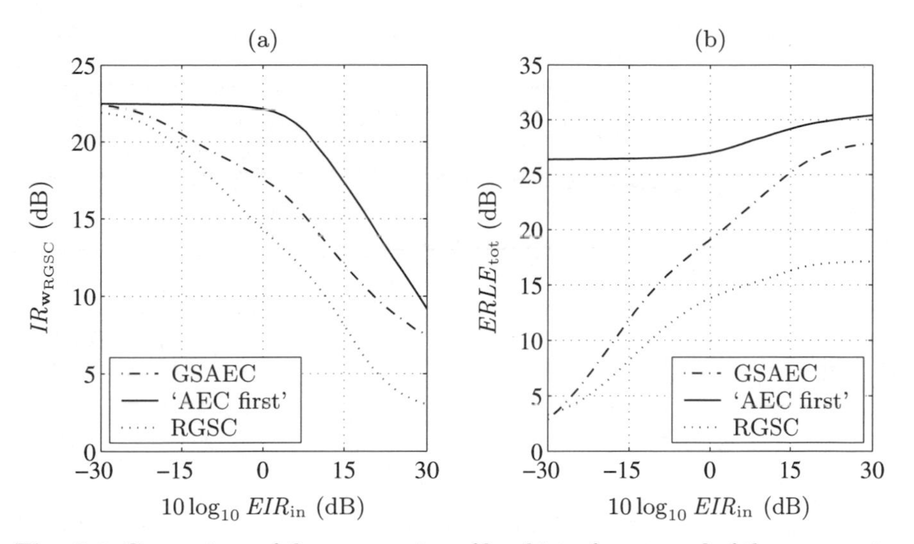

Fig. 6.6. Comparison of the suppression of local interference and of the suppression of acoustic echoes for RGSC alone, 'AEC first', and GSAEC after convergence of the adaptive filters; multimedia room with $T_{60} = 250\,\mathrm{ms}$, $M = 4$

first' (Fig. 6.3) except that $ERLE_{\mathrm{tot}}(k)$ of GSAEC is limited to 22.5 dB while $ERLE_{\mathrm{AEC}}(k)$ increases up to 31 dB after convergence of the AEC. This means that acoustic echoes are efficiently cancelled in the reference path, but that leakage of the acoustic echoes is not completely prevented by the interference canceller due to the limited number of degrees of freedom. However, $IR_{\mathbf{w}_{\mathrm{RGSC}}}(k)$ and $ERLE_{\mathrm{tot}}(k)$ are 2.5 dB and 7 dB greater than for the RGSC alone ($t = 0$).

In summary, the experimental results show that positive synergies can be obtained between acoustic echo cancellation and beamforming with GSAEC. Most synergies relative to the RGSC are obtained for predominant acoustic echoes. In this case, the performance of GSAEC is close to the performance of 'AEC first'.

Equal Propagation Properties

Now, we illustrate the behavior of GSAEC for identical propagation characteristics of acoustic echoes and of local interference. 'Identical propagation characteristics' means that the wave field of acoustic echoes and of local interference is identical. Although this is an unrealistic (extreme) case, this gives further insights into the behavior of GSAEC, since the case with common zeroes of the room impulse responses w.r.t. acoustic echoes and w.r.t. local interferers is covered. For simplicity, we assume WSS signals, and we use the DTFT-domain description after Chap. 4.4.2 and Chap. 5.3 as a starting point. First, we derive the optimum weight vector $\mathbf{w}^{(3)}_{\mathrm{f,a,o}}(\omega)$ of GSAEC for equal

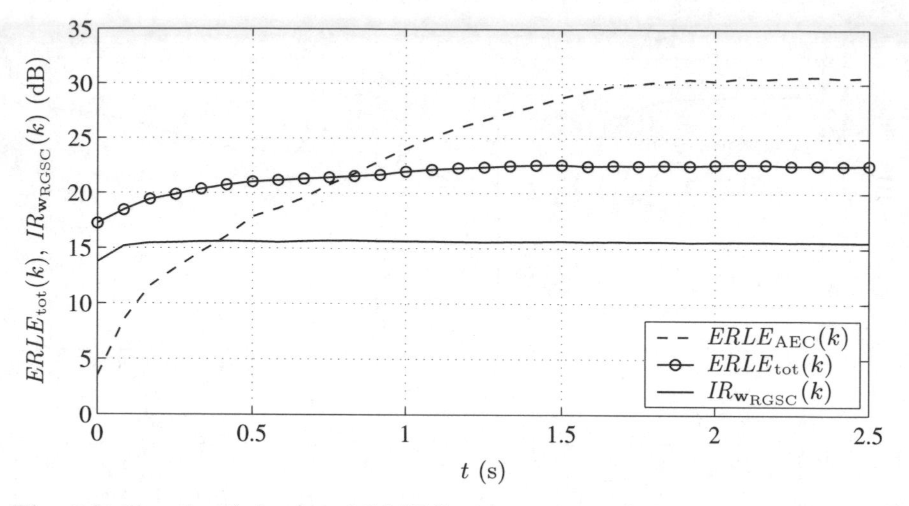

Fig. 6.7. Transient behavior of GSAEC with presence of acoustic echoes and local interference; multimedia room with $T_{60} = 250\,\mathrm{ms}$, $M = 4$

wave fields, i.e., for equal spatial coherence functions. Second, we assume an identity blocking matrix, $\mathbf{B}_{\mathrm{f}}(\omega) = \mathbf{I}_{M \times M}$, which allows to simplify $\mathbf{w}^{(3)}_{\mathrm{f,a,o}}(\omega)$ for a better illustration. We finish by some examples for $\mathbf{B}_{\mathrm{f}}(\omega) = \mathbf{I}_{M \times M}$ and for $\mathbf{B}_{\mathrm{f}}(\omega) \neq \mathbf{I}_{M \times M}$.

Derivation of the Optimum Interference Canceller

We assume that the desired signal is not active and that sensor noise is not present. Local interference and acoustic echoes should be mutually uncorrelated. The sensor signals can then be divided into local interference components and into echo components with different PSDs, $S_{n_{\mathrm{c}}n_{\mathrm{c}}}(\omega)$ and $S_{n_{\mathrm{ls}}n_{\mathrm{ls}}}(\omega)$, respectively, and with equal spatial coherence matrices $\mathbf{\Gamma}_{\mathbf{n}_{\mathrm{c}}\mathbf{n}_{\mathrm{c}}}(\omega)$. The spatio-spectral correlation matrix of the sensor signals (4.38) can thus be written with (4.16) and with (4.36) as

$$\mathbf{S}_{\mathbf{xx}}(\omega) = \left(S_{n_{\mathrm{c}}n_{\mathrm{c}}}(\omega) + S_{n_{\mathrm{ls}}n_{\mathrm{ls}}}(\omega)\right) \mathbf{\Gamma}_{\mathbf{n}_{\mathrm{c}}\mathbf{n}_{\mathrm{c}}}(\omega) \,. \tag{6.5}$$

For modeling the AEC, we use a frequency-dependent gain function $0 \leq G^{(3)}_{\mathbf{a}}(\omega) \leq 1$ after the quiescent weight vector in the reference path of the beamformer. For acoustic echoes, the factor $G^{(3)}_{\mathbf{a}}(\omega)$ is equivalent to $ERLE_{\mathrm{f,AEC}}(\omega)$. (See Fig. 6.8.) For local interference, $G^{(3)}_{\mathbf{a}}(\omega) = 1$.

According to (4.151), the optimum weight vector $\mathbf{w}^{(3)}_{\mathrm{f,a,o}}(\omega)$ is equivalent to the matrix-vector product of the inverse of the spatio-spectral correlation matrix of the output signals of the blocking matrix and the spatio-spectral correlation vector between the output signal of the echo canceller and the output signals of the blocking matrix. Using (6.5) and the model $G^{(3)}_{\mathbf{a}}(\omega)$ of the AEC in (4.151), we obtain $\mathbf{w}^{(3)}_{\mathrm{f,a,o}}(\omega)$ as

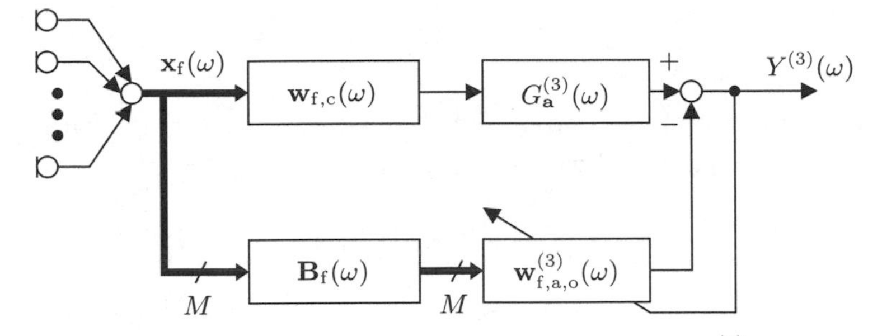

Fig. 6.8. Modeling of the AEC as a gain function $G_\mathbf{a}^{(3)}(\omega)$

$$\begin{aligned}\mathbf{w}_{\mathrm{f,a,o}}^{(3)}(\omega) = & \frac{S_{n_\mathrm{c}n_\mathrm{c}}(\omega) + G_\mathbf{a}^{(3)}(\omega)S_{n_\mathrm{ls}n_\mathrm{ls}}(\omega)}{S_{n_\mathrm{c}n_\mathrm{c}}(\omega) + S_{n_\mathrm{ls}n_\mathrm{ls}}(\omega)} \times \\ & \times \left(\mathbf{B}_\mathrm{f}^H(\omega)\mathbf{\Gamma}_{\mathbf{n}_\mathrm{c}\mathbf{n}_\mathrm{c}}(\omega)\mathbf{B}_\mathrm{f}(\omega)\right)^+ \mathbf{B}_\mathrm{f}^H(\omega)\mathbf{\Gamma}_{\mathbf{n}_\mathrm{c}\mathbf{n}_\mathrm{c}}(\omega)\mathbf{w}_{\mathrm{f,c}}(\omega)\,. \end{aligned} \quad (6.6)$$

With the previous assumptions, the weight vector $\mathbf{w}_{\mathrm{f,a,o}}^{(3)}(\omega)$ is thus equivalent to $\mathbf{w}_{\mathrm{f,a,o}}(\omega)$ of the RGSC, except for the fractional weighting factor, which is equal to one for the RGSC. We assume first that the quiescent weight vector is not optimum for the given wave field. Then, the sidelobe cancelling path is active ($\mathbf{w}_{\mathrm{f,a,o}}^{(3)}(\omega) \neq \mathbf{0}_{M\times 1}$). For $G_\mathbf{a}^{(3)}(\omega) \neq 1$, it follows that the AEC disturbs the RGSC and that neither acoustic echoes nor local interference can be completely suppressed, independently of the spatial coherence matrix of the interference at the sensors. Second, if the quiescent weight vector is optimum for the given wave field, then, $\mathbf{w}_{\mathrm{f,a,o}}^{(3)}(\omega) = \mathbf{0}_{M\times 1}$. The suppression of local interference is equal to the suppression of $\mathbf{w}_{\mathrm{f,c}}(\omega)$. Acoustic echoes are suppressed according to the cascade of $\mathbf{w}_{\mathrm{f,c}}(\omega)$ and of $G_\mathbf{a}^{(3)}(\omega)$.

Identity Blocking Matrix and Special Cases

For simplifying (6.6), we assume that the blocking matrix $\mathbf{B}_\mathrm{f}(\omega)$ is the identity matrix. We thus assume the blocking matrix of the RGSC without activity of the desired signal at frequency ω, i.e., $S_{dd}(\omega) = 0$. For $\mathbf{B}_\mathrm{f}(\omega) = \mathbf{I}_{M\times M}$, we obtain from (6.6)

$$\mathbf{w}_{\mathrm{f,a,o}}^{(3)}(\omega) = \frac{S_{n_\mathrm{c}n_\mathrm{c}}(\omega) + G_\mathbf{a}^{(3)}(\omega)S_{n_\mathrm{ls}n_\mathrm{ls}}(\omega)}{S_{n_\mathrm{c}n_\mathrm{c}}(\omega) + S_{n_\mathrm{ls}n_\mathrm{ls}}(\omega)}\mathbf{w}_{\mathrm{f,c}}(\omega)\,. \quad (6.7)$$

With the ratio of the PSDs of acoustic echoes and local interference,

$$EIR_{\mathrm{f,in}}(\omega) = \frac{S_{n_\mathrm{ls}n_\mathrm{ls}}(\omega)}{S_{n_\mathrm{c}n_\mathrm{c}}(\omega)}\,, \quad (6.8)$$

(6.7) can be written as

$$\mathbf{w}_{\mathrm{f,a,o}}^{(3)}(\omega) = \frac{1 + G_{\mathbf{a}}^{(3)}(\omega) EIR_{\mathrm{f,in}}(\omega)}{1 + EIR_{\mathrm{f,in}}(\omega)} \mathbf{w}_{\mathrm{f,c}}(\omega) \,. \tag{6.9}$$

The PSD $S_{yy}^{(3)}(\omega)$ of the output signal of GSAEC then reads with (4.148), with $\mathbf{B}_{\mathrm{f}}(\omega) = \mathbf{I}_{M \times M}$, and with $\mathbf{w}_{\mathrm{f,a,o}}^{(3)}(\omega)$ as

$$S_{yy}^{(3)}(\omega) = \left(\mathbf{w}_{\mathrm{f,c}}(\omega) - \mathbf{w}_{\mathrm{f,a,o}}^{(3)}(\omega) \right)^H \mathbf{S}_{\mathbf{xx}}(\omega) \left(\mathbf{w}_{\mathrm{f,c}}(\omega) - \mathbf{w}_{\mathrm{f,a,o}}^{(3)}(\omega) \right) \,. \tag{6.10}$$

It follows by replacing in (6.10) $\mathbf{S}_{\mathbf{xx}}(\omega)$ by (6.5) and $\mathbf{w}_{\mathrm{f,a,o}}^{(3)}(\omega)$ by (6.9) the expression

$$\begin{aligned} S_{yy}^{(3)}(\omega) = & \left[S_{n_c n_c}(\omega) + G_{\mathbf{a}}^{(3)}(\omega) S_{n_{ls} n_{ls}}(\omega) \right] \mathbf{w}_{\mathrm{f,c}}^H(\omega) \mathbf{\Gamma}_{\mathbf{n}_c \mathbf{n}_c}(\omega) \mathbf{w}_{\mathrm{f,c}}(\omega) + \\ & + \left[S_{n_c n_c}(\omega) + S_{n_{ls} n_{ls}}(\omega) \right] \frac{\left| 1 + G_{\mathbf{a}}^{(3)}(\omega) EIR_{\mathrm{f,in}}(\omega) \right|^2}{\left(1 + EIR_{\mathrm{f,in}}(\omega)\right)^2} \times \\ & \times \mathbf{w}_{\mathrm{f,c}}^H(\omega) \mathbf{\Gamma}_{\mathbf{n}_c \mathbf{n}_c}(\omega) \mathbf{w}_{\mathrm{f,c}}(\omega) + \\ & \left[S_{n_c n_c}(\omega) + G_{\mathbf{a}}^{(3)}(\omega) S_{n_{ls} n_{ls}}(\omega) \right] \frac{1 + G_{\mathbf{a}}^{(3)}(\omega) EIR_{\mathrm{f,in}}(\omega)}{1 + EIR_{\mathrm{f,in}}(\omega)} \times \\ & \times \mathbf{w}_{\mathrm{f,c}}^H(\omega) \mathbf{\Gamma}_{\mathbf{n}_c \mathbf{n}_c}(\omega) \mathbf{w}_{\mathrm{f,c}}(\omega) \,. \end{aligned} \tag{6.11}$$

We now split $S_{yy}^{(3)}(\omega)$ into a term that covers all components of the local interference and into a term for acoustic echo signals. This yields after some rearrangements the PSDs

$$\begin{aligned} S_{y_{\mathbf{n}_c} y_{\mathbf{n}_c}}^{(3)}(\omega) = & S_{n_c n_c}(\omega) EIR_{\mathrm{f,in}}^2(\omega) \frac{\left| 1 - G_{\mathbf{a}}^{(3)}(\omega) \right|^2}{\left(1 + EIR_{\mathrm{f,in}}(\omega)\right)^2} \times \\ & \times \mathbf{w}_{\mathrm{f,c}}^H(\omega) \mathbf{\Gamma}_{\mathbf{n}_c \mathbf{n}_c}(\omega) \mathbf{w}_{\mathrm{f,c}}(\omega), \end{aligned} \tag{6.12}$$

$$\begin{aligned} S_{y_{\mathbf{n}_{ls}} y_{\mathbf{n}_{ls}}}^{(3)}(\omega) = & S_{n_{ls} n_{ls}}(\omega) \frac{\left| 1 - G_{\mathbf{a}}^{(3)}(\omega) \right|^2}{\left(1 + EIR_{\mathrm{f,in}}(\omega)\right)^2} \times \\ & \times \mathbf{w}_{\mathrm{f,c}}^H(\omega) \mathbf{\Gamma}_{\mathbf{n}_c \mathbf{n}_c}(\omega) \mathbf{w}_{\mathrm{f,c}}(\omega) \,, \end{aligned} \tag{6.13}$$

at the output of the RGSC w.r.t. local interference and w.r.t. acoustic echoes, respectively. Finally, the suppression of acoustic echoes, $ERLE_{\mathrm{f,tot}}(\omega)$ and the suppression of local interference $IR_{\mathrm{f},\mathbf{w}_{\mathrm{RGSC}}}(\omega)$ are obtained by assuming narrowband signals with frequency ω, by inverting (6.12) and (6.13), and by multiplying both sides of (6.12) and of (6.13) with $S_{n_c n_c}(\omega)$ and with $S_{n_{ls} n_{ls}}(\omega)$, respectively. That is,

$$IR_{\mathrm{f},\mathbf{w}_{\mathrm{RGSC}}}(\omega) = \frac{\left(1 + EIR_{\mathrm{f,in}}(\omega)\right)^2}{EIR_{\mathrm{f,in}}^2(\omega) \left| 1 - G_{\mathbf{a}}^{(3)}(\omega) \right|^2 \mathbf{w}_{\mathrm{f,c}}^H(\omega) \mathbf{\Gamma}_{\mathbf{n}_c \mathbf{n}_c}(\omega) \mathbf{w}_{\mathrm{f,c}}(\omega)} \tag{6.14}$$

$$ERLE_{\mathrm{f,tot}}(\omega) = \frac{\left(1 + EIR_{\mathrm{f,in}}(\omega)\right)^2}{\left|1 - G_{\mathbf{a}}^{(3)}(\omega)\right|^2 \mathbf{w}_{\mathrm{f,c}}^H(\omega)\boldsymbol{\Gamma}_{\mathbf{n}_c\mathbf{n}_c}(\omega)\mathbf{w}_{\mathrm{f,c}}(\omega)} . \tag{6.15}$$

Note that $(\mathbf{w}_{\mathrm{f,c}}^H(\omega)\boldsymbol{\Gamma}_{\mathbf{n}_c\mathbf{n}_c}(\omega)\mathbf{w}_{\mathrm{f,c}}(\omega))^{-1}$ is equivalent to the interference rejection $IR_{\mathrm{f},\mathbf{w}_c}(\omega)$ of the quiescent weight vector (see (4.83) for $\boldsymbol{\Gamma}_{\mathbf{n}_w\mathbf{n}_w}(\omega) = \mathbf{0}_{M\times M}$). Generally, since $\mathbf{B}_{\mathrm{f}}(\omega) = \mathbf{I}_{M\times M}$, complete suppression of acoustic echoes and of local interference is possible for $G_{\mathbf{a}}^{(3)}(\omega) = 1$, i.e., if the AEC is inefficient, or for $EIR_{\mathrm{f,in}}(\omega) \in \{0, \infty\}$, i.e., if only local interference or if only acoustic echoes are present. Otherwise, the conflict for the weight vector of the interference canceller prevents complete suppression of acoustic echoes and of local interference. If the echo suppression increases, the interference rejection decreases and vice versa. The worst case is obtained for $EIR_{\mathrm{f,in}}(\omega) = 1$, $G_{\mathbf{a}}^{(3)}(\omega) = 0$, i.e., acoustic echoes and local interferers have the same power level, and the AEC completely cancels acoustic echoes. It yields for (6.14) and for (6.15) the minimum

$$IR_{\mathrm{f},\mathbf{w}_{\mathrm{RGSC}}}(\omega) = ERLE_{\mathrm{f,tot}}(\omega) = \frac{4}{\mathbf{w}_{\mathrm{f,c}}^H(\omega)\boldsymbol{\Gamma}_{\mathbf{n}_c\mathbf{n}_c}(\omega)\mathbf{w}_{\mathrm{f,c}}(\omega)} , \tag{6.16}$$

which corresponds to 6 dB gain relative to $IR_{\mathrm{f},\mathbf{w}_c}(\omega)$.

Examples

For illustration, we evaluate $IR_{\mathrm{f},\mathbf{w}_{\mathrm{RGSC}}}(\omega)$ and $ERLE_{\mathrm{f,tot}}(\omega)$ of GSAEC for diffuse wave fields as a function of frequency and as a function of $EIR_{\mathrm{f,in}}(\omega)$ for the sensor array with $M = 4$ as defined in App. B. We assume that the AEC suppresses acoustic echoes by $ERLE_{\mathrm{f,AEC}}(\omega) = 25$ dB.

For the scenario with $\mathbf{B}_{\mathrm{f}}(\omega) = \mathbf{I}_{M\times M}$ (without presence of the desired signal) after (6.14) and (6.15), the RGSC completely suppresses acoustic echoes and local interference. Figure 6.9a shows that high suppression of the local interference is only possible if local interference predominates ($10\log_{10} EIR_{\mathrm{f,in}}(\omega) < -20$ dB). For predominant acoustic echoes, the interference rejection $IR_{\mathrm{f},\mathbf{w}_{\mathrm{RGSC}}}(\omega)$ is low, and acoustic echoes are efficiently suppressed (Fig. 6.9b). Relative to an AEC alone, the combination of AEC and RGSC only provides greater $ERLE_{\mathrm{f,tot}}(\omega)$ if local interference can be neglected. In our example, we have $ERLE_{\mathrm{f,tot}}(\omega) > ERLE_{\mathrm{f,AEC}}(\omega)$ for $10\log_{10} EIR_{\mathrm{f,in}}(\omega) = 20$ dB at frequencies $f > 2.3$ kHz.

Now, the blocking matrix is defined by (5.51) so that $\mathbf{B}_{\mathrm{f}}(\omega) \neq \mathbf{I}_{M\times M}$. The results are depicted in Fig. 6.10. $IR_{\mathrm{f},\mathbf{w}_{\mathrm{RGSC}}}(\omega)$ and $ERLE_{f,\mathbf{w}_{\mathrm{RGSC}}}(\omega)$ of the RGSC alone are equal to the directivity index $DI(\omega)$ of the RGSC. (See Sect. 4.2.3.) For comparison of GSAEC and RGSC, $DI(\omega)$ is plotted in Fig. 6.10b. In Fig. 6.10a, $DI(\omega)$ overlaps with $IR_{\mathrm{f},\mathbf{w}_{\mathrm{RGSC}}}(\omega)$ for $EIR_{\mathrm{f,in}}(\omega) = 0.01$. Figure 6.10a shows that presence of the AEC reduces $IR_{\mathrm{f},\mathbf{w}_{\mathrm{RGSC}}}(\omega)$ by less than 3 dB for $f < 1$ kHz relative to the RGSC for $10\log_{10} EIR_{\mathrm{f,in}}(\omega) =$

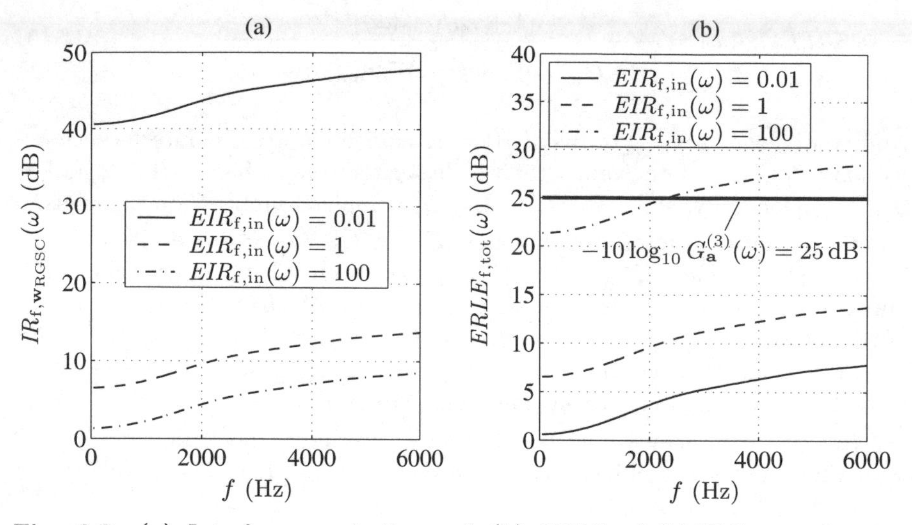

Fig. 6.9. **(a)** Interference rejection and **(b)** ERLE of GSAEC as a function of frequency and of $EIR_{\mathrm{f,in}}(\omega)$ for diffuse wave fields for the blocking matrix $\mathbf{B}_{\mathrm{f}}(\omega) = \mathbf{I}_{M\times M}$; sensor array with $M = 4$ according to App. B, uniformly weighted beamformer $\mathbf{w}_{\mathrm{c}} = 1/M\mathbf{1}_{M\times 1}$

20 dB. For higher frequencies and smaller $EIR_{\mathrm{f,in}}(\omega)$, nearly the same performance is obtained as for the RGSC. Figure 6.10b shows that $ERLE_{\mathrm{f,tot}}(\omega)$ is improved relative to the RGSC for $10\log_{10} EIR_{\mathrm{f,in}}(\omega) > -20\,\mathrm{dB}$ over the entire frequency range. For $f > 1.2\,\mathrm{kHz}$, $ERLE_{\mathrm{f,tot}}(\omega) > ERLE_{\mathrm{f,AEC}}(\omega)$, so that leakage of acoustic echoes through the sidelobe cancelling path of the RGSC can be prevented. These observations can be explained by the fact that diffuse wave fields become spatially uncorrelated for wavelengths which are small relative to the array aperture, so that the uniformly weighted beamformer becomes optimum, and $\mathbf{w}_{\mathrm{f,a,o}}^{(3)}(\omega)$ becomes zero.

In summary, the analysis of GSAEC for identical wave fields of acoustic echoes and of local interference shows that positive synergies can be obtained for GSAEC especially for strong acoustic echoes despite of the conflict for the interference canceller. The results for identical wave fields and different wave fields of acoustic echoes and of local interference are thus similar.

6.3.2 AEC Combined with the Interference Canceller (GEIC)

The analysis of GSAEC has shown that the independent adaptation of the AEC in the reference path does not increase the number of spatial degrees of freedom for the interference canceller, since acoustic echoes are not cancelled in the sidelobe cancelling path before they are seen by the interference canceller. Joint adaptation of the interference canceller and of the AEC resolves this problem as we will see in the following. In Fig. 6.11, the GSAEC is redrawn

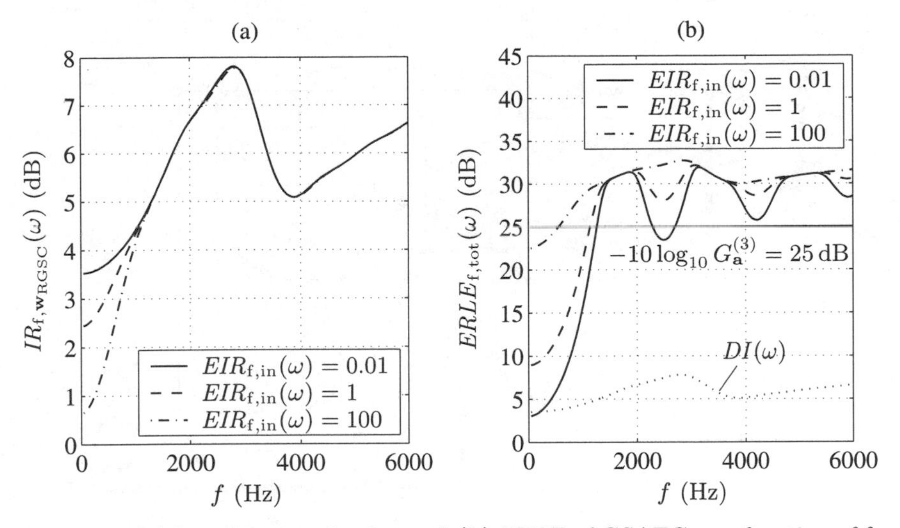

Fig. 6.10. **(a)** Interference rejection and **(b)** ERLE of GSAEC as a function of frequency and of $EIR_{\mathrm{f,in}}(\omega)$ for diffuse wave fields for the blocking matrix after (5.51); sensor array with $M = 4$ according to App. B, uniformly weighted beamformer $\mathbf{w}_{\mathrm{c}} = 1/M \mathbf{1}_{M\times 1}$

with the filters of the AEC moved parallel to the interference canceller. Both units thus use the same error signal, i.e., the output signal of the RGSC, for optimization so that the loudspeaker signals can be interpreted as additional channels of the sidelobe cancelling path. This combination is denoted by GEIC in the following. The stacked weight vector of AEC and interference canceller is defined as

$$\mathbf{w}_{\mathrm{a,s}}^{(4)}(k) = \left(\mathbf{w}_{\mathrm{a,s}}^{T}(k)\,,\mathbf{a}^{(4)T}(k)\right)^{T} . \tag{6.17}$$

First, we describe a necessary precondition for using GEIC. Second, we study the advantages of this structure compared to the previously described approaches. Third, we derive the optimum weight vector of the interference canceller combined with the AEC. Finally, the performance of the combined system is illustrated by some experiments.

Implications for the AEC

The weight vector of the AEC, $\mathbf{a}^{(4)}(k)$, should be adapted simultaneously with $\mathbf{w}_{\mathrm{a,s}}(k)$, and the length $N_{\mathbf{a}}$ of the filters $\mathbf{a}^{(4)}(k)$ should be equal to the length N of the filters $\mathbf{w}_{\mathrm{a,s}}(k)$. With these assumptions, the rate of convergence of the AEC is equal to the rate of convergence of the interference canceller, and tracking of the time-variance of the beamformer is assured. Due to the limited length of the AEC filters, it cannot be expected for converged adaptive filters that the suppression of acoustic echoes of GEIC is as high as for 'AEC first',

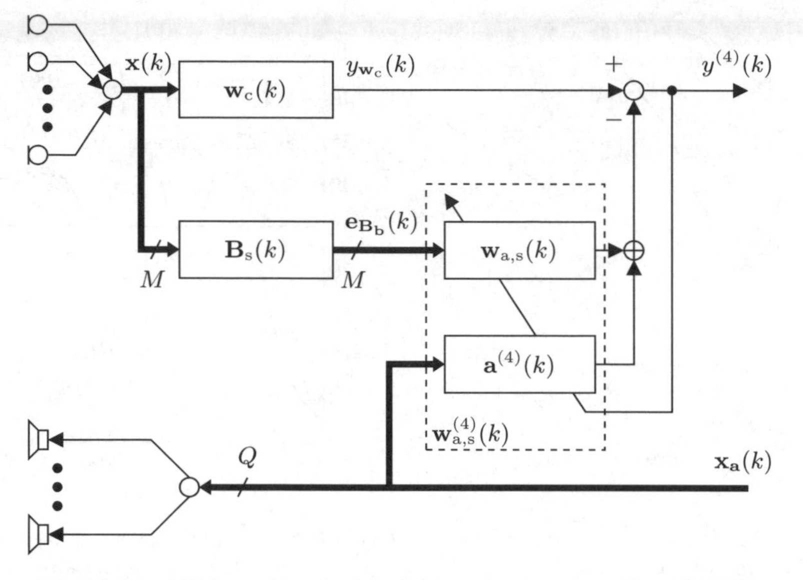

Fig. 6.11. AEC combined with the interference canceller (GEIC)

where longer filters for the AEC are possible. The maximum suppression of acoustic echoes, which would be obtained for stationary conditions, can hardly be obtained in turn. However, in many situations, the echo paths are strongly time-varying, which means that long adaptive filters are rarely completely converged. It thus depends on the time-variance of the acoustic environment if the longer adaptive filters of 'AEC first' yield better echo suppression, or if the same performance can be obtained with shorter adaptive filters of GEIC.

Advantages of the Joint Adaptation of AEC and Interference Canceller

If the length of $\mathbf{a}^{(4)}(k)$ is chosen appropriately, a better tracking of time-varying local interference and of time-varying acoustic echoes is obtained with GEIC relative to the previously described systems. The joint degrees of freedom of the beamformer and of the AEC are always available supposed that an appropriate adaptation algorithm is used. This is assured by the adaptation strategy of the RGSC (Sect. 5.6) when it is applied for the adaptation of $\mathbf{w}_{\mathrm{a,s}}^{(4)}(k)$. This adaptation strategy allows to update $\mathbf{B}_{\mathrm{s}}(k)$ and $\mathbf{w}_{\mathrm{a,s}}^{(4)}(k)$ during double-talk between desired signal, local interference, and acoustic echoes. Therefore, non-stationarity can be optimally tracked, which leads to maximum interference rejection and maximum ERLE. This maximum tracking capability cannot be obtained with the other combinations of beamformer and AEC, since the beamformer and the AEC are adapted independently there: The AEC can only be updated if presence of local interference does not lead to instabilities of the adaptation algorithm of the AEC. This is the case for competing burst signals as, e.g., speech or music. If the echo paths change during these periods, the AEC is inefficient, and only the degrees of freedom of the beamformer are available.

As a consequence of the joint adaptation of beamformer and AEC, the AEC does not require a separate adaptation control unit. However, it depends on the complexity of the adaptation algorithm for the stacked weight vector $\mathbf{w}_{\mathrm{a,s}}^{(4)}(k)$ with $M+Q$ input channels whether the computational complexity can be reduced with GEIC compared to the other combinations of AEC and RGSC.

Optimum Weight Vector of the Interference Canceller Combined with the AEC

Next, we derive the LS solution of the stacked weight vector $\mathbf{w}_{\mathrm{a,s}}^{(4)}(k)$ in order to illustrate the conditions for which the AEC filters operate independently of the RGSC, and, thus, do not 'see' the time-variance of the RGSC. The output signal $y^{(4)}(k)$ of GEIC reads with (5.29)

$$y^{(4)}(k) = \mathbf{w}_{\mathrm{c}}^T(k)\mathbf{J}_{MN^3\times MN}^{(1)T}\mathbf{x}_{\mathrm{ss}}(k) - \mathbf{w}_{\mathrm{a,s}}^T(k)\mathbf{B}_{\mathrm{ss}}^T(k)\mathbf{x}_{\mathrm{ss}}(k) - \mathbf{a}^{(4)T}(k)\mathbf{x}_{\mathbf{a}}(k)\,, \tag{6.18}$$

where $\mathbf{x}_{\mathbf{a}}(k)$ is defined as

$$\mathbf{x}_{\mathbf{a}}(k) = \left(\mathbf{x}_{\mathbf{a},0}^T(k)\,, \mathbf{x}_{\mathbf{a},1}^T(k)\,, \ldots, \mathbf{x}_{\mathbf{a},Q-1}^T(k)\right)^T\,, \tag{6.19}$$

$$\mathbf{x}_{\mathbf{a},q}(k) = \left(x_{\mathbf{a},q}(k),\, x_{\mathbf{a},q}(k-1),\, \ldots,\, x_{\mathbf{a},q}(k-N_{\mathbf{a}}+1)\right)^T\,. \tag{6.20}$$

Using the vector of sensor signals $\mathbf{x}_{\mathrm{ss}}(k)$ and the vector of loudspeaker signals $\mathbf{x}_{\mathbf{a}}(k)$ as input signals of a MISO system, we can write (6.18) in block-matrix form as

$$\begin{aligned} y^{(4)}(k) = &\left[\begin{pmatrix} \mathbf{J}_{MN^3\times MN}^{(1)} & \mathbf{0}_{MN^3\times QN_{\mathbf{a}}} \\ \mathbf{0}_{QN_{\mathbf{a}}\times MN} & \mathbf{0}_{QN_{\mathbf{a}}\times QN_{\mathbf{a}}} \end{pmatrix} \begin{pmatrix} \mathbf{w}_{\mathrm{c}}(k) \\ \mathbf{0}_{QN_{\mathbf{a}}\times 1} \end{pmatrix} - \right. \\ &\left. - \begin{pmatrix} \mathbf{B}_{\mathrm{ss}}(k) & \mathbf{0}_{MN^3\times QN_{\mathbf{a}}} \\ \mathbf{0}_{QN_{\mathbf{a}}\times MN} & \mathbf{I}_{QN_{\mathbf{a}}\times QN_{\mathbf{a}}} \end{pmatrix} \begin{pmatrix} \mathbf{w}_{\mathrm{a,s}}(k) \\ \mathbf{a}^{(4)}(k) \end{pmatrix}\right]^T \begin{pmatrix} \mathbf{x}_{\mathrm{ss}}(k) \\ \mathbf{x}_{\mathbf{a}}(k) \end{pmatrix}. \end{aligned} \tag{6.21}$$

Let the data matrix w.r.t. the loudspeaker signals be defined as

$$\mathbf{X}_{\mathbf{a}}(k) = \left(\mathbf{x}_{\mathbf{a}}(k),\, \mathbf{x}_{\mathbf{a}}(k+1),\, \ldots,\, \mathbf{x}_{\mathbf{a}}(k+K-1)\right)\,. \tag{6.22}$$

Then, identification of (6.21) with the quantities in (5.29) allows to write the optimum weight vector $\mathbf{w}_{\mathrm{a,s,o}}^{(4)}(k)$ in the same way as (5.36). After some rearrangements, we can simplify $\mathbf{w}_{\mathrm{a,s,o}}^{(4)}(k)$ as follows:

$$\begin{pmatrix} \mathbf{w}_{\mathrm{a,s,o}}(k) \\ \mathbf{a}_{\mathrm{o}}^{(4)}(k) \end{pmatrix} = \begin{pmatrix} \mathbf{B}_{\mathrm{ss}}^T(k)\mathbf{\Phi}_{\mathbf{xx},\mathrm{ss}}(k)\mathbf{B}_{\mathrm{ss}}(k) & \mathbf{B}_{\mathrm{ss}}^T(k)\mathbf{X}_{\mathrm{ss}}(k)\mathbf{X}_{\mathbf{a}}^T(k) \\ \mathbf{X}_{\mathbf{a}}(k)\mathbf{X}_{\mathrm{ss}}^T(k)\mathbf{B}_{\mathrm{ss}}(k) & \mathbf{X}_{\mathbf{a}}(k)\mathbf{X}_{\mathbf{a}}^T(k) \end{pmatrix}^+ \times$$
$$\times \begin{pmatrix} \mathbf{B}_{\mathrm{ss}}^T(k)\mathbf{\Phi}_{\mathbf{xx},\mathrm{ss}}(k)\mathbf{J}_{MN^3\times MN}^{(1)}\mathbf{w}_{\mathrm{c}}(k) \\ \mathbf{X}_{\mathbf{a}}(k)\mathbf{X}_{\mathrm{ss}}^T(k)\mathbf{J}_{MN^3\times MN}^{(1)}\mathbf{w}_{\mathrm{c}}(k) \end{pmatrix}. \tag{6.23}$$

The first term in parentheses on the right side of (6.23) is the pseudoinverse of the sample correlation matrix of the stacked input signal vector of $\mathbf{w}_{\mathrm{a,s,o}}^{(4)}(k)$. The second term is the sample correlation matrix of the output signal of the quiescent weight vector and the input signals of $\mathbf{w}_{\mathrm{a,s,o}}^{(4)}(k)$. We see that $\mathbf{a}_{\mathrm{o}}^{(4)}(k)$ is independent of the output signals of the blocking matrix only if $\mathbf{B}_{\mathrm{ss}}(k)$ cancels acoustic echoes. The sample correlation matrix is then block-diagonal. The optimum weight vector $\mathbf{a}_{\mathrm{o}}^{(4)}(k)$ is equivalent to the optimum AEC filters $\mathbf{a}_{\mathrm{o}}^{(3)}(k)$ of GSAEC, and the adaptive weight vector $\mathbf{w}_{\mathrm{a,s,o}}(k)$ is equivalent to the adaptive weight vector of the RGSC according to (5.36). Since acoustic echoes are assumed to be in the null space of the blocking matrix, $\mathbf{w}_{\mathrm{a,s,o}}(k)$ is independent of acoustic echoes.

Experiments

We illustrate the properties of GEIC for WSS signals and time-invariant environments and for transient conditions.

Stationary Conditions

For studying the behavior of GEIC for temporally WSS signals and for a time-invariant environment, $IR_{\mathbf{w}_{\mathrm{RGSC}}}$ and $ERLE_{\mathrm{tot}}$ are illustrated in Fig. 6.12a and in Fig. 6.12b, respectively, as a function of EIR_{in}. The experimental setup is the same as in Sect. 6.2.1. The same number of filter coefficients is used for the filters of the AEC as for the filters of the interference canceller, i.e., $N = N_{\mathbf{a}} = 256$. For comparing GEIC with the previously discussed combinations of AEC and GSC, $IR_{\mathbf{w}_{\mathrm{RGSC}}}$ and $ERLE_{\mathrm{tot}}$ for GSAEC, 'AEC first', and RGSC from Fig. 6.6 are repeated. It can be seen that GEIC outperforms the interference rejection capability of GSAEC since the number of degrees of freedom of the interference canceller is increased compared to GSAEC by the number of loudspeaker channels. However, $ERLE_{\mathrm{tot}}$ of GEIC is less than for GSAEC for large EIR_{in} since $ERLE_{\mathrm{AEC}}$ is limited by the small number of filter coefficients of the AEC. For stationary conditions, the GSAEC can be outperformed by GEIC w.r.t. $IR_{\mathbf{w}_{\mathrm{RGSC}}}$ but not w.r.t. $ERLE_{\mathrm{tot}}$.

Transient Conditions

Here, GEIC is studied for transient signals relative to 'AEC first'. The experimental setup of Sect. 6.2.2 is used. For GEIC, we use $N = N_{\mathbf{a}} = 256$. For 'AEC first', we use $N = 256$, $N_{\mathbf{a}} = 1024$. The results are depicted in Fig. 6.13. Figures 6.13a–c show the desired signal, one acoustic echo signal, and the signal of

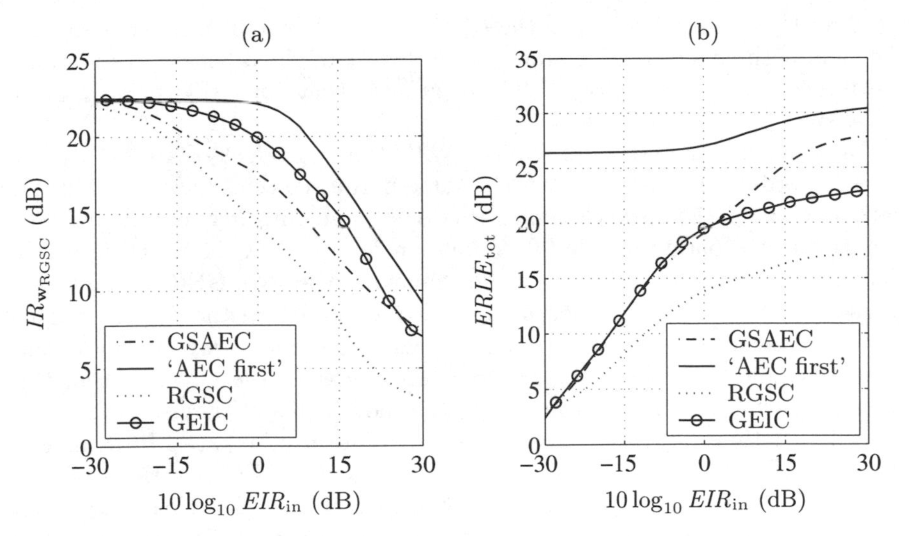

Fig. 6.12. Comparison of the interference rejection and of the echo suppression for RGSC alone, 'AEC first', GSAEC, and GEIC after convergence of the adaptive filters; experimental setup according to Sect. 6.2.1, multimedia room with $T_{60} = 250$ ms, $M = 4$, $N = 256$, $N_{\mathbf{a}} = 256$ (GEIC), $N_{\mathbf{a}} = 256$ ('AEC first')

the local interferer, respectively. For simulating a time-varying echo path, the loudspeaker positions are moved at $t = 12.5$ s from the two endfire positions $\theta = 0$ and $\theta = \pi$ to $\theta = \pi/18$ and $\theta = \pi - \pi/18$, respectively. Figure 6.13d shows the segmentation of the time axis into 6 phases and the adaptation of the modules. The echo suppression and the interference rejection of GEIC and of 'AEC first' are illustrated in Fig. 6.13e and in Fig. 6.13f, respectively. Figure 6.13g and Fig. 6.13h show the residual acoustic echoes and the residual local interference at the output of GEIC and at the output of 'AEC first', respectively. In Phase I, the blocking matrix is adapted, acoustic echoes and local interference is not present. In Phase II, only acoustic echoes are active so that the AEC in 'AEC first' can adapt without being impaired by presence of local interference. The echo suppression is more than 30 dB. For GEIC the echo suppression is limited to about 23 dB due to the limited length of the AEC filters. From Phase III to Phase V, acoustic echoes and local interference are active. In Phase III, the interference rejection of 'AEC first' is greater than for GEIC, since the interference canceller converges slowlier for GEIC than for 'AEC first' due to the larger number of filter coefficients. At the beginning of Phase IV, the echo paths are changed while acoustic echoes and local interference are present. The AEC of 'AEC first' cannot be adapted because of the double-talk situation with local interference, and the AEC is inefficient in turn. Only the degrees of freedom of the RGSC are available so that the echo suppression drops by 20 dB and the interference rejection reduces by 5 dB. Only when acoustic echoes alone are present the AEC can be adapted

(Phase VI). For GEIC, however, the AEC can be adapted continuously so that the echo path change can be immediately tracked. The echo suppression and the interference rejection of GEIC are greater than for 'AEC first' in such situations.

In summary, the GEIC provides – with the equivalent of an AEC for a single recording channel – at least the same performance as 'AEC first' in situations with continuously changing echo paths and with frequent double-talk situations. A separate double-talk detection for the AEC is not required. It thus depends on the application if the echo suppression of GEIC is sufficient. Consider for example the microphone array mounted on the top of or integrated in the screen of a laptop computer with the user sitting in front of the computer, and a pair of loudspeakers placed next to the screen. Then, the echo paths change continuously due to the movements of the user. The echo canceller continuously has to track the changes of the echo paths. If the user is often speaking or local interference is often active, the AEC of 'AEC first' can often not be adapted and may be often inefficient in turn. For GEIC, however, the AEC can be adapted continuously so that the changes of the echo paths can be better tracked.

6.4 Discussion

In this chapter, we have discussed four principle methods for integrating acoustic echo cancellation into beamformers. Though the combination of AECs with fixed beamformers is studied, too, we put more focus on combinations of AECs and adaptive realizations of data-dependent optimum beamformers. The results are summarized in Table 6.1. While the structure with the AEC following the adaptive beamformer ('beamformer first') can hardly be applied, the usage of the other structures depends on the acoustic conditions and on the available computational resources.

'AEC first' provides maximum echo suppression and interference rejection if the acoustic echo paths do not continuously change and if frequent double-talk between desired signal, local interference, and acoustic echoes is not expected, since the AEC can only be adapted if acoustic echoes predominate. Continuously changing echo paths cannot be tracked during double-talk so that the performance reduces in strongly time-varying acoustic environments. The computational complexity may be prohibitively high, since at least the filtering and the filter coefficient update are required for each sensor channel.

The GSAEC with the AEC in the reference path of a GSC-like structure requires the AEC only for a single recording channel. Thus, computational complexity is considerably reduced relative to 'AEC first'. GSAEC has almost the same echo suppression capability as 'AEC first' for predominant acoustic echoes. However, theoretical studies and experimental results showed that the performance of this system is limited since acoustic echoes are still present in the sidelobe cancelling path while they may be efficiently cancelled by the

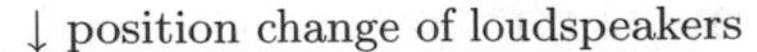

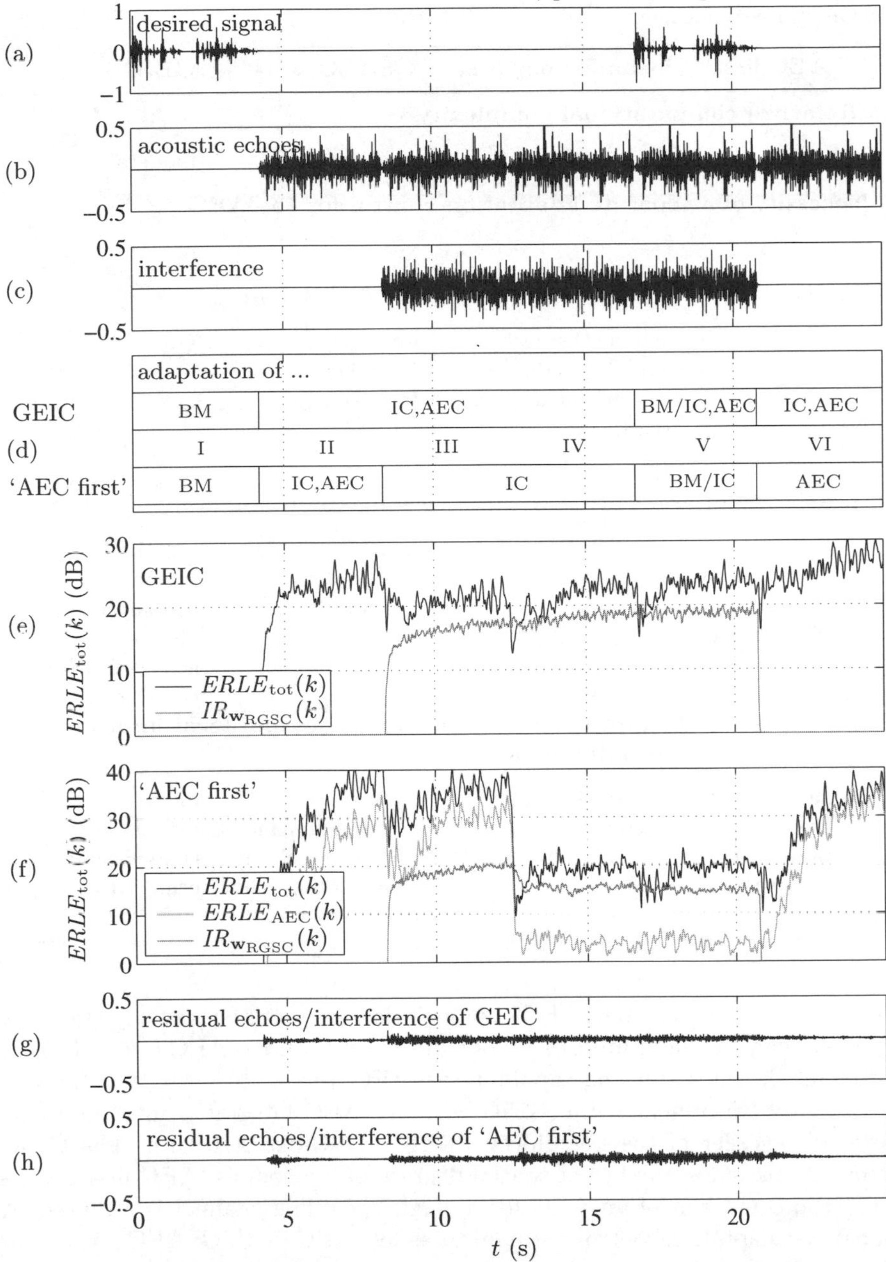

Fig. 6.13. Transient behavior of GEIC and of 'AEC first' with presence of acoustic echoes, local interference, and desired signal; multimedia room with $T_{60} = 250\,\mathrm{ms}$, $M = 4$

Table 6.1. Overview of the properties of the combinations of acoustic echo cancellation and beamforming

'AEC first'	'beamformer first'	GSAEC	GEIC
(Relative) computational complexity			
high	moderate	moderate	moderate
Necessity of a separate adaptation control for the AEC			
yes	yes	yes	no
Possibility of exploitation of all degrees of freedom			
yes	only at the cost of highly reduced tracking or for WSS conditions	only for WSS local interference	yes
Tracking capability (possibility of application to time-varying environments)			
high, if echo paths are slowly time-varying	very little	high, if echo paths are slowly time-varying	very high, at the cost of less suppression of acoustic echoes for WSS conditions
Improvement of suppression of acoustic echoes and local interference relative to beamforming alone			
high for slowly time-varying echo paths	only for special cases	high for predominant echoes and slowly time-varying echo paths	high for strongly time-varying echo paths

AEC in the reference path. Echo suppression and interference rejection are greater than for the beamformer alone but lower than for 'AEC first'. GSAEC provides the same tracking capability as 'AEC first'.

Finally, we presented the GEIC with the AEC integrated into the interference canceller of the beamformer with a GSC-like structure. The GEIC provides the same number of spatial degrees of freedom as 'AEC first', while only the equivalent of an AEC for a single recording channel is required. A separate adaptation control mechanism as for 'AEC first', GSAEC, or 'beamformer first' is not necessary, since the AEC is adapted simultaneously with the interference canceller. The AEC can be adapted even during continuous double-talk of local interference and acoustic echoes. Therefore, optimum tracking capability of the AEC is obtained with GEIC. This is especially advantageous in applications with time-varying acoustic echo paths if continuous double-talk is expected. However, the filters for the AEC should not be longer than the filters of the interference canceller. Therefore, the echo suppression after convergence of the AEC is limited.

Audio examples which illustrate the performance of the studied systems can be found at www.wolfgangherbordt.de/micarraybook/.

7

Efficient Real-Time Realization of an Acoustic Human/Machine Front-End

In the previous chapters, we discussed options for data-dependent optimum beamforming for acoustic human/machine front-ends on cost-sensitive platforms of limited dimension. The main difficulties for realizing adaptive data-dependent optimum beamforming were studied, which resulted in an attractive solution for practical audio signal acquisition systems using a robust generalized sidelobe canceller with spatio-temporal constraints. Various techniques for combining the RGSC with multi-channel acoustic echo cancellation as a complementary speech enhancement technique for full-duplex applications were analyzed.

For a seamless audio interface realized as an embedded application or as a PC-based system, the cost has to be negligible in comparison to the total system cost. The cost for the audio capture includes both hardware and software. For the beamforming algorithms, the number of sensors should be minimized in order to keep the amount of additional hardware components low. Considering, however, the continuously increasing processing power on all computing platforms, sophisticated algorithms can be considered.

In this chapter, we present a concept for realizing adaptive beamforming integrating multi-channel acoustic echo cancellation based on the RGSC, which reconciles the need for low computational complexity with versatility and robustness for real-world scenarios. Especially, the advantages of the RGSC to deal with (a) mixtures of desired speech and non-stationary directional or diffuse interference and (b) with time-varying reverberant acoustic environments are preserved. The efficient realization of the AEC is discussed for the example of a stereophonic system [BM00]. The extension to more than two reproduction channels is presented in [BBK03].

The RGSC was first published using the LMS algorithm in the time domain for adaptation [HSH99]. This realization does not resolve the adaptation problems of the blocking matrix and of the interference canceller for realistic conditions with non-stationary signals and time-varying acoustic environments. Here, we use multi-channel DFT-domain adaptive filtering with RLS-like characteristics [BM00, BBK03] as a computationally efficient means to meet the requirements of both RGSC and multi-channel acoustic echo cancellation. The de-crosscorrelating property of multi-channel DFT-domain adaptive filtering is required by the AEC for maximum convergence speed in the presence of mutually strongly correlated loudspeaker signals [BM00, BBK03]. The direct access to distinct frequency bins of the DFT is exploited for the frequency-selective adaptation of the blocking matrix and of the interference canceller of the RGSC. Compared to the unconstrained optimization criterion in the DFT domain leading to the MIMO adaptive filter [BM00, BBK03], we are using constrained optimization in the DFT domain for formulating the update equation. In [RE00], constrained single-channel adaptive filtering in the DFT domain using penalty functions is obtained from an optimization criterion with penalty terms in the discrete time domain by transforming the resulting update equation into the DFT domain. With our approach, by starting from an optimization criterion with penalty function in the DFT domain, the constrained MIMO adaptive filter in the DFT domain follows directly. This method will be applied in order to add diagonal loading to the RLS-like algorithm in the DFT domain for maximum robustness against cancellation of the desired signal [HK01b, HK02b].

This chapter is organized as follows: Section 7.1 introduces adaptive filtering in the DFT domain for MIMO systems. This generic MIMO adaptation algorithm is applied in Sect. 7.2 for realizing the RGSC combined with multi-channel acoustic echo cancellation in the DFT domain. The AEC module is described for the example of 'AEC first'. The other combinations of RGSC and AEC can be obtained by integrating the AEC into the RGSC at the desired position. Section 7.3 concludes this chapter by experimental results which confirm that our approach meets well the practical requirements of real-world scenarios. Especially, the capability of the RGSC to deal with non-stationary signals and time-varying environments is illustrated, and the RGSC is applied as a front-end to an automatic speech recognizer.

7.1 Multi-channel Block-Adaptive Filtering in the Discrete Fourier Transform (DFT) Domain

The integrated system of RGSC and AEC, which will be presented in this chapter, is solely based on efficient frequency-domain adaptive filtering using the overlap-save method [PM96]. In the following, we give a compact formulation of a generic adaptive filter structure with M input channels and P output channels as shown in Fig. 7.1 [BBK03, HBK03a]. This formalism will then be applied in Sect. 7.2 to our combination of RGSC and AEC. As it turns out, this formulation supports a systematic transformation of the entire structure (Chaps. 5, 6) into the DFT domain, and leads to the desired effects, such as solution of the adaptation problem and robustness improvement of the RGSC and accounting of the mutual correlation between the loudspeaker signals of the AEC module. Note that the application of the overlap-save method using

DFTs requires block processing of the input and output data streams. In the following, we derive the algorithm for a block length K that is equal to the filter length N for formal simplicity, which yields maximum efficiency. However, to keep the processing delay short and to preserve optimum tracking behavior, the data blocks are overlapped by the overlap factor α in our realization [SP90, BD91, BD92, MAG95, PM94]. Moreover, we consider multi-channel frequency-domain adaptive filters for constrained optimization with penalty functions for the derivation. A more general treatment of this class of adaptive algorithms (without penalty terms) including an in-depths convergence analysis can be found in [BBK03]. Introductions to (single-channel) DFT-domain adaptive filtering can be found in, e.g., [Shy92, Hay96].

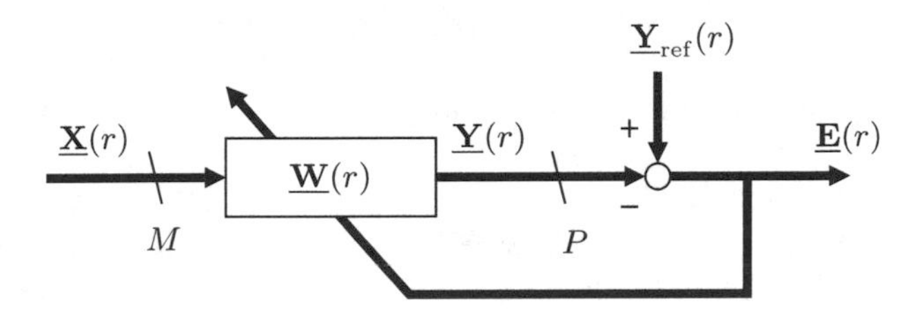

Fig. 7.1. Adaptive MIMO filtering in the DFT domain [BBK03, HBK03a]

7.1.1 Optimization Criterion

To obtain a MIMO algorithm in the DFT domain, we start from the block-error signal $\mathbf{E}(k) = \mathbf{Y}_{\mathrm{ref}}(k) - \mathbf{X}^T(k)\mathbf{W}(k)$ in the discrete time domain as defined in (3.13) with $K = N$. First, a block-overlap factor $\alpha = N/R$, where R is the number of 'new' samples per block, is introduced into $\mathbf{E}(k)$. This replaces the discrete time index k by the block time index r, which is related to k by $k := rR$. Then, we write the convolutional product $\mathbf{X}^T(k)\mathbf{W}(k)$ as a fast convolution in the DFT domain using the overlap-save method [PM96]. With the DFT matrix $\mathbf{F}_{2N\times 2N}$ of size $2N \times 2N$, the data matrix $\mathbf{X}(k)$ is transformed into a block-diagonal matrix $\underline{\mathbf{X}}(r)$ of size $2N \times 2MN$:

$$\underline{\mathbf{X}}(r) = \left(\underline{\mathbf{X}}_0(r), \underline{\mathbf{X}}_1(r), \dots, \underline{\mathbf{X}}_{M-1}(r)\right) , \tag{7.1}$$

where

$$\underline{\mathbf{X}}_m(r) = \mathrm{diag}\left\{ \mathbf{F}_{2N\times 2N} \begin{pmatrix} x_m(rR-N) \\ x_m(rR-N+1) \\ \vdots \\ x_m(rR+N-1) \end{pmatrix} \right\} . \tag{7.2}$$

The MIMO system $\mathbf{W}(rR)$ after (3.1) with filters $\mathbf{w}_{m,p}(rR)$ of length N is transformed into the DFT domain as

$$\underline{\mathbf{W}}(r) = \mathrm{diag}\left\{\mathbf{F}_{2N\times 2N}\mathbf{W}^{10}_{2N\times N}, \ldots, \mathbf{F}_{2N\times 2N}\mathbf{W}^{10}_{2N\times N}\right\} \mathbf{W}(rR)\,, \tag{7.3}$$

where the windowing matrix

$$\mathbf{W}^{10}_{2N\times N} = \left(\mathbf{I}_{N\times N}, \mathbf{0}_{N\times N}\right)^T \tag{7.4}$$

appends N zeroes to the coefficient vectors $\mathbf{w}_{m,p}(rR)$ in order to prevent circular convolution in the DFT domain. Using fast convolution, we obtain for the block error matrix $\mathbf{E}(rR)$ after (3.13) for $K = N$ the expression

$$\mathbf{E}(rR) = \mathbf{Y}_{\mathrm{ref}}(rR) - \mathbf{W}^{01}_{N\times 2N}\,\mathbf{F}^{-1}_{2N\times 2N}\,\underline{\mathbf{X}}(r)\,\underline{\mathbf{W}}(r)\,, \tag{7.5}$$

where the windowing matrix

$$\mathbf{W}^{01}_{N\times 2N} = \left(\mathbf{0}_{N\times N}, \mathbf{I}_{N\times N}\right)^T \tag{7.6}$$

extracts the block of size $N \times P$ of output signals $\mathbf{Y}(rR)$ from the matrix $\mathbf{F}^{-1}_{2N\times 2N}\,\underline{\mathbf{X}}(r)\,\underline{\mathbf{W}}(r)$ of size $2N \times P$.

Next, the optimization criterion is defined in the DFT domain similar to (3.7). Multiplying (7.5) by $\mathbf{F}_{2N\times 2N}\mathbf{W}^{01}_{2N\times N}$, where $\mathbf{W}^{01}_{2N\times N} := (\mathbf{W}^{01}_{N\times 2N})^T$ we get the block-error signal matrix in the DFT domain:

$$\underline{\mathbf{E}}(r) = \underline{\mathbf{Y}}_{\mathrm{ref}}(r) - \mathbf{G}^{01}_{2N\times 2N}\,\underline{\mathbf{X}}(r)\,\underline{\mathbf{W}}(r)\,, \tag{7.7}$$

where

$$\begin{aligned}\underline{\mathbf{E}}(r) &= \mathbf{F}_{2N\times 2N}\mathbf{W}^{01}_{2N\times N}\mathbf{E}(rR) \\ &= \mathbf{F}_{2N\times 2N}\begin{pmatrix}\mathbf{0}_{N\times 1} & \cdots & \mathbf{0}_{N\times 1} \\ \mathbf{e}_0(rR) & \cdots & \mathbf{e}_{P-1}(rR)\end{pmatrix}, \end{aligned}\tag{7.8}$$

$$\begin{aligned}\underline{\mathbf{Y}}_{\mathrm{ref}}(r) &= \mathbf{F}_{2N\times 2N}\mathbf{W}^{01}_{2N\times N}\mathbf{Y}_{\mathrm{ref}}(rR) \\ &= \mathbf{F}_{2N\times 2N}\begin{pmatrix}\mathbf{0}_{N\times 1} & \cdots & \mathbf{0}_{N\times 1} \\ \mathbf{y}_{\mathrm{ref},0}(rR) & \cdots & \mathbf{y}_{\mathrm{ref},P-1}(rR)\end{pmatrix}, \end{aligned}\tag{7.9}$$

$$\mathbf{G}^{01}_{2N\times 2N} = \mathbf{F}_{2N\times 2N}\mathbf{W}^{01}_{2N\times 2N}\mathbf{F}^{-1}_{2N\times 2N}\,, \tag{7.10}$$

$$\mathbf{W}^{01}_{2N\times 2N} = \mathbf{W}^{01}_{2N\times N}\mathbf{W}^{01}_{N\times 2N} = \begin{pmatrix}\mathbf{0}_{N\times N} & \mathbf{0}_{N\times N} \\ \mathbf{0}_{N\times N} & \mathbf{I}_{N\times N}\end{pmatrix}, \tag{7.11}$$

and where

$$\mathbf{e}_p(k) = \left(e_p(k),\, e_p(k+1),\, \ldots,\, e_p(k+N-1)\right)^T\,, \tag{7.12}$$

$$\mathbf{y}_{\mathrm{ref},p}(k) = \left(y_{\mathrm{ref},p}(k),\, y_{\mathrm{ref},p}(k+1),\, \ldots,\, y_{\mathrm{ref},p}(k+N-1)\right)^T\,. \tag{7.13}$$

The cost function $\underline{\xi}(r)$ in the DFT domain is obtained by recursively averaging the accumulated sum $\mathrm{tr}\{\underline{\mathbf{E}}^H(r)\underline{\mathbf{E}}(r)\}$ of DFT frequency bins over the block time r. That is,

$$\underline{\xi}(r) = (1-\beta)\sum_{i=0}^{r}\beta^{r-i}\,\left\|\underline{\mathbf{E}}(i)\right\|_2^2\,. \tag{7.14}$$

The exponential forgetting factor $0 < \beta < 1$ includes past blocks of input data into the optimization criterion and allows for tracking of time-varying second-order signal statistics. The criterion (7.14) in the DFT domain is analogous to the one leading to the RLS algorithm [Hay96] in the time domain. The advantage of using (7.14) is to take advantage of the fast Fourier transform (FFT) [OS75] in order to have low-complexity adaptive filters.

Constraints may be incorporated into the cost function $\underline{\xi}(r)$ by adding penalty terms to $\underline{\xi}(r)$ (7.14) to penalize $\underline{\xi}(r)$ for violating given constraints [Fle87, RE00]. The new cost function $\underline{\xi}_{\mathrm{c}}(r)$ can be formulated as

$$\underline{\xi}_{\mathrm{c}}(r) := \underline{\xi}(r) + \sigma \sum_{i=0}^{C'-1} \left|\max\{c_i(\underline{\mathbf{W}}(r)),\, 0\}\right|^2 , \tag{7.15}$$

where $c_i(\underline{\mathbf{W}}(r))$ defines the i-th penalty function for a constraint that should be satisfied by the adaptive filter matrix $\underline{\mathbf{W}}(r)$. C' is the number of penalty functions. The penalty function for the i-th constraint is zero if $c_i(\underline{\mathbf{W}}(r)) < 0$, i.e., if the i-th constraint is fulfilled. The penalty function is equal to the squared sum of the constraints weighted by σ if the constraints are violated.

7.1.2 Adaptive Algorithm

From the cost functions $\underline{\xi}(r)$ and $\underline{\xi}_{\mathrm{c}}(r)$, we can now derive the recursive adaptation algorithm in the DFT domain. First, we consider an unconstrained adaptive algorithm using $\underline{\xi}(r)$ after (7.14). Second, the unconstrained algorithm is extended to the constrained case using (7.15).

Adaptation Without Penalty Functions

An RLS-like adaptation algorithm can be derived in a straightforward way from the normal equation that is obtained by setting the gradient of (7.14) w.r.t. $\underline{\mathbf{W}}^*(r)$ equal to zero. According to [Hay96], and by noting that

$$(\mathbf{G}^{01}_{2N\times 2N})^H \mathbf{G}^{01}_{2N\times 2N} = \mathbf{G}^{01}_{2N\times 2N} , \tag{7.16}$$

$$(\mathbf{G}^{01}_{2N\times 2N})^H \underline{\mathbf{Y}}_{\mathrm{ref}}(r) = \underline{\mathbf{Y}}_{\mathrm{ref}}(r) , \tag{7.17}$$

we have for the gradient the expression

$$\nabla_{\underline{\mathbf{W}}(r)} \underline{\xi}(r) = 2 \frac{\partial \underline{\xi}(r)}{\partial \underline{\mathbf{W}}^*(r)} \tag{7.18}$$

$$= 2(1-\beta) \sum_{i=0}^{r} \beta^{r-i} \Big(-\underline{\mathbf{X}}^H(i) \underline{\mathbf{Y}}_{\mathrm{ref}}(i) + + \underline{\mathbf{X}}^H(i) \mathbf{G}^{01}_{2N\times 2N} \underline{\mathbf{X}}(i) \underline{\mathbf{W}}(r) \Big) . \tag{7.19}$$

Setting this gradient equal to zero, we obtain the normal equation in analogy to (3.34)

$$\hat{\underline{\mathbf{S}}}_{\mathbf{xx}}(r)\,\underline{\mathbf{W}}(r) = \hat{\underline{\mathbf{S}}}_{\mathbf{xy}}(r)\,, \tag{7.20}$$

where

$$\begin{aligned}\hat{\underline{\mathbf{S}}}_{\mathbf{xx}}(r) &= (1-\beta)\sum_{i=0}^{r}\beta^{r-i}\,\underline{\mathbf{X}}^H(i)\mathbf{G}^{01}_{2N\times 2N}\underline{\mathbf{X}}(i)\\ &= \beta\,\hat{\underline{\mathbf{S}}}_{\mathbf{xx}}(r-1) + (1-\beta)\,\underline{\mathbf{X}}^H(r)\mathbf{G}^{01}_{2N\times 2N}\underline{\mathbf{X}}(r)\,, \end{aligned}\tag{7.21}$$

$$\begin{aligned}\hat{\underline{\mathbf{S}}}_{\mathbf{xy}}(r) &= (1-\beta)\sum_{i=0}^{r}\beta^{r-i}\underline{\mathbf{X}}^H(i)\underline{\mathbf{Y}}_{\mathrm{ref}}(i)\\ &= \beta\,\hat{\underline{\mathbf{S}}}_{\mathbf{xy}}(r-1) + (1-\beta)\,\underline{\mathbf{X}}^H(r)\underline{\mathbf{Y}}_{\mathrm{ref}}(r)\,. \end{aligned}\tag{7.22}$$

Note that $\hat{\underline{\mathbf{S}}}_{\mathbf{xx}}(r)$ in (7.21) is an exponentially-windowed average of instantaneous estimates $\underline{\mathbf{X}}^H(r)\mathbf{G}^{01}_{2N\times 2N}\underline{\mathbf{X}}(r)$ of the (spatio-spectral) correlation matrix $\mathbf{S}_{\mathbf{xx}}(\omega)$ at block time r in the DFT domain. Accordingly, (7.22) can be interpreted as an exponentially-windowed average of instantaneous estimates $\underline{\mathbf{X}}^H(r)\underline{\mathbf{Y}}_{\mathrm{ref}}(r)$ of the correlation between the input signals $\mathbf{x}(k)$ (2.30) and the reference signals $\mathbf{y}_{\mathrm{ref}}(k)$ (3.4) in the DFT domain.

The iterative algorithm, i.e., the recursive update of the coefficient matrix $\underline{\mathbf{W}}(r)$ is directly derived from (7.20)–(7.22). In the recursive equation (7.22), we replace $\hat{\underline{\mathbf{S}}}_{\mathbf{xy}}(r)$ and $\hat{\underline{\mathbf{S}}}_{\mathbf{xy}}(r-1)$ by formulating (7.20) in terms of block time indices r and $r-1$, respectively. We next eliminate $\hat{\underline{\mathbf{S}}}_{\mathbf{xx}}(r-1)$ from the resulting equation using (7.21). Reintroducing the error signal vector (7.7) with $\underline{\mathbf{W}}(r)$ replaced by $\underline{\mathbf{W}}(r-1)$, we obtain the iterative algorithm[1]

$$\underline{\mathbf{E}}(r) = \underline{\mathbf{Y}}_{\mathrm{ref}}(r) - \mathbf{G}^{01}_{2N\times 2N}\underline{\mathbf{X}}(r)\underline{\mathbf{W}}(r-1)\,, \tag{7.23}$$

$$\underline{\mathbf{W}}(r) = \underline{\mathbf{W}}(r-1) + (1-\beta)\hat{\underline{\mathbf{S}}}^{-1}_{\mathbf{xx}}(r)\underline{\mathbf{X}}^H(r)\underline{\mathbf{E}}(r)\,. \tag{7.24}$$

The matrix $\hat{\underline{\mathbf{S}}}_{\mathbf{xx}}(r)$ is estimated by (7.21). The adaptation algorithm that consists of (7.21), (7.23), and (7.24) is equivalent to the RLS algorithm in the sense that its mean-squared error convergence is independent of the condition number of the input spatio-spectral correlation matrix. To reduce the computational complexity of the adaptation algorithm, it is shown in [BBK03] that the matrix $\mathbf{G}^{01}_{2N\times 2N}$ in (7.21) can be well approximated by a diagonal matrix: $\mathbf{G}^{01}_{2N\times 2N} \approx \mathbf{I}_{2N\times 2N}/2$. Using this approximation and introducing a diagonal $2MN\times 2MN$ matrix $\boldsymbol{\mu}(r)$, which contains frequency-dependent stepsizes $\mu(r,n)$, $n = 0, 1, \ldots, 2N-1$, on the main diagonal, into (7.24), we may rewrite (7.21) and (7.24) as

[1] Note that $\underline{\mathbf{W}}(r)$ at block time r is not available when the error signal vector is calculated at block time r. Therefore, $\underline{\mathbf{W}}(r)$ is replaced by $\underline{\mathbf{W}}(r-1)$ in (7.7).

$$\hat{\underline{\mathbf{S}}}'_{\mathbf{xx}}(r) = \beta\hat{\underline{\mathbf{S}}}'_{\mathbf{xx}}(r-1) + (1-\beta)\underline{\mathbf{X}}^H(r)\underline{\mathbf{X}}(r)\,, \tag{7.25}$$

$$\underline{\mathbf{W}}(r) = \underline{\mathbf{W}}(r-1) + (1-\beta)\boldsymbol{\mu}(r)\hat{\underline{\mathbf{S}}}'^{-1}_{\mathbf{xx}}(r)\underline{\mathbf{X}}^H(r)\underline{\mathbf{E}}(r)\,. \tag{7.26}$$

Prior to inversion of $\hat{\underline{\mathbf{S}}}'_{\mathbf{xx}}(r)$ in (7.26), a proper regularization by adding a suitable diagonal matrix (e.g., [Tre02, BBK03]) is important to ensure robust convergence behavior. Equations (7.23), (7.25), and (7.26) define an unconstrained frequency-domain adaptive algorithm, since circular convolution that may be introduced by the multiplications in the DFT domain on the right side of the update equation (7.26) is not inhibited due to the approximation $\mathbf{G}^{01}_{2N\times 2N} \approx \mathbf{I}_{2N\times 2N}/2$. Multiplication of the diagonal matrix $\boldsymbol{\mu}(r)$ from the left side with a constraint matrix $\mathbf{G}^{10}_{2MN\times 2MN}$,

$$\mathbf{G}^{10}_{2MN\times 2MN} = \operatorname{diag}\left\{\mathbf{G}^{10}_{2N\times 2N}, \mathbf{G}^{10}_{2N\times 2N}, \ldots, \mathbf{G}^{10}_{2N\times 2N}\right\}, \tag{7.27}$$

$$\mathbf{G}^{10}_{2N\times 2N} = \mathbf{F}_{2N\times 2N}\mathbf{W}^{2N\times 2N}_{10}\mathbf{F}^{-1}_{2N\times 2N}, \tag{7.28}$$

eliminates these circular convolution effects.[2] The constraint $\mathbf{G}^{10}_{2MN\times 2MN}$ is required if the energy of the components introduced by the circular convolution is not negligible and if the artifacts are annoying for the human listener or disturb the application. The convergence speed is slightly increased compared to the unconstrained frequency-domain adaptive filter [SP90] at the cost of an increased computational complexity since one additional FFT and one inverse FFT (IFFT) are required for each input channel for calculating the constraint matrix $\mathbf{G}^{10}_{2MN\times 2MN}$. Instead of this intuitive formulation, this constrained version of the DFT-domain adaptive filter can be rigorously derived as in [BM00, BBK03].

Adaptation with Penalty Functions

The constrained form of the generic algorithm using penalty functions is obtained by setting the gradient of (7.15) w.r.t. $\underline{\mathbf{W}}^*(r)$ equal to zero. Following the derivation without penalty functions, we obtain after combining (7.22) with the normal equation (7.20), with (7.21), and with (7.7) the expression

$$\hat{\underline{\mathbf{S}}}_{\mathbf{xx}}(r)\underline{\mathbf{W}}(r)+\underline{\mathbf{\Psi}}(r) = \hat{\underline{\mathbf{S}}}_{\mathbf{xx}}(r)\underline{\mathbf{W}}(r-1)+(1-\beta)\underline{\mathbf{X}}^H(r)\underline{\mathbf{E}}(r)+\beta\underline{\mathbf{\Psi}}(r-1)\,, \tag{7.29}$$

where

$$\underline{\mathbf{\Psi}}(r) = \sigma\nabla_{\underline{\mathbf{W}}(r)}\sum_{i=0}^{C'-1}\left|\max\{c_i(\underline{\mathbf{W}}(r)),\,0\}\right|^2\,. \tag{7.30}$$

Next, we replace in $\underline{\mathbf{\Psi}}(r)$ on the left side of (7.29) the block time r by $r-1$, since $\underline{\mathbf{\Psi}}(r)$ depends on $\underline{\mathbf{W}}(r)$, which is not available at block time r. Solving

[2] The constraints against circular convolution should not be confused with the constraints added by the penalty terms in (7.15).

(7.29) for $\underline{\mathbf{W}}(r)$ and introducing the matrix of stepsizes $\boldsymbol{\mu}(r)$, the update equation of the constrained adaptation algorithm can be formulated as

$$\underline{\mathbf{W}}(r) = \underline{\mathbf{W}}(r-1) + (1-\beta)\boldsymbol{\mu}(r)\hat{\underline{\mathbf{S}}}_{\mathbf{xx}}^{-1}(r)[\underline{\mathbf{X}}^H(r)\underline{\mathbf{E}}(r) - \underline{\boldsymbol{\Psi}}(r-1)]\,. \quad (7.31)$$

7.2 RGSC Combined with Multi-channel Acoustic Echo Cancellation in the DFT Domain

In this section, we present a DFT-domain realization of the RGSC combined with multi-channel AEC. Systematic application of multi-channel DFT domain adaptive filtering yields a system that exploits the advantages of multi-channel DFT-domain adaptive filtering while preserving positive synergies between RGSC and AEC. Especially, (1) mutual correlation between the loudspeaker signals and between the sensor signals is taken into account for fast convergence of the AEC and of the RGSC, respectively. (2), Adaptation problems of the sidelobe cancelling path of the RGSC are effectively resolved. (3) Computational complexity is minimized for efficient implementation of the integrated system on low-cost PC platforms or in embedded systems [HYBK02]. An alternative realization in subbands is presented in [NFB02]. However, the adaptation problems of the sidelobe cancelling path are not adressed.

Section 7.2.1 describes the RGSC in the DFT domain. Section 7.2.2 explains the realization of the multi-channel AEC. In Sect. 7.2.3, the proposed algorithm is summarized and the complexity is compared with a realization in the time domain.

7.2.1 RGSC in the DFT Domain

We transform the RGSC according to Fig. 5.1 into the DFT domain using the FFT. It was explained in Chap. 5 that alternating adaptation of the blocking matrix and of the interference canceller in the eigenspace of the desired signal and in the eigenspace of interference-plus-noise is necessary to obtain maximum suppression of interferers with minimum distortion of the desired signal, respectively. However, the complexity of diagonalizing the estimates of spatio-temporal correlation matrices of the sensor signals might exceed the available computational resources on cost-sensitive platforms. Therefore, we propose the DFT for an approximate diagonalization of the sample spatio-temporal correlation matrix of the sensor signals [Gra72]. This transforms the classification of eigenvectors of the sample spatio-temporal correlation matrix of the sensor signals to a corresponding classification of the DFT bins and simplifies a practical realization of the RGSC. We will use overlapping input data blocks for optimum tracking behavior of the adaptive filters and for reducing the processing delay for delay-sensitive applications. The computational complexity is roughly multiplied by the overlap factor α. If $\alpha = 1$, the number of FFTs can be further reduced relative to the realization given here. (See [HK01a, HK01c, HK02b].) For generality, we do not consider this special case in this framework.

Quiescent Weight Vector

The quiescent weight vector steers the sensor array to the position of the desired source and enhances the desired signal relative to interference-plus-noise. A reference signal is thus produced for the adaptation of the sidelobe cancelling path. The more interference-plus-noise is suppressed by the quiescent weight vector, the less is the influence of the adaptive sidelobe cancelling path, and the higher is the robustness of the RGSC against artifacts due to the adaptation process. Therefore, it is advantageous to introduce information about the wave field of interference-plus-noise into the quiescent weight vector for optimum suppression of (time-invariant) interference-plus-noise. For diffuse wave fields, data-dependent beamformers which are optimum for these wave fields can be used [HK02b]. Additionally, they provide better separation at low frequencies than simple windowed beamformers [CZK87, SBB93a, SBB93b, SR93, BS01]. For applications where small variations of the DOA of the signal of the desired source should not be tracked by the quiescent weight vector, beamformer designs which widen the main-lobe of the beampattern can be used. (See, e.g., Sect. 4.3 and [HK02b]).

For a DFT-domain processing, the sensor signals are transformed into the DFT domain according to (7.1) and (7.2) as[3]

$$\underline{\mathbf{X}}_{\mathbf{w}_c}(r) = \left(\underline{\mathbf{X}}_{\mathbf{w}_c,0}(r), \underline{\mathbf{X}}_{\mathbf{w}_c,1}(r), \ldots, \underline{\mathbf{X}}_{\mathbf{w}_c,M-1}(r)\right) , \quad (7.32)$$

$$\underline{\mathbf{X}}_{\mathbf{w}_c,m}(r) = \mathrm{diag}\left\{\mathbf{F}_{2N\times 2N}\mathbf{x}_{\mathbf{w}_c,m}(rR)\right\} , \quad (7.33)$$

$$\mathbf{x}_{\mathbf{w}_c,m}(k) = (x_m(k-N),\, x_m(k-N+1),\, \ldots,\, x_m(k+N-1))^T . \quad (7.34)$$

Then, a vector $\mathbf{y}_{\mathbf{w}_c}(rR)$ of size $N \times 1$ is formed of the output signal $y_{\mathbf{w}_c}(k) = \mathbf{w}_c^T \mathbf{x}(k)$ of the fixed beamformer after (4.140) as

$$\mathbf{y}_{\mathbf{w}_c}(k) = (y_{\mathbf{w}_c}(k),\, y_{\mathbf{w}_c}(k+1),\, \ldots,\, y_{\mathbf{w}_c}(k+N-1))^T . \quad (7.35)$$

Defining the quiescent weight vector $\mathbf{w}_c$ in the DFT domain as the vector $\underline{\mathbf{w}}_c$ of size $2MN \times 1$ analogously to (7.3), we can write $\mathbf{y}_{\mathbf{w}_c}(rR)$ as

$$\mathbf{y}_{\mathbf{w}_c}(rR) = \mathbf{W}^{01}_{N\times 2N}\, \mathbf{F}^{-1}_{2N\times 2N}\, \underline{\mathbf{X}}_{\mathbf{w}_c}(r)\, \underline{\mathbf{w}}_c . \quad (7.36)$$

[3] Although (7.1), (7.2) and (7.32), (7.33), are identical, respectively, a new matrix $\underline{\mathbf{X}}_{\mathbf{w}_c}(r)$ is introduced in order to avoid confusion for 'AEC first' where the input signals of the fixed beamformer are not the sensor signals, but the output signals of the AEC.

Blocking Matrix

The blocking matrix consists of adaptive filters between the output of the quiescent weight vector and the sensor channels: It adaptively subtracts the desired signal from the sidelobe cancelling path in order to prevent the desired signal to be cancelled by the interference canceller. The M-channel output of the blocking matrix, ideally, contains interference and noise components, which are used in the interference canceller to form an estimate of the interference-plus-noise contained in the reference path.

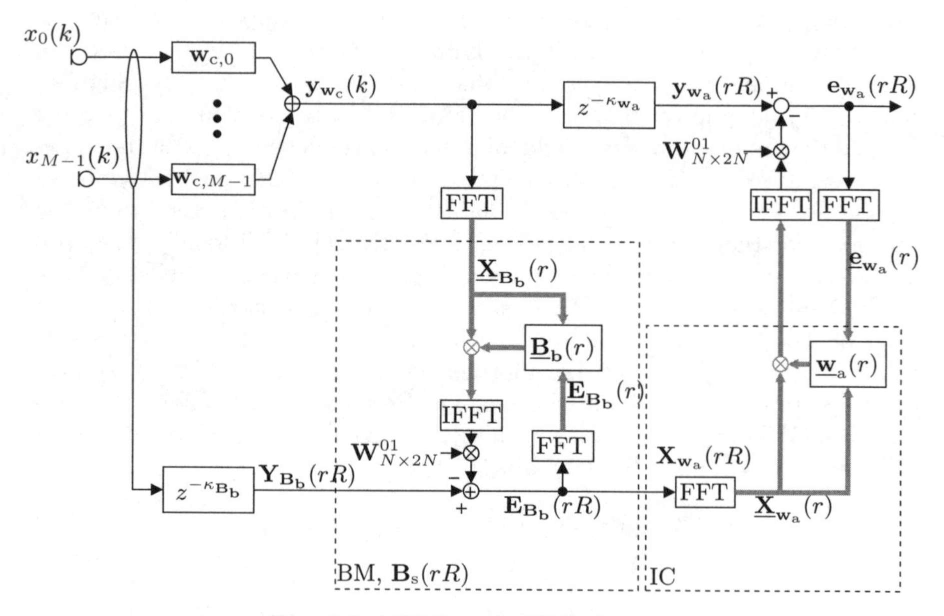

Fig. 7.2. RGSC in the DFT domain

In Fig. 7.2, the RGSC is depicted for DFT-domain processing. Identifying the blocking matrix with a SIMO adaptive filter, we can rigorously apply the generic adaptive filter structure after Fig. 7.1. According to (7.2), we capture the last $2N$ samples of the output signal $y_{\mathbf{w}_c}(k)$ of the fixed beamformer in a vector, and obtain after transformation into the DFT domain:

$$\underline{\mathbf{X}}_{\mathbf{B}_b}(r) = \mathrm{diag}\left\{\mathbf{F}_{2N\times 2N}\mathbf{x}_{\mathbf{B}_b}(rR)\right\} , \tag{7.37}$$

$$\mathbf{x}_{\mathbf{B}_b}(k) = \left(y_{\mathbf{w}_c}(k-N),\, y_{\mathbf{w}_c}(k-N+1),\, \ldots,\, y_{\mathbf{w}_c}(k+N-1)\right)^T . \tag{7.38}$$

With the adaptive filters of the blocking matrix $\mathbf{B}_b(rR)$ after (5.12) written in the DFT domain analogously to (7.3) as a $2N \times M$ matrix $\underline{\mathbf{B}}_b(r)$, the $N \times M$ block error matrix $\mathbf{E}_{\mathbf{B}_b}(rR)$ after (5.14) for $K = N$ is obtained analogously to (7.5) as

$$\mathbf{E}_{\mathbf{B}_b}(rR) = \mathbf{Y}_{\mathbf{B}_b}(rR) - \mathbf{W}^{01}_{N\times 2N}\,\mathbf{F}^{-1}_{2N\times 2N}\,\underline{\mathbf{X}}_{\mathbf{B}_b}(r)\,\underline{\mathbf{B}}_b(r-1) . \tag{7.39}$$

The $N \times M$ data matrix of the sensor signals is defined analogously to (3.12) as

$$\mathbf{Y}_{\mathbf{B}_\mathrm{b}}(k) = \left(\mathbf{y}_{\mathbf{B}_\mathrm{b},0}(k),\, \mathbf{y}_{\mathbf{B}_\mathrm{b},1}(k),\, \ldots,\, \mathbf{y}_{\mathbf{B}_\mathrm{b},M-1}(k)\right)^T, \tag{7.40}$$

with

$$\mathbf{y}_{\mathbf{B}_\mathrm{b},m}(k) = \begin{pmatrix} x_m(k-\kappa_{\mathbf{B}_\mathrm{b}}) \\ x_m(k-\kappa_{\mathbf{B}_\mathrm{b}}+1) \\ \vdots \\ x_m(k-\kappa_{\mathbf{B}_\mathrm{b}}+N-1) \end{pmatrix}, \tag{7.41}$$

where we introduced the delays $\kappa_{\mathbf{B}_\mathrm{b}}$ for causality of the blocking matrix filters. Let $\underline{\mathbf{E}}_{\mathbf{B}_\mathrm{b}}(r)$ be defined as

$$\underline{\mathbf{E}}_{\mathbf{B}_\mathrm{b}}(r) = \mathbf{F}_{2N\times 2N}\mathbf{W}^{01}_{2N\times N}\mathbf{E}_{\mathbf{B}_\mathrm{b}}(rR), \tag{7.42}$$

the update equation for $\underline{\mathbf{B}}_\mathrm{b}(r)$ reads according to (7.26) as[4]:

$$\underline{\mathbf{B}}_\mathrm{b}(r) = \underline{\mathbf{B}}_\mathrm{b}(r-1) + (1-\beta_{\mathbf{B}_\mathrm{b}})\mathbf{G}^{10}_{2N\times 2N}\boldsymbol{\mu}_{\mathbf{B}_\mathrm{b}}(r)\hat{\underline{\mathbf{S}}}^{-1}_{\mathbf{x}_{\mathbf{B}_\mathrm{b}}\mathbf{x}_{\mathbf{B}_\mathrm{b}}}(r)\underline{\mathbf{X}}^H_{\mathbf{B}_\mathrm{b}}(r)\underline{\mathbf{E}}_{\mathbf{B}_\mathrm{b}}(r). \tag{7.43}$$

Note that the circular convolution constraints $\mathbf{G}^{10}_{2N\times 2N}$ are required, since the impulse responses of optimum blocking matrix filters are generally much longer than the length of the adaptive filters (Chap. 5). The $2N \times 2N$ diagonal matrix $\underline{\boldsymbol{\mu}}_{\mathbf{B}_\mathrm{b}}(r)$ captures frequency-dependent stepsizes $\mu_{\mathbf{B}_\mathrm{b}}(r,n)$, $n = 0, 1, \ldots, 2N-1$, on the main diagonal, which control the adaptation of the blocking matrix. Typically, the stepsizes are switched between 0 and a frequency-independent constant value $\mu_{\mathbf{B}_\mathrm{b},0} \neq 0$ for stopping and continuing the adaptation. The factor $0 < \beta_{\mathbf{B}_\mathrm{b}} < 1$ is the exponential forgetting factor. The diagonal matrix $\hat{\underline{\mathbf{S}}}_{\mathbf{x}_{\mathbf{B}_\mathrm{b}}\mathbf{x}_{\mathbf{B}_\mathrm{b}}}(r)$ of size $2N \times 2N$ with the recursively averaged PSD of the output signal of the fixed beamformer is obtained according to (7.25) as

$$\hat{\underline{\mathbf{S}}}_{\mathbf{x}_{\mathbf{B}_\mathrm{b}}\mathbf{x}_{\mathbf{B}_\mathrm{b}}}(r) = \beta_{\mathbf{B}_\mathrm{b}}\hat{\underline{\mathbf{S}}}_{\mathbf{x}_{\mathbf{B}_\mathrm{b}}\mathbf{x}_{\mathbf{B}_\mathrm{b}}}(r-1) + (1-\beta_{\mathbf{B}_\mathrm{b}})\underline{\mathbf{X}}^H_{\mathbf{B}_\mathrm{b}}(r)\underline{\mathbf{X}}_{\mathbf{B}_\mathrm{b}}(r). \tag{7.44}$$

Interference Canceller

The interference canceller adaptively subtracts the signal components from the reference path which are correlated with the output signals of the blocking matrix. Since, for non-stationary signals, it cannot always be assured that the output signals of the blocking matrix do not contain components of the desired signal, a quadratic constraint is applied to the adaptive filters of the

[4]For some cases, it might be useful to use coefficient constraints for improving robustness against cancellation of desired signal by the blocking matrix as proposed in [HSH99]. These coefficient constraints can be added to the DFT-domain adaptive filters according to [HK01b, HK02b].

interference canceller in order to improve robustness against cancellation of the desired signal (Chap. 4.4.2).

The input data matrix of the MISO system with M input channels that describes the interference canceller is given in the DFT domain by

$$\underline{\mathbf{X}}_{\mathbf{w}_\mathrm{a}}(r) = \left(\underline{\mathbf{X}}_{\mathbf{w}_\mathrm{a},0}(r), \underline{\mathbf{X}}_{\mathbf{w}_\mathrm{a},1}(r), \ldots, \underline{\mathbf{X}}_{\mathbf{w}_\mathrm{a},M-1}(r)\right), \tag{7.45}$$

where

$$\underline{\mathbf{X}}_{\mathbf{w}_\mathrm{a},m}(r) = \mathrm{diag}\left\{\mathbf{F}_{2N\times 2N}\mathbf{x}_{\mathbf{w}_\mathrm{a},m}(rR)\right\}, \tag{7.46}$$

and where

$$\mathbf{x}_{\mathbf{w}_\mathrm{a},m}(k) = \begin{pmatrix} e_{\mathbf{B}_\mathrm{b},m}(k-N) \\ e_{\mathbf{B}_\mathrm{b},m}(k-N+1) \\ \vdots \\ e_{\mathbf{B}_\mathrm{b},m}(k+N-1) \end{pmatrix}, \tag{7.47}$$

analogously to (7.37) and (7.38), respectively.

Transformation of the adaptive filters $\mathbf{w}_{\mathrm{a,s}}(rR)$ of the interference canceller into the DFT domain according to (7.3) yields the stacked $2MN \times 1$ vector $\underline{\mathbf{w}}_\mathrm{a}(r)$. The block error vector of the interference canceller then reads

$$\mathbf{e}_{\mathbf{w}_\mathrm{a}}(rR) = \mathbf{y}_{\mathbf{w}_\mathrm{a}}(rR) - \mathbf{W}^{01}_{N\times 2N}\mathbf{F}^{-1}_{2N\times 2N}\underline{\mathbf{X}}_{\mathbf{w}_\mathrm{a}}(r)\underline{\mathbf{w}}_\mathrm{a}(r-1), \tag{7.48}$$

where $\mathbf{y}_{\mathbf{w}_\mathrm{a}}(rR)$ is equal to the data block $\mathbf{y}_{\mathbf{w}_\mathrm{c}}(rR)$ after (7.35) delayed by the synchronization delay $\kappa_{\mathbf{w}_\mathrm{a}}$,

$$\mathbf{y}_{\mathbf{w}_\mathrm{a}}(rR) = \mathbf{y}_{\mathbf{w}_\mathrm{c}}(rR - \kappa_{\mathbf{w}_\mathrm{a}}), \tag{7.49}$$

and where the vector $\mathbf{e}_{\mathbf{w}_\mathrm{a}}(k)$ contains a block of N samples of the output signal $y_{\mathbf{w}_\mathrm{RGSC}}(k)$ of the RGSC:

$$\mathbf{e}_{\mathbf{w}_\mathrm{a}}(k) = \left(y_{\mathbf{w}_\mathrm{RGSC}}(k),\, y_{\mathbf{w}_\mathrm{RGSC}}(k+1),\, \ldots,\, y_{\mathbf{w}_\mathrm{RGSC}}(k+N-1)\right)^T. \tag{7.50}$$

Let the block error matrix be defined in the DFT domain as

$$\underline{\mathbf{e}}_{\mathbf{w}_\mathrm{a}}(r) = \mathbf{F}_{2N\times 2N}\mathbf{W}^{01}_{2N\times N}\mathbf{e}_{\mathbf{w}_\mathrm{a}}(rR), \tag{7.51}$$

then, the multi-channel filter update equation with penalty vector $\underline{\boldsymbol{\psi}}(r-1)$ can be written (see (7.31)) as

$$\begin{aligned}\underline{\mathbf{w}}_\mathrm{a}(r) &= \underline{\mathbf{w}}_\mathrm{a}(r-1) + (1-\beta_{\mathbf{w}_\mathrm{a}})\times \\ &\times \mathbf{G}^{10}_{2MN\times 2MN}\boldsymbol{\mu}_{\mathbf{w}_\mathrm{a}}(r)\hat{\underline{\mathbf{S}}}^{-1}_{\mathbf{x}_{\mathbf{w}_\mathrm{a}}\mathbf{x}_{\mathbf{w}_\mathrm{a}}}(r)\left(\underline{\mathbf{X}}^H_{\mathbf{w}_\mathrm{a}}(r)\underline{\mathbf{e}}_{\mathbf{w}_\mathrm{a}}(r) - \underline{\boldsymbol{\psi}}(r-1)\right). \end{aligned} \tag{7.52}$$

The diagonal matrix $\boldsymbol{\mu}_{\mathbf{w}_\mathrm{a}}(r)$ of size $2MN \times 2MN$ captures frequency-dependent stepsizes $\mu_{\mathbf{w}_\mathrm{a}}(r,n)$, $n = 0, 1, \ldots, 2N-1$, on the main diagonal for controlling the adaptation of the interference canceller. As for the blocking matrix, the stepsizes are, generally, switched between 0 and $\mu_{\mathbf{w}_\mathrm{a},0} \neq 0$ for stopping and continuing the adaptation, depending on the decision of the adaptation control. The factor $0 < \beta_{\mathbf{w}_\mathrm{a}} < 1$ is an exponential forgetting factor. Circular convolution is prevented by the matrix $\mathbf{G}^{10}_{2MN\times 2MN}$, which is necessary since the impulse responses of the optimum filters of the interference canceller are generally much longer than the adaptive filters (Chap. 5). The CPSD matrix $\hat{\underline{\mathbf{S}}}_{\mathbf{x}_{\mathbf{w}_\mathrm{a}}\mathbf{x}_{\mathbf{w}_\mathrm{a}}}(r)$ of size $2MN \times 2MN$ is computed according to (7.25) as

$$\hat{\underline{\mathbf{S}}}_{\mathbf{x}_{\mathbf{w}_\mathrm{a}}\mathbf{x}_{\mathbf{w}_\mathrm{a}}}(r) = \beta_{\mathbf{w}_\mathrm{a}}\hat{\underline{\mathbf{S}}}_{\mathbf{x}_{\mathbf{w}_\mathrm{a}}\mathbf{x}_{\mathbf{w}_\mathrm{a}}}(r-1) + (1-\beta_{\mathbf{w}_\mathrm{a}})\underline{\mathbf{X}}^H_{\mathbf{w}_\mathrm{a}}(r)\underline{\mathbf{X}}_{\mathbf{w}_\mathrm{a}}(r)\,. \tag{7.53}$$

Note that, compared to the PSD matrix $\hat{\underline{\mathbf{S}}}_{\mathbf{x}_{\mathrm{B_b}}\mathbf{x}_{\mathrm{B_b}}}(r)$ of the input signals of the blocking matrix, the CPSD matrix $\hat{\underline{\mathbf{S}}}_{\mathbf{x}_{\mathbf{w}_\mathrm{a}}\mathbf{x}_{\mathbf{w}_\mathrm{a}}}(r)$ is not diagonal due to multi-channel input. The CPSDs of the input signals are thus taken into account for maximum convergence speed of the adaptive filters [BM00, BBK03].[5] From (7.52), it can be seen that this requires the inversion of the $2MN \times 2MN$ CPSD matrix $\hat{\underline{\mathbf{S}}}_{\mathbf{x}_{\mathbf{w}_\mathrm{a}}\mathbf{x}_{\mathbf{w}_\mathrm{a}}}(r)$. For efficiently computing the inverse, the block-diagonal structure of $\hat{\underline{\mathbf{S}}}_{\mathbf{x}_{\mathbf{w}_\mathrm{a}}\mathbf{x}_{\mathbf{w}_\mathrm{a}}}(r)$ (see (7.46) and (7.53)) can be exploited. It allows to decompose $\hat{\underline{\mathbf{S}}}_{\mathbf{x}_{\mathbf{w}_\mathrm{a}}\mathbf{x}_{\mathbf{w}_\mathrm{a}}}(r)$ into $2N$ CPSD matrices of size $M \times M$ w.r.t. each DFT bin, which, then, only requires the computation of $2N$ inverses of $M \times M$ matrices. Furthermore, application of the matrix inversion lemma [Hay96] avoids the matrix inversion completely. More details on the efficient realization of these multi-channel DFT-domain adaptive filters can be found in [BBK03]. In order to further reduce the complexity, the CPSD matrix can be replaced by its main diagonal, which replaces in (7.53) the update term $\underline{\mathbf{X}}^H_{\mathbf{w}_\mathrm{a}}(r)\underline{\mathbf{X}}_{\mathbf{w}_\mathrm{a}}(r)$ by its main diagonal, $\mathrm{diag}\{\underline{\mathbf{X}}^H_{\mathbf{w}_\mathrm{a}}(r)\underline{\mathbf{X}}_{\mathbf{w}_\mathrm{a}}(r)\}$. The matrix inverses can thus be completely avoided at the cost of reduced convergence speed.

The constraint equation for incorporating diagonal loading into the update equation (7.52) can be formulated in the DFT domain according to Sects. 4.4.2, 5.6.2 as

$$c(\underline{\mathbf{w}}_\mathrm{a}(r)) = \|\underline{\mathbf{w}}_\mathrm{a}(r)\|_2^2 - T_0 \overset{!}{<} 0\,, \tag{7.54}$$

where T_0 is the limit of the L_2-norm of the filter coefficients [RE00]. The penalty vector $\underline{\psi}(r)$ is obtained from (7.54) with (7.30) as

$$\begin{aligned}\underline{\psi}(r) &= \sigma\nabla_{\underline{\mathbf{w}}^*_\mathrm{a}(r)}|\max\{c(\underline{\mathbf{w}}_\mathrm{a}(r)),\,0\}|^2\\ &= 4\sigma\max\{c(\underline{\mathbf{w}}_\mathrm{a}(r)),\,0\}\underline{\mathbf{w}}_\mathrm{a}(r)\end{aligned} \tag{7.55}$$

[RE00]. Finally, the r-th block of length R of the output signal $y_{\mathbf{w}_{\mathrm{RGSC}}}(k)$ of the RGSC is given by the last R samples of the vector $\mathbf{e}_{\mathbf{w}_\mathrm{a}}(rR)$ according to (7.50).

[5]The consideration of CPSDs of the input signals of the interference canceller for maximum convergence speed is also discussed in [AC93] for an orthogonalization using the discrete cosine transform or in, e.g., [Tre02] for the RLS algorithm in the time domain.

Adaptation Control

The RGSC adaptation strategy can be summarized for the DFT-domain implementation as follows: The fixed beamformer cannot produce an estimate of the desired signal that is free of interference-plus-noise. Therefore, the blocking matrix should only be adapted when the SINR is high in order to prevent suppression of interference and noise by the blocking matrix. Interference and noise components that are suppressed by the blocking matrix cannot be cancelled by the interference canceller, and, thus, leak to the output of the RGSC via the fixed beamformer. The blocking matrix does, generally, not produce an estimate of interference-plus-noise at its output that is completely free of the desired signal. Therefore, the interference canceller should only be adapted if the SINR is low in order to prevent cancellation of the desired signal. For a full-band adaptation control, all adaptation is stopped during double-talk of desired signal and interference-plus-noise as a consequence. Since the interference canceller cannot be adapted together with the blocking matrix, the interference canceller cannot compensate for the blocking matrix variations during adaptation of the blocking matrix over the entire frequency range. The interference canceller is thus inefficient during adaptation of the blocking matrix, which leads to emerging interference and noise until the interference canceller reconverges. With this full-band adaptation strategy, the suppression of interference-plus-noise is reduced to that of the fixed beamformer during activity of the desired signal, and, only, during presence of interference-plus-noise alone, a high suppression of interference-plus-noise can be obtained with the sidelobe cancelling path.

Obviously, the eigenspace adaptation method described in Sect. 5.6 maximizes the efficiency of the sidelobe cancelling path. For our DFT-domain implementation of the RGSC, the classification of eigenvalues and eigenvectors translates to a classification of DFT frequency bins according to the bin-wise SINR, $SINR_{\mathrm{f,in}}(r,n)$.[6] If $SINR_{\mathrm{f,in}}(r,n)$ is high or low, the blocking matrix or the interference canceller is updated, respectively. For realizing this adaptation strategy, thus, a method is required, which estimates $SINR_{\mathrm{f,in}}(r,n)$ and which puts decision thresholds on the estimated bin-wise SINR, or which detects activity of the desired signal only, activity of interference only, and double-talk in the DFT bins.

One possible solution to this problem is described in App. A and in [HTK03]. In our scenario, we assume that the desired source is located within the mainlobe of the fixed beamformer. This suggests to exploit the directivity of the fixed beamformer for the classification of the DFT bins:[7] The PSD of the output signal of the fixed beamformer is an estimate of the PSD of the

[6]The SINR $SINR_{\mathrm{f,in}}(r,n)$ as a function of the block time r and of the DFT bin n is obtained according to (A.51) (Sect. A.3.5) as the ratio of recursively averaged periodograms of the desired signal and of interference-plus-noise at the sensors.

[7]In [HSH99], this principle is used for controlling the adaptation of a time-domain realization of the GSC.

desired signal. A complementary fixed beamformer yields an estimate of the PSD of the interference. A biased estimate of $SINR_{\mathrm{f,in}}(r,n)$ is thus obtained by the ratio of these PSDs. In App. A, it is shown that this estimate of the SINR at the sensors is biased, and that the bias depends on the spatial coherence matrices w.r.t. the desired signal and w.r.t. interference-plus-noise, and on the SINR at the sensors. Since the spatial coherence functions strongly depend on frequency and on the acoustic environment, it is not possible to put fixed thresholding on this biased SINR estimate for classification of the DFT bins. Instead, we propose to track the minima and the maxima of the biased SINR estimate, and to adapt the blocking matrix or the interference canceller whenever the biased SINR estimate is maximum or minimum, respectively. For tracking the maxima and the minima of the biased estimate of $SINR_{\mathrm{f,in}}(r,n)$, the minimum statistics after [Mar01a] can be used.[8]

Figure 7.3 shows an example for the behavior of a bin-wise adaptation control compared to a full-band decision. The adaptation controls are idealized in a way that the bin-wise decision is based on $SINR_{\mathrm{f,in}}(r,n)$.[9] For the DFT-bin-wise adaptation control, the blocking matrix and the interference canceller are adapted if $SINR_{\mathrm{f,in}}(r,n)$ is greater than 15 dB and smaller than −15 dB, respectively. For the full-band decision, the blocking matrix and the interference canceller are adapted if the full-band SINR, $SINR_{\mathrm{in}}(rR)$, is greater than 10 dB and smaller than −10 dB, respectively. Figures 7.3a, b show the desired signal and the interference signal at the $M/2$-th microphone, respectively. In Fig. 7.3c, $SINR_{\mathrm{f,in}}(r,n)$ is depicted. Figures 7.3d, e illustrate the decision of the bin-wise adaptation control and of the full-band adaptation control, respectively. By comparing Fig. 7.3c and Fig. 7.3d, we see that the bin-wise decision allows more precise and more frequent adaptation of the blocking matrix and of the interference canceller than the full-band approach. Because of this behavior, the tracking problem of the sidelobe cancelling path can be resolved, as will be shown in Sect. 7.3 by experimental results.

7.2.2 Combination with the AEC

The AEC can be represented by a MIMO system with Q input (=loudspeaker) channels. The number of output channels depends on the position of the AEC relative to the RGSC. See Figs. 6.2, 6.5, and 6.11 for 'AEC first', for 'beamforming first', for GSAEC, and for GEIC, respectively. For a unified view of these systems, we first discuss the realization of a multi-channel AEC

[8]This classification mechanism can be equally used as a basis for removing the bias from the bin-wise SINR estimate. More details on this bias-free estimation of the DFT bin-wise SINR can be found in App. A.

[9]For these figures, we use these 'idealized' decisions for the adaptation controls for a better illustration. For the real adaptation control, the regions of adaptation of the blocking matrix and of the interference canceller in Fig. 7.3d are less contiguous. See also Sect. 7.3 for a comparison of the real adaptation control with the 'idealized' adaptation control.

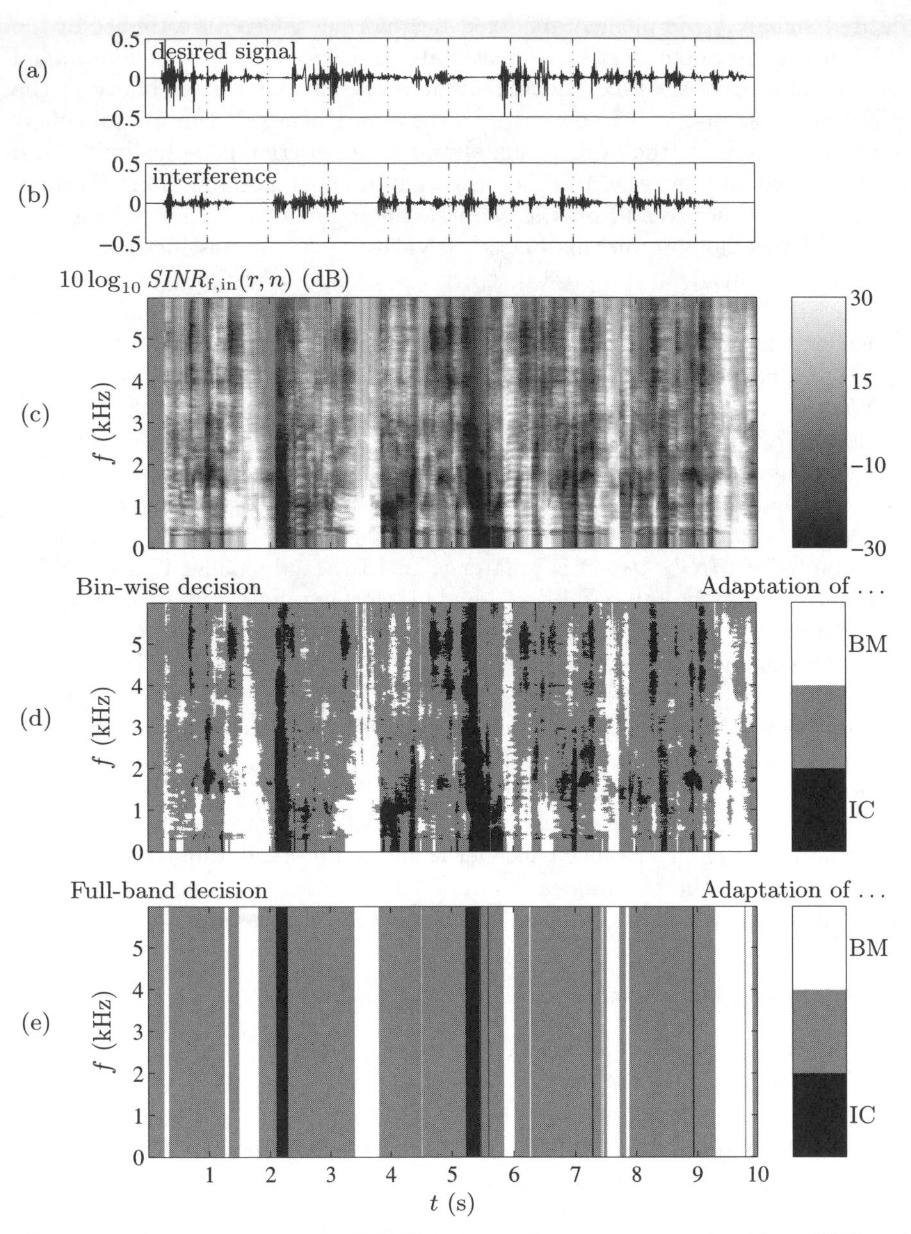

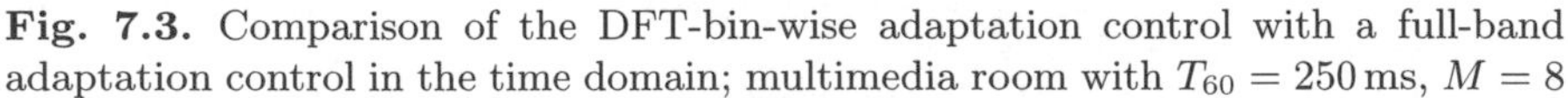

Fig. 7.3. Comparison of the DFT-bin-wise adaptation control with a full-band adaptation control in the time domain; multimedia room with $T_{60} = 250\,\mathrm{ms}$, $M = 8$

for multiple recording channels according to Fig. 6.1. Second, the integration of the AEC into the interference canceller (GEIC) is described. The algorithms for the other combinations may be obtained from 'AEC first' by adjusting the number M of the recording channels, by choosing the desired inputs and outputs for the AEC, and by using appropriate synchronization delays. The adaptation control for the AEC is not considered within this context. Various step-size controls can be found in, e.g., [MPS00, GB00].

'AEC First', 'Beamformer First', and GSAEC

The DFT length for the AEC is $2N_{\mathbf{a}}$, yielding the DFT matrix $\mathbf{F}_{2N_{\mathbf{a}}\times 2N_{\mathbf{a}}}$, with typically $N_{\mathbf{a}} > N$. The block overlap factor for the AEC is denoted as $\alpha_{\mathbf{a}}$, where typically $\alpha_{\mathbf{a}} \geq \alpha$ for maximum speed of convergence. Therefore, the block of 'new' samples $R' := N_{\mathbf{a}}/\alpha_{\mathbf{a}}$ for the AEC is generally different from the block of 'new' samples for the RGSC, which means that the block time index r', $r'R' = k$, for the AEC is different from the block time r for the RGSC. This has to be considered when synchronizing AEC and RGSC [HBK03a]. The block diagram of the AEC for a single recording channel is depicted in Fig. 7.4.

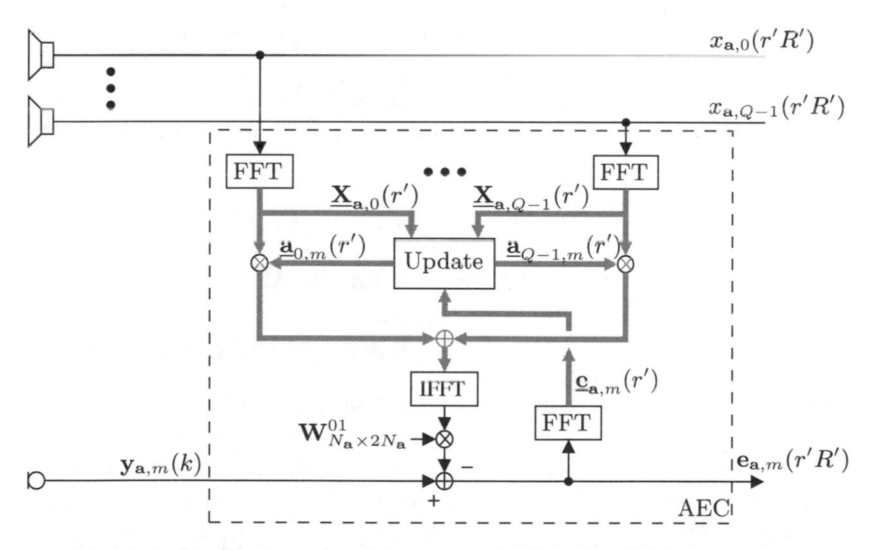

Fig. 7.4. Realization of multi-channel AEC in the DFT domain

For de-crosscorrelating the loudspeaker signals, the loudspeaker signals $x_{\mathbf{a},q}(k)$ are assumed to be pre-processed by a simple inaudible nonlinearity [BMS98]. The pre-processed loudspeaker signals $x_{\mathbf{a},q}(k)$ are transformed into the DFT domain according to (7.1) and (7.2). It yields

$$\underline{\mathbf{X}}_{\mathbf{a}}(r') = \left(\underline{\mathbf{X}}_{\mathbf{a},0}(r'), \underline{\mathbf{X}}_{\mathbf{a},1}(r'), \ldots, \underline{\mathbf{X}}_{\mathbf{a},Q-1}(r')\right), \qquad (7.56)$$

where

$$\underline{\mathbf{X}}_{\mathbf{a},q}(r') = \operatorname{diag}\left\{ \mathbf{F}_{2N_\mathbf{a}\times 2N_\mathbf{a}} \begin{pmatrix} x_{\mathbf{a},q}(r'R' - N_\mathbf{a}) \\ x_{\mathbf{a},q}(r'R' - N_\mathbf{a} + 1) \\ \dots \\ x_{\mathbf{a},q}(r'R' + N_\mathbf{a} - 1) \end{pmatrix} \right\} . \tag{7.57}$$

According to (7.3), the adaptive filters $[\mathbf{A}^{(1)}(r'R')]_{qN_\mathbf{a},m} = \mathbf{a}^{(1)}_{q,m}(r'R')$ of the AEC after (6.3) are transformed into the DFT domain, yielding the coefficient matrix $[\underline{\mathbf{A}}^{(1)}(r')]_{2qN_\mathbf{a},m} = \underline{\mathbf{a}}^{(1)}_{q,m}(r')$ of size $2N_\mathbf{a}Q \times M$. Let the block error matrix $\mathbf{E}_\mathbf{a}(r'R')$ of size $N_\mathbf{a} \times M$ be determined in the discrete time domain according to (7.5) as

$$\mathbf{E}_\mathbf{a}(r'R') = \mathbf{Y}_\mathbf{a}(r'R') - \mathbf{W}^{01}_{N_\mathbf{a}\times 2N_\mathbf{a}} \mathbf{F}^{-1}_{2N_\mathbf{a}\times 2N_\mathbf{a}} \underline{\mathbf{X}}_\mathbf{a}(r') \underline{\mathbf{A}}^{(1)}(r'-1) , \tag{7.58}$$

where

$$\mathbf{E}_\mathbf{a}(k) = (\mathbf{e}_{\mathbf{a},0}(k), \mathbf{e}_{\mathbf{a},1}(k), \dots, \mathbf{e}_{\mathbf{a},M-1}(k)) , \tag{7.59}$$

$$\mathbf{e}_{\mathbf{a},m}(k) = (e_{\mathbf{a},m}(k), e_{\mathbf{a},m}(k+1), \dots, e_{\mathbf{a},m}(k+N_\mathbf{a}-1))^T , \tag{7.60}$$

$$\mathbf{Y}_\mathbf{a}(k) = (\mathbf{y}_{\mathbf{a},0}(k), \mathbf{y}_{\mathbf{a},1}(k), \dots, \mathbf{y}_{\mathbf{a},M-1}(k)) , \tag{7.61}$$

$$\mathbf{y}_{\mathbf{a},m}(k) = (x_m(k), x_m(k+1), \dots, x_m(k+N_\mathbf{a}-1))^T , \tag{7.62}$$

and where $\mathbf{W}^{01}_{N_\mathbf{a}\times 2N_\mathbf{a}}$ is given by (7.6) with N replaced by $N_\mathbf{a}$. After transformation into the DFT domain according to (7.7), we obtain the block error signal in the DFT domain as

$$\underline{\mathbf{E}}_\mathbf{a}(r') = \mathbf{F}_{2N_\mathbf{a}\times 2N_\mathbf{a}} \mathbf{W}^{01}_{2N_\mathbf{a}\times N_\mathbf{a}} \mathbf{E}_\mathbf{a}(r'R') . \tag{7.63}$$

This allows to write the update equation for the AEC as

$$\underline{\mathbf{A}}^{(1)}(r') = \underline{\mathbf{A}}^{(1)}(r'-1) + \mu_\mathbf{a}(1-\beta_\mathbf{a}) \hat{\underline{\mathbf{S}}}^{-1}_{\mathbf{x}_\mathbf{a}\mathbf{x}_\mathbf{a}}(r') \underline{\mathbf{X}}^H_\mathbf{a}(r') \underline{\mathbf{E}}_\mathbf{a}(r') , \tag{7.64}$$

where $\mu_\mathbf{a}$ is the step-size parameter. Note that $\mu_\mathbf{a}$ is independent of frequency, which results from the fact that present adaptation control mechanisms for AECs provide decisions for the whole frequency range only. The factor $0 < \beta_\mathbf{a} < 1$ is the exponential forgetting factor. The spatio-spectral correlation matrix $\hat{\underline{\mathbf{S}}}_{\mathbf{x}_\mathbf{a}\mathbf{x}_\mathbf{a}}(r')$ is the recursively averaged CPSD matrix of the loudspeaker signals:

$$\hat{\underline{\mathbf{S}}}_{\mathbf{x}_\mathbf{a}\mathbf{x}_\mathbf{a}}(r') = \beta_\mathbf{a} \hat{\underline{\mathbf{S}}}_{\mathbf{x}_\mathbf{a}\mathbf{x}_\mathbf{a}}(r'-1) + (1-\beta_\mathbf{a}) \underline{\mathbf{X}}^H_\mathbf{a}(r') \underline{\mathbf{X}}_\mathbf{a}(r') . \tag{7.65}$$

With the inverse of the CPSD matrix $\hat{\underline{\mathbf{S}}}_{\mathbf{x}_\mathbf{a}\mathbf{x}_\mathbf{a}}(r')$ in the update equation, the mutual correlation of the loudspeaker channels is explicitly taken into account in order to assure high convergence speed, which is nearly independent of the condition number of $\hat{\underline{\mathbf{S}}}_{\mathbf{x}_\mathbf{a}\mathbf{x}_\mathbf{a}}(r')$. The efficient realization of the matrix inverse is discussed for the stereophonic case in [BM00], for more than two reproduction channels in [BBK03].

The constraints for preventing circular convolution are not necessarily required for the update of the AEC filters – in contrast to the blocking matrix and to the interference canceller. Since room impulse responses can be assumed to decay nearly exponentially (Chap. 2), circular convolution can be neglected if the AEC filters are chosen sufficiently long. Generally, this is the case for assuring high suppression of acoustic echoes.

One block of length R' for the m-th output signal of the AEC is given by the last R' samples of the m-th column of the block error signal $\mathbf{E}_{\mathbf{a}}(r'R')$.

GEIC

For optimum synergies between beamforming and acoustic echo cancellation for GEIC, the convergence speed of the AEC and of the interference canceller must be comparable. Therefore, we consider only the case with $N = N_{\mathbf{a}}$ and $\alpha = \alpha_{\mathbf{a}}$. The integration of the AEC into the interference canceller (GEIC, Fig. 6.11) is then obtained straightforwardly from the interference canceller (Sect. 7.2.1): The MISO system of the interference canceller $\underline{\mathbf{w}}_{\mathrm{a}}(r)$ is complemented by the adaptive filters $\mathbf{a}^{(4)}(rR)$ for the AEC in the DFT domain, $\underline{\mathbf{a}}^{(4)}(r)$ and the data matrix $\underline{\mathbf{X}}_{\mathbf{a}}(r)$ of the loudspeaker signals is stacked to the input data matrix $\underline{\mathbf{X}}_{\mathbf{w}_{\mathrm{a}}}(r)$ after (7.45) correspondingly. The block error vector and the filter update equation follow directly from (7.48) and from (7.52), respectively. The AEC filters $\underline{\mathbf{a}}^{(4)}(r)$ are excluded from the quadratic constraint. The loudspeaker signals should be synchronized with the echo signals at the output of the blocking matrix in order to remove unnecessary delays which may reduce the echo suppression. The signal levels of the loudspeaker signals should be adjusted to the level of the sensor signals in order to avoid high condition numbers of the correlation matrix of the input signals of the interference canceller, which may lead to instabilities. As already discussed in Chap. 6, an additional adaptation control for the AEC is not required.

7.2.3 Real-Time Algorithm and Computational Complexity

The real-time algorithm is now summarized. We compare the computational complexity of the block-based DFT-domain realization with a straightforward non-block full-band time-domain implementation with the LMS algorithm [HSH99]. Although the requirements of the RGSC and of the multi-channel AEC cannot be met by a simple full-band LMS-based realization, the comparison illustrates that the proposed DFT-domain system is even more efficient than realizations with the very simple LMS algorithm.

In Tables 7.1–7.5, an overview of the signal processing for 'AEC first' is given for the stereophonic case ($Q = 2$). A typical parameter setup for $f_{\mathrm{s}} = 12\,\mathrm{kHz}$ for a moderately reverberant environment is indicated. The number of real multiplications per output sample (NRMs) is examined for all modules. The NRM for the other combinations of RGSC and AEC can easily be calculated using the analysis in Tables 7.1–7.5 as a basis. The NRMs for all combinations are summarized in Fig. 7.5. The DFT is carried out by the radix-2 algorithm for real-valued time-domain sequences. The number of real multiplications for one FFT (IFFT) is equal to $2N \log_2 2N$. Symmetry of real-valued time-domain sequences in the DFT domain is considered for minimizing the computational load. The possibility to exploit the block overlapping for increasing the efficiency of the FFT is not taken into account. (See, e.g., [Mar71, Hol87, Sch99, BBK03].) Exploiting the symmetry of the DFT of real-valued sequences, and considering that one complex multiplication requires 4 real multiplications, the Hadamard product [Bel72] for fast convolution consists of $4N - 2$ real scalar multiplications.

Table 7.1. Summary of the r'-th iteration of the AEC of 'AEC first'; typical parameters for $f_\mathrm{s} = 12\,\mathrm{kHz}$: $N_\mathbf{a} = 1024$, $\alpha_\mathbf{a} = 8$, $\beta_\mathbf{a} = 0.97$, $\mu_\mathbf{a} = 0.8$

	Acoustic echo cancellation
	$q,\, q_1,\, q_2 = 0,\, 1;\; Q = 2$
1	FFT of Q loudspeaker signals $x_{\mathbf{a},q}(k)$, yielding $\underline{\mathbf{X}}_\mathbf{a}(r')$, (7.56), (7.57)
2	$\mathbf{E}_\mathbf{a}(r'R') = \mathbf{Y}_\mathbf{a}(r'R') - \mathbf{W}^{01}_{N_\mathbf{a} \times 2N_\mathbf{a}} \mathbf{F}^{-1}_{2N_\mathbf{a} \times 2N_\mathbf{a}} \underline{\mathbf{X}}_\mathbf{a}(r') \underline{\mathbf{A}}^{(1)}(r'-1)$, (7.58)
3	$\underline{\mathbf{E}}_\mathbf{a}(r') = \mathbf{F}_{2N_\mathbf{a} \times 2N_\mathbf{a}} \mathbf{W}^{01}_{2N_\mathbf{a} \times N_\mathbf{a}} \mathbf{E}_\mathbf{a}(r'R')$, (7.63)
4	*Calculation of* $\underline{\mathbf{K}}(r') = \hat{\underline{\mathbf{S}}}^{-1}_{\mathbf{x}_\mathbf{a}\mathbf{x}_\mathbf{a}}(r') \underline{\mathbf{X}}^H_\mathbf{a}(r')$ *(7.64)*
4a	$\hat{\underline{\mathbf{S}}}_{\mathbf{x}_{\mathbf{a},q_1}\mathbf{x}_{\mathbf{a},q_2}}(r') = \beta_\mathbf{a} \hat{\underline{\mathbf{S}}}_{\mathbf{x}_{\mathbf{a},q_1}\mathbf{x}_{\mathbf{a},q_2}}(r'-1) + (1-\beta_\mathbf{a}) \underline{\mathbf{X}}^*_{\mathbf{a},q_1}(r') \underline{\mathbf{X}}_{\mathbf{a},q_2}(r')$
	with $\hat{\underline{\mathbf{S}}}_{\mathbf{x}_{\mathbf{a},0}\mathbf{x}_{\mathbf{a},1}}(r') = \hat{\underline{\mathbf{S}}}^*_{\mathbf{x}_{\mathbf{a},1}\mathbf{x}_{\mathbf{a},0}}(r')$
4b	$\hat{\underline{\mathbf{S}}}_q(r') = \hat{\underline{\mathbf{S}}}_{\mathbf{x}_{\mathbf{a},q}\mathbf{x}_{\mathbf{a},q}}(r') [\mathbf{I}_{2N_\mathbf{a} \times 2N_\mathbf{a}} -$
	$- \hat{\underline{\mathbf{S}}}^*_{\mathbf{x}_{\mathbf{a},0}\mathbf{x}_{\mathbf{a},1}}(r') \hat{\underline{\mathbf{S}}}_{\mathbf{x}_{\mathbf{a},0}\mathbf{x}_{\mathbf{a},1}}(r') \{\hat{\underline{\mathbf{S}}}_{\mathbf{x}_{\mathbf{a},0}\mathbf{x}_{\mathbf{a},0}}(r') \hat{\underline{\mathbf{S}}}_{\mathbf{x}_{\mathbf{a},1}\mathbf{x}_{\mathbf{a},1}}(r')\}^{-1}]^{-1}$
4c	$\underline{\mathbf{K}}_0(r') = \hat{\underline{\mathbf{S}}}^{-1}_0(r') [\underline{\mathbf{X}}^*_{\mathbf{a},0}(r') - \hat{\underline{\mathbf{S}}}_{\mathbf{x}_{\mathbf{a},0}\mathbf{x}_{\mathbf{a},1}}(r') (\hat{\underline{\mathbf{S}}}_{\mathbf{x}_{\mathbf{a},1}\mathbf{x}_{\mathbf{a},1}}(r'))^{-1} \underline{\mathbf{X}}^*_{\mathbf{a},1}(r')]$
4d	$\underline{\mathbf{K}}_1(r') = \hat{\underline{\mathbf{S}}}^{-1}_1(r') [\underline{\mathbf{X}}^*_{\mathbf{a},1}(r') - \hat{\underline{\mathbf{S}}}^*_{\mathbf{x}_{\mathbf{a},0}\mathbf{x}_{\mathbf{a},1}}(r') (\hat{\underline{\mathbf{S}}}_{\mathbf{x}_{\mathbf{a},0}\mathbf{x}_{\mathbf{a},0}}(r'))^{-1} \underline{\mathbf{X}}^*_{\mathbf{a},0}(r')]$
	$\underline{\mathbf{K}}(r') = (\underline{\mathbf{K}}^T_0(r'),\, \underline{\mathbf{K}}^T_1(r'))^T$
5	$\underline{\mathbf{A}}^{(1)}(r') = \underline{\mathbf{A}}^{(1)}(r'-1) + \mu_\mathbf{a}(1-\beta_\mathbf{a}) \underline{\mathbf{K}}(r') \underline{\mathbf{E}}_\mathbf{a}(r')$, (7.64)
	$NRM_\mathbf{a} = \alpha_\mathbf{a}[(2M+Q)(2N_\mathbf{a} \log_2 2N_\mathbf{a}) + (13N_\mathbf{a} - 7)Q + 24N_\mathbf{a} - 6 + 2MN_\mathbf{a}Q]/N_\mathbf{a}$

Table 7.2. Summary of the fixed beamformer for the r-th iteration of 'AEC first'

	Fixed beamformer
6	FFT of M output signals of the AEC, yielding $\underline{\mathbf{X}}_{\mathbf{w}_\mathrm{c}}(r)$
7	$y_{\mathbf{w}_\mathrm{c}}(rR) = \mathbf{W}^{01}_{N \times 2N} \mathbf{F}^{-1}_{2N \times 2N} \underline{\mathbf{X}}_{\mathbf{w}_\mathrm{c}}(r)\, \underline{\mathbf{w}}_\mathrm{c}(r)$, (7.36)
	$NRM_{\mathbf{w}_\mathrm{c}} = \alpha[(M+1)(2N \log_2 2N) + M(4N-2)]/N$

Table 7.3. Summary of the adaptation control for the r-th iteration of 'AEC first'; typical parameters for $f_s = 12\,\text{kHz}$, $N = 256$, $\alpha = 2$: $D = 96$, $\Delta\hat{\Upsilon}_{\text{th,d}} = 0.1$, $\Delta\hat{\Upsilon}_{\text{th,n}} = 0.3$, $\Upsilon_{0,\text{d}} = 1.0$, $\Upsilon_{0,\text{n}} = 2.0$, $\beta_{\mathbf{w}_\text{c}} = 0.58$

	Adaptation control
	$n = 0, 1, \ldots, N$
8	FFT of the output signal of the fixed beamformer, yielding $\underline{\mathbf{X}}_{\mathbf{B}_\text{b}}(r)$, (7.37), (7.38)
9	$\hat{\underline{\mathbf{S}}}_{\mathbf{y}_{\mathbf{w}_\text{c}}\mathbf{y}_{\mathbf{w}_\text{c}}}(r) = \beta_{\mathbf{w}_\text{c}}\hat{\underline{\mathbf{S}}}_{\mathbf{y}_{\mathbf{w}_\text{c}}\mathbf{y}_{\mathbf{w}_\text{c}}}(r-1) + (1-\beta_{\mathbf{w}_\text{c}})\underline{\mathbf{X}}^H_{\mathbf{B}_\text{b}}(r)\underline{\mathbf{X}}_{\mathbf{B}_\text{b}}(r)$
10	$\bar{\underline{\mathbf{Y}}}_{\mathbf{w}_\text{c},m}(r) = \underline{\mathbf{X}}_{\mathbf{w}_\text{c},m}(r - \frac{\kappa_{\mathbf{w}_\text{c}}}{R}) - \underline{\mathbf{X}}_{\mathbf{B}_\text{b}}(r)$
11	$\hat{\underline{\mathbf{S}}}_{\bar{\mathbf{y}}_{\mathbf{w}_\text{c}}\bar{\mathbf{y}}_{\mathbf{w}_\text{c}}}(r) = \beta_{\mathbf{w}_\text{c}}\hat{\underline{\mathbf{S}}}_{\bar{\mathbf{y}}_{\mathbf{w}_\text{c}}\bar{\mathbf{y}}_{\mathbf{w}_\text{c}}}(r-1) + (1-\beta_{\mathbf{w}_\text{c}})\times$ $\times\frac{1}{M}\sum_{m=0}^{M-1}\bar{\underline{\mathbf{Y}}}^*_{\mathbf{w}_\text{c},m}(r)\bar{\underline{\mathbf{Y}}}_{\mathbf{w}_\text{c},m}(r)$
12	$\hat{\underline{\mathbf{\Upsilon}}}(r) = (\hat{\underline{\mathbf{S}}}_{\bar{\mathbf{y}}_{\mathbf{w}_\text{c}}\bar{\mathbf{y}}_{\mathbf{w}_\text{c}}}(r))^{-1}\hat{\underline{\mathbf{S}}}_{\mathbf{y}_{\mathbf{w}_\text{c}}\mathbf{y}_{\mathbf{w}_\text{c}}}(r)$ Elements on the main diagonal of $\hat{\underline{\mathbf{\Upsilon}}}(r)$ denoted by $\hat{\Upsilon}(r,n)$
13	*Track minima of* $1/\hat{\Upsilon}(r,n)$ *over a window of length* D IF $\lvert 1/\hat{\Upsilon}(r,n) - \hat{\Upsilon}_{\text{th,d}}(r-1,n)\rvert \leq \hat{\Upsilon}_{\text{th,d}}(r-1,n)\Delta\hat{\Upsilon}_{\text{th,d}}$ $\quad\hat{\Upsilon}_{\text{th,d}}(r,n) = \hat{\Upsilon}(r,n)$ $\quad$Replace all stored values of $\hat{\Upsilon}_\text{d}(r-i,n)$, $i = 0, 1, \ldots, D$, by $\hat{\Upsilon}_{\text{th,d}}(r,n)$ ELSE $\quad$Find $\hat{\Upsilon}_{\text{th,d}}(r,n)$, the minimum of $\hat{\Upsilon}_\text{d}(r-i,n)$, $i = 0, 1, \ldots, D$ $\quad\hat{\Upsilon}_\text{d}(r,n) = \hat{\Upsilon}(r,n)$
14	*Track minima of* $\hat{\Upsilon}(r,n)$ *over a window of length* D IF $\lvert\hat{\Upsilon}(r,n) - \hat{\Upsilon}_{\text{th,n}}(r-1,n)\rvert \leq \hat{\Upsilon}_{\text{th,n}}(r-1,n)\Delta\hat{\Upsilon}_{\text{th,n}}$ $\quad\hat{\Upsilon}_{\text{th,n}}(r,n) = \hat{\Upsilon}(r,n)$ $\quad$Replace all stored values of $\hat{\Upsilon}_\text{n}(r-i,n)$, $i = 0, 1, \ldots, D$, by $\hat{\Upsilon}_{\text{th,n}}(r,n)$ ELSE $\quad$Find $\hat{\Upsilon}_{\text{th,n}}(r,n)$, the minimum of $\hat{\Upsilon}_\text{n}(r-i,n)$, $i = 0, 1, \ldots, D$ $\quad\hat{\Upsilon}_\text{n}(r,n) = \hat{\Upsilon}(r,n)$
15	*Adaptation control for the blocking matrix* IF $\hat{\Upsilon}(r,n) \leq \Upsilon_{0,\text{d}}\hat{\Upsilon}_{\text{th,d}}(r,n)$ $\quad\mu_{\mathbf{B}_\text{b}}(r,n) = \mu_{\mathbf{B}_\text{b},0}$ ELSE $\quad\mu_{\mathbf{B}_\text{b}}(r,n) = 0$
16	*Adaptation control for the interference canceller* IF $\hat{\Upsilon}(r,n) \leq \Upsilon_{0,\text{n}}\hat{\Upsilon}_{\text{th,n}}(r,n)$ $\quad\mu_{\mathbf{w}_\text{a}}(r,n) = \mu_{\mathbf{w}_\text{a},0}$ ELSE $\quad\mu_{\mathbf{w}_\text{a}}(r,n) = 0$
	$NRM_{\text{AC}} = \alpha[(M+1)2N\log_2 2N + 2MN + 10N]/N$

Table 7.4. Summary of the blocking matrix for the r-th iteration of 'AEC first'; typical parameters for $f_s = 12\,\text{kHz}$: $N = 256$, $\alpha = 2$, $\beta_{\mathbf{B}_b} = 0.95$, $\mu_{\mathbf{B}_b,0} = 1.0$

	Blocking matrix
17	$\mathbf{E}_{\mathbf{B}_b}(rR) = \mathbf{Y}_{\mathbf{B}_b}(rR) - \mathbf{W}^{01}_{N\times 2N}\mathbf{F}^{-1}_{2N\times 2N}\underline{\mathbf{X}}_{\mathbf{B}_b}(r)\underline{\mathbf{B}}_b(r-1)$, (7.39)
18	$\underline{\mathbf{E}}_{\mathbf{B}_b}(r) = \mathbf{F}_{2N\times 2N}\mathbf{W}^{01}_{2N\times N}\mathbf{E}_{\mathbf{B}_b}(rR)$, (7.42)
19	$\hat{\underline{\mathbf{S}}}_{\mathbf{x}_{\mathbf{B}_b}\mathbf{x}_{\mathbf{B}_b}}(r) = \beta_{\mathbf{B}_b}\hat{\underline{\mathbf{S}}}_{\mathbf{x}_{\mathbf{B}_b}\mathbf{x}_{\mathbf{B}_b}}(r-1) + (1-\beta_{\mathbf{B}_b})\underline{\mathbf{X}}^H_{\mathbf{B}_b}(r)\underline{\mathbf{X}}_{\mathbf{B}_b}(r)$, (7.44)
20	$\underline{\mathbf{B}}_b(r) = \underline{\mathbf{B}}_b(r-1) +$ $+(1-\beta_{\mathbf{B}_b})\mathbf{G}^{10}_{2N\times 2N}\boldsymbol{\mu}_{\mathbf{B}_b}(r)\hat{\underline{\mathbf{S}}}^{-1}_{\mathbf{x}_{\mathbf{B}_b}\mathbf{x}_{\mathbf{B}_b}}(r)\underline{\mathbf{X}}^H_{\mathbf{B}_b}(r)\underline{\mathbf{E}}_{\mathbf{B}_b}(r)$, (7.43)
	$NRM_{\mathbf{B}_b} = \alpha[(3M+1)(2N\log_2 2N) + 2M(4N-2) + 5N - 3]/N$

Table 7.5. Summary of the interference canceller for the r-th iteration of 'AEC first'; typical parameters for $f_s = 12\,\text{kHz}$: $N = 256$, $\alpha = 2$, $\beta_{\mathbf{w}_a} = 0.97$, $\mu_{\mathbf{w}_a,0} = 0.2$

	Interference canceller
21	FFT of the blocking matrix output signals, yielding $\underline{\mathbf{X}}_{\mathbf{w}_a}(r)$, (7.45)
22	$\mathbf{e}_{\mathbf{w}_a}(rR) = \mathbf{y}_{\mathbf{w}_a}(rR) - \mathbf{W}^{01}_{N\times 2N}\mathbf{F}^{-1}_{2N\times 2N}\underline{\mathbf{X}}_{\mathbf{w}_a}(r)\underline{\mathbf{w}}_a(r-1)$, (7.48)
23	$\underline{\mathbf{e}}_{\mathbf{w}_a}(r) = \mathbf{F}_{2N\times 2N}\mathbf{W}^{01}_{2N\times N}\mathbf{e}_{\mathbf{w}_a}(rR)$
24	$\hat{\underline{\mathbf{S}}}_{\mathbf{x}_{\mathbf{w}_a}\mathbf{x}_{\mathbf{w}_a}}(r) = \beta_{\mathbf{w}_a}\hat{\underline{\mathbf{S}}}_{\mathbf{x}_{\mathbf{w}_a}\mathbf{x}_{\mathbf{w}_a}}(r-1) + (1-\beta_{\mathbf{w}_a})\mathrm{diag}\{\underline{\mathbf{X}}^H_{\mathbf{w}_a}(r)\underline{\mathbf{X}}_{\mathbf{w}_a}(r)\}$, (7.53)
25	$\underline{\mathbf{w}}_a(r) = \underline{\mathbf{w}}_a(r-1) + (1-\beta_{\mathbf{w}_a})\times$ $\times\mathbf{G}^{10}_{2MN\times 2MN}\boldsymbol{\mu}_{\mathbf{w}_a}(r)\hat{\underline{\mathbf{S}}}^{-1}_{\mathbf{x}_{\mathbf{w}_a}\mathbf{x}_{\mathbf{w}_a}}(r)[\underline{\mathbf{X}}^H_{\mathbf{w}_a}(r)\underline{\mathbf{e}}_{\mathbf{w}_a}(r) - \underline{\boldsymbol{\psi}}(r-1)]$, (7.52)
26	$c(\underline{\mathbf{w}}_a(r)) = \Vert\underline{\mathbf{w}}_a(r)\Vert^2_F - T_0$, (7.54)
27	$\underline{\boldsymbol{\psi}}(r) = 4\sigma\max\{c(\underline{\mathbf{w}}_a(r)), 0\}\underline{\mathbf{w}}_a(r)$, (7.55)
	$NRM_{\mathbf{w}_a} = \alpha[(3M+2)(2N\log_2 2N) + 2M(5N-1) + 9N - 3]/N$

Figure 7.5 examines the NRM of the efficient real-time system (solid lines) compared to an implementation in the time domain using the simple LMS algorithm [HSH99] (dashed lines). We choose $\alpha = 2$, $\alpha_{\mathbf{a}} = 8$, $N_{\mathbf{a}} = 1024$, which is typical for our real-time system in moderately reverberant environments. We see that for $N = 256$, the RGSC ('$*$') in the DFT domain requires $\approx 75\%$ less NRMs than the RGSC in the time domain. By combining the RGSC with the AEC, the reduction of the number of real multiplications is even higher because of the longer AEC filters and because of the increasing efficiency of DFT-domain processing with increasing filter length. The complexity of GEIC is not illustrated in Fig. 7.5. The NRMs of GEIC as a function of $N := N_{\mathbf{a}}$ is very close to GSAEC.

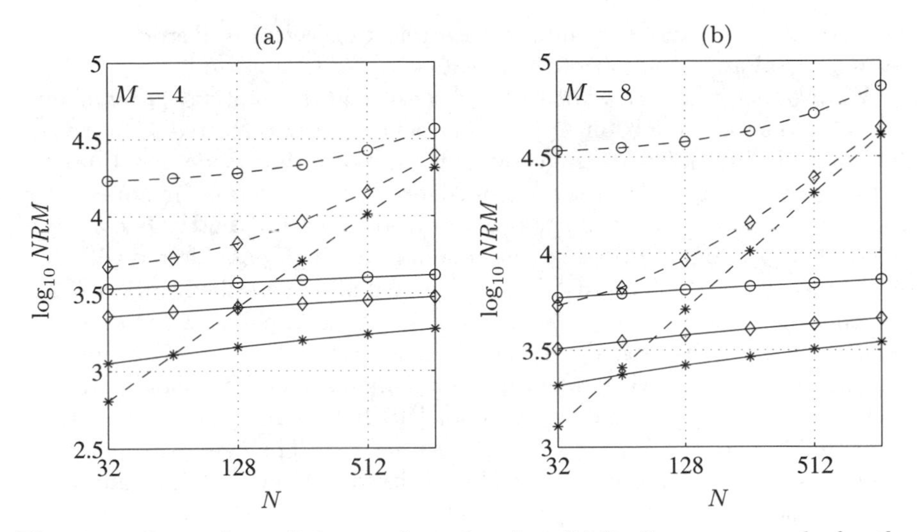

Fig. 7.5. Comparison of the number of real multiplications per sample for the efficient real-time system in the DFT domain (solid lines) with a realization using the LMS algorithm in the time domain (dashed lines): '∘': 'AEC first', '⋄': GSAEC, '∗': RGSC; $\alpha = 2$, $\alpha_\mathrm{a} = 8$, $N_\mathrm{a} = 1024$

7.3 Experimental Results

In this section, we evaluate the performance of the RGSC experimentally. In Sect. 7.3.1, the advantage of the DFT-bin-wise adaptation relative to a full-band adaptation is illustrated. Section 7.3.2 compares the RGSC with a GSC that uses a fixed blocking matrix with one spatial distortionless constraint instead of the time-varying spatio-temporal constraints. In Sect. 7.3.3, the proposed adaptation control of the RGSC is compared to an 'ideal' adaptation control, where the SINR at the sensors is assumed to be known. Finally, in Sect. 7.3.4, the RGSC is applied as a front-end for an automatic speech recognizer for verifying the usability of the proposed system for such applications. The combination of the RGSC with acoustic echo cancellation is not studied here, since the characteristics were already illustrated experimentally in Chap. 6.

7.3.1 Comparison of the DFT-Bin-Wise Adaptation with a Full-Band Adaptation

We compare the RGSC using the DFT-bin-wise adaptation control with the RGSC using a full-band adaptation based on fixed thresholding for illustrating the advantages of the proposed system. The DFT-bin-wise adaptation of the RGSC, as summarized in Table 7.3, allows a better tracking of time-varying signals and of time-varying acoustic environments than adaptation in the full-band. This experiment will show that the DFT-bin-wise adaptation is even indispensable to assure high interference rejection with low distortion of the desired signal at the output of the RGSC.

We use the sensor array with $M = 8$ sensors in the multimedia room with $T_{60} = 250$ ms as described in App. B. The sensor array is steered to broadside direction ($\theta = \pi/2$). The desired source is located in broadside direction at a distance of $d = 60$ cm from the array center. The interfering (male) speaker is located at $\theta_{\mathrm{n}} = 0$ at a distance of 1.2 m from the array center. The average SINR at the sensors is 0 dB. The sensor signals are high-pass-filtered with cut-off frequency 200 Hz. The RGSC is implemented as described in Sect. 7.2.3. For the adaptation in the full-band, the tracking mechanism for detecting presence of the desired signal only and presence of interference only (Table 7.3) is removed and replaced by fixed frequency-independent thresholding of the decision variable $\hat{\underline{\Upsilon}}(r, n)$ as suggested in [HSH99]. This removes steps 13-15 in Table 7.3, and sets $\hat{\underline{\Upsilon}}_{\mathrm{th,d}}(r, n) = \hat{\underline{\Upsilon}}_{\mathrm{th,n}}(r, n) = 1$. The fixed thresholds are chosen as $\Upsilon_{0,\mathrm{d}} = 3.5$ and $\Upsilon_{0,\mathrm{n}} = 3$. The fixed beamformer is a uniformly weighted beamformer ($\mathbf{w}_{\mathrm{c}} = 1/M \mathbf{1}_{M \times 1}$).

The results are shown in Fig. 7.6. The desired signal and the interference signal at the sensor $M/2$ are depicted in Fig. 7.6a and Fig. 7.6b, respectively. We notice the nearly continuous double-talk of desired speaker and interferer. Figure 7.6c illustrates the cancellation of the desired signal by the blocking matrix for both adaptation controls. It may be seen that $DC_{\mathbf{B}_{\mathrm{s}}}(k)$ for the adaptation in the full-band is almost as high as for the DFT-bin-wise adaptation. This can be explained by the parameter setting of the full-band adaptation: The thresholds are chosen such that $DC_{\mathbf{B}_{\mathrm{s}}}(k)$ is maximized in order to maximally prevent cancellation of the desired signal at the output of the RGSC. However, as can be seen in Fig. 7.6d, distortion of the desired signal $DC_{\mathbf{w}_{\mathrm{RGSC}}}(k)$ cannot entirely prevented. While $DC_{\mathbf{w}_{\mathrm{RGSC}}}(k)$ for the DFT-bin-wise adaptation is less than 2 dB, $DC_{\mathbf{w}_{\mathrm{RGSC}}}(k)$ is down to -5 dB. The negative sign of $DC_{\mathbf{w}_{\mathrm{RGSC}}}(k)$ for the fullband adaptation results from instabilities of the adaptive filters, since adaptation of the interference canceller during double-talk of desired speaker and interference cannot be completely prevented by the simple adaptation control in the full-band. Adaptation of the interference canceller during double-talk may lead to instabilities since the desired signal which is efficiently suppressed by the blocking matrix is still present in the reference path of the beamformer, and, thus, disturbs the adaptation of the interference canceller. Figure 7.6e illustrates the interference rejection $IR_{\mathbf{w}_{\mathrm{RGSC}}}(k)$ of the RGSC with the two adaptation mechanisms and the interference rejection $IR_{\mathbf{w}_{\mathrm{c}}}(k)$ of the fixed beamformer. We see that the interference rejection of the RGSC for the full-band adaptation is only little higher than $IR_{\mathbf{w}_{\mathrm{c}}}(k)$, while high interference rejection is obtained with the RGSC using the DFT-bin-wise adaptation. As expected for the full-band adaptation, the interference canceller cannot track the time-variance of the PSDs of the sensor signals and of the blocking matrix. As a consequence, the sidelobe cancelling path is almost inefficient. This is reflected by the residual interference signals at the output of the RGSC in Figs. 7.6f, g. While for the full-band adapta-

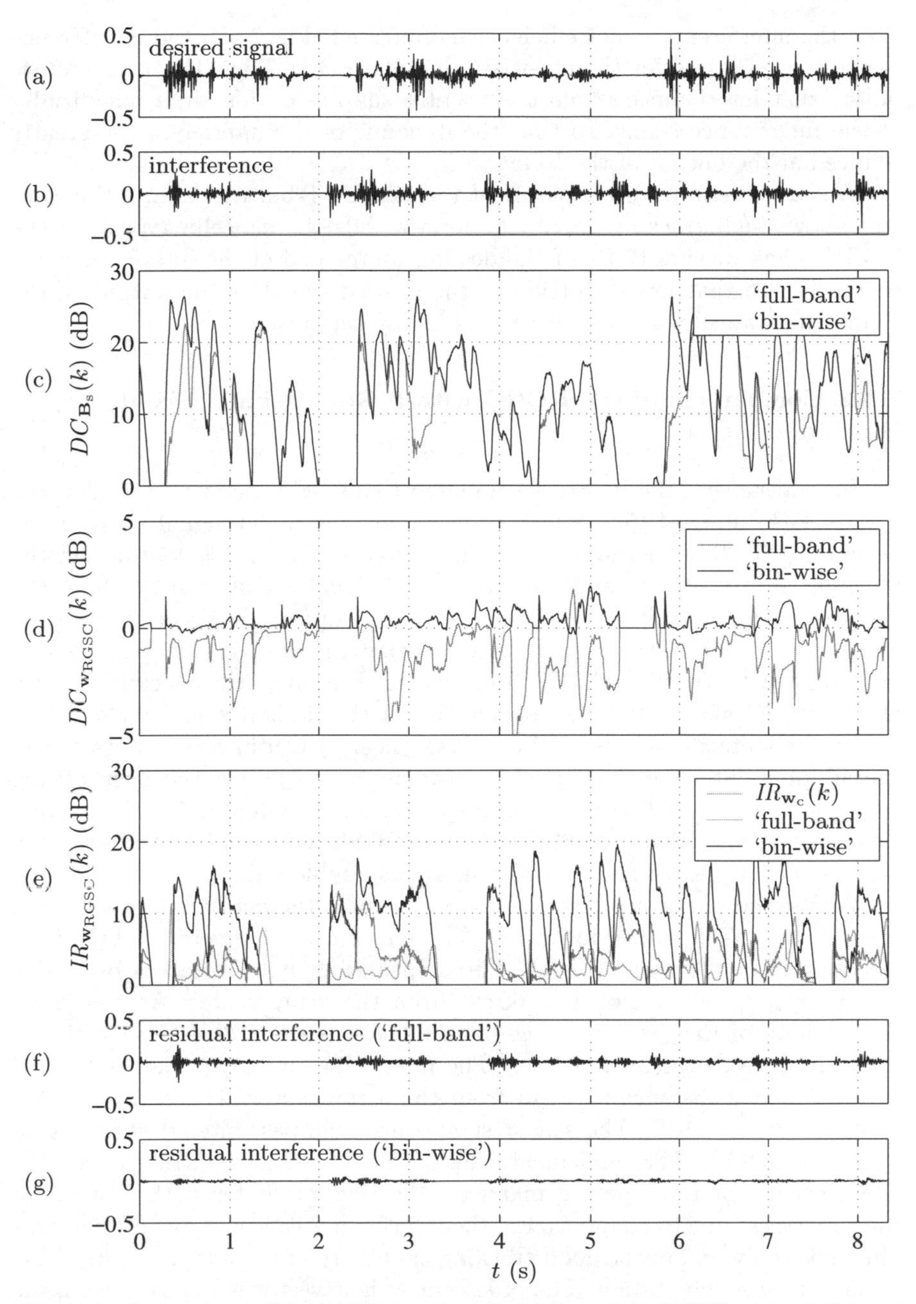

Fig. 7.6. Comparison of the DFT bin-wise adaptation of the RGSC with a full-band adaptation; multimedia room with $T_{60} = 250$ ms, $M = 8$

tion, the interference is not efficiently suppressed (Fig. 7.6f), the interference is highly suppressed for the bin-wise adaptation (Fig. 7.6g). Furthermore, we notice that interference is most efficiently suppressed for large magnitudes of the interference signal, so that the dynamic of the interference is greatly reduced at the output of the RGSC.

We conclude from this experiment that the DFT-bin-wise adaptation control allows high tracking capability for the sidelobe cancelling path of the RGSC. Time-varying PSDs of the desired source and of the interference can be tracked so that low distortion of the desired signal at the output of the RGSC is obtained while interference is highly suppressed.

7.3.2 Comparison of the RGSC with a GSC Using a Fixed Blocking Matrix

In this experiment, the RGSC is compared to a GSC which uses the blocking matrix $\mathbf{B}(k)$ after (4.152), which cancels only the direct signal path of the desired signal. Reverberation w.r.t. the desired signal thus leaks through the blocking matrix. We expect that the desired signal is distorted by the GSC. For the RGSC, however, the blocking matrix suppresses direct signal path and reverberation of the desired signal so that the desired signal should not be distorted by the RGSC. The adaptive realization of the blocking matrix of the RGSC allows to track movements of the desired source. The DFT-bin-wise adaptation of the RGSC allows (nearly) continuous adaptation of the adaptive filters so that speaker movements can be tracked even during continuous double-talk of the desired speaker and interferers. For illustrating this tracking capability of the DFT-domain RGSC, we consider a time-varying position of the desired source and continuous double-talk.

We use the sensor array with $M = 8$ sensors in the multimedia room with $T_{60} = 250$ ms as described in App. B. The sensor array is steered to broadside direction ($\theta = \pi/2$). The (male) desired speaker is first located in broadside direction at a distance of $d = 60$ cm from the array center. At $t = 8.6$ s, the position of the desired source is switched to $\theta_{\mathrm{d}} = \pi/2 + 2\pi/9$ with the same distance to the array center. The interfering (male) speaker is located at $\theta_{\mathrm{n}} = 0$ at a distance of 1.2 m from the array center. The average SINR at the sensors is 0 dB. The sensor signals are high-pass-filtered with cut-off frequency 200 Hz. The implementation of the RGSC is the same as for the GSC, except for the blocking matrices. We thus profit for both systems of the robustness due to adaptation of the interference canceller for interference-plus-noise only and of the good tracking capability of the adaptive filters due to the bin-wise adaptation. The quiescent weight vector is realized using the wideband Dolph-Chebyshev design which is presented in Sect. 4.3 and which is illustrated in Fig. 4.10 in order to widen the peak-to-zero distance of the fixed beamformer. The parameter setup is the same for the RGSC and for the GSC. (See Sect. 7.2.3.)

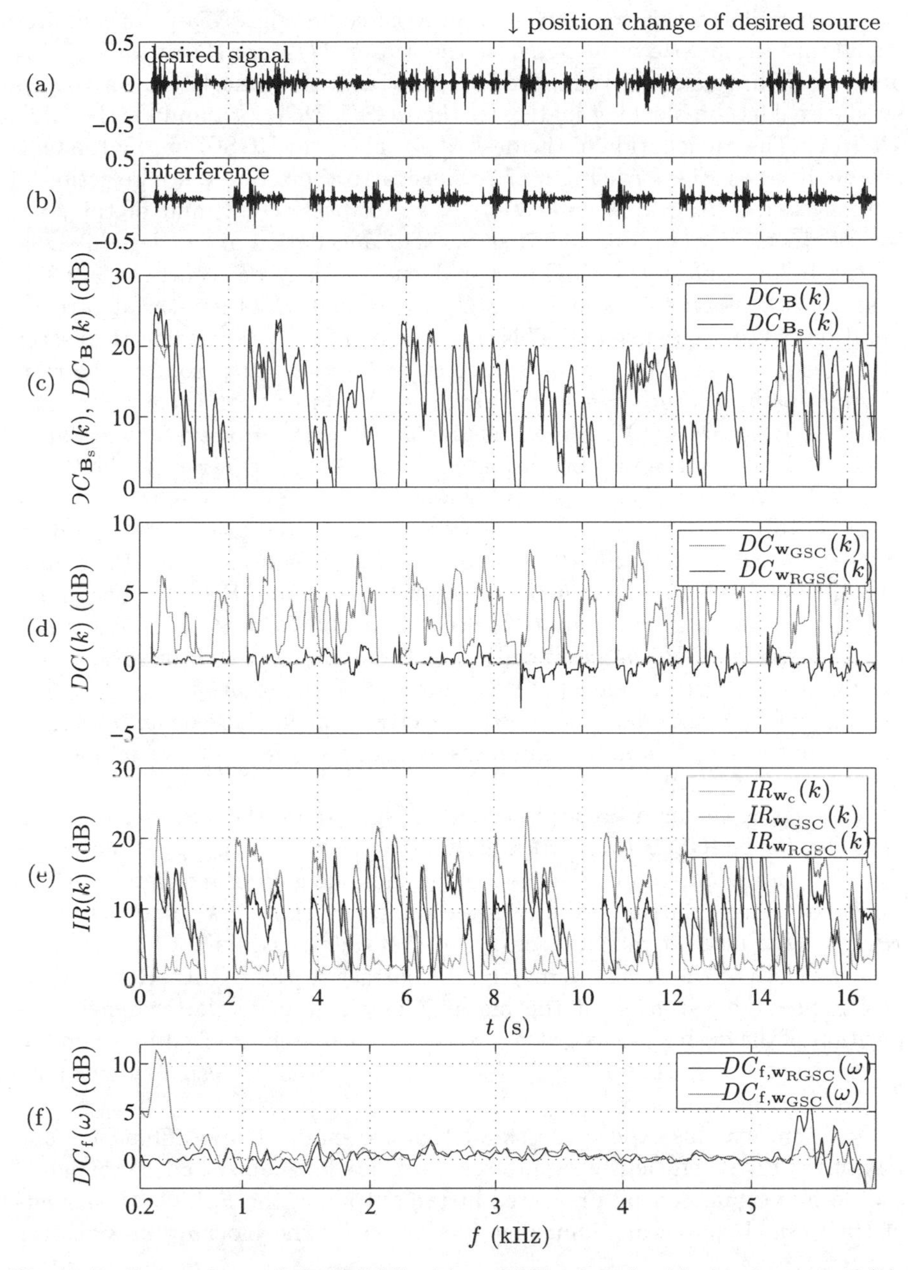

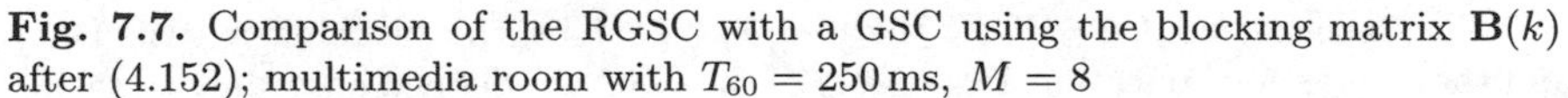

Fig. 7.7. Comparison of the RGSC with a GSC using the blocking matrix $\mathbf{B}(k)$ after (4.152); multimedia room with $T_{60} = 250\,\mathrm{ms}$, $M = 8$

The results of this experiment are depicted in Figs. 7.7a-f. The desired signal and the interference signal at the sensor $M/2$ are shown in Fig. 7.7a and Fig. 7.7b, respectively. Figure 7.7c illustrates the cancellation of the desired signal by the blocking matrix for the RGSC, $DC_{\mathbf{B}_s}(k)$, and for the GSC, $DC_{\mathbf{B}}(k)$. The cancellation of the desired signal by the RGSC and by the GSC are depicted in Fig. 7.7d. Figure 7.7e illustrates the interference rejection of both systems, $IR_{\mathbf{w}_{\mathrm{RGSC}}}(k)$ and $IR_{\mathbf{w}_{\mathrm{GSC}}}(k)$, respectively[10], and of the fixed beamformer, $IR_{\mathbf{w}_c}(k)$. Figure 7.7f shows the cancellation of the desired signal by the RGSC and by the GSC over frequency. It may be seen from Fig. 7.7c that $DC_{\mathbf{B}}(k)$ is almost as high as $DC_{\mathbf{B}_s}(k)$ though $\mathbf{B}(k)$ lets through reverberation of the desired signal. This can be explained by the geometrical setup of this experiment: The desired source is close to the array so that the ratio of direct-path to reverberation is high. Since the direct path is efficiently cancelled by $\mathbf{B}(k)$, $DC_{\mathbf{B}}(k)$ is almost as high as $DC_{\mathbf{B}_s}(k)$. However, cancellation of the desired signal cannot be prevented by the GSC, in contrast to the RGSC (Fig. 7.7d). While $|DC_{\mathbf{w}_{\mathrm{RGSC}}}(k)|$ is below 1 dB most of the time, $|DC_{\mathbf{w}_{\mathrm{GSC}}}(k)|$ is up to 5 dB. The change of the position of the desired source at $t = 8.6$ s is tracked immediately by the RGSC. For the GSC, the desired signal is mainly cancelled at frequencies $f < 600$ Hz, as can be seen from Fig. 7.7f. For higher frequencies, the performance of the RGSC and of the GSC is similar. This can be explained by the combination of three effects: (a) Reverberation is strongest at low frequencies. (b) Robustness of the adaptation control is lowest at low frequencies due to the little directivity of the fixed beamformer. (c) data-dependent optimum beamformers are most sensitive at low frequencies [GM55, CZK86, SR93].

Comparing the interference rejection in Fig. 7.7e for the fixed beamformer, $IR_{\mathbf{w}_c}(k)$, for the GSC, $IR_{\mathbf{w}_{\mathrm{GSC}}}(k)$, and for the RGSC, $IR_{\mathbf{w}_{\mathrm{RGSC}}}(k)$, we notice that for both adaptive beamformers a considerable gain of interference rejection is obtained relative to the fixed beamformer. $IR_{\mathbf{w}_{\mathrm{GSC}}}(k)$ is greater than $IR_{\mathbf{w}_{\mathrm{RGSC}}}(k)$, since the interference canceller of the GSC does not have to track time-variance of the blocking matrix, in contrast to the RGSC. However, we notice that the variation of the blocking matrix due to the change of the position of the desired source at $t = 8.6$ s is immediately tracked by the interference canceller so that $IR_{\mathbf{w}_{\mathrm{RGSC}}}(k)$ does not decrease perceptibly when the desired speaker is moving.

In summary, this experiment shows that the spatio-temporal blocking matrix of the RGSC efficiently suppresses the desired signal so that distortion of the desired signal can be prevented at the output of the RGSC. Movements of the desired source are immediately tracked by the blocking matrix even

[10]The cancellation $DC_{\mathbf{B}}(k)$ of the desired signal by the blocking matrix $\mathbf{B}(k)$ and the interference rejection $IR_{\mathbf{w}_{\mathrm{GSC}}}(k)$ of the GSC are defined according to (5.58) and (5.59) with $\mathbf{B}_s(k)$ replaced by $\mathbf{B}(k)$, respectively. Estimates of $DC_{\mathbf{B}}(k)$ and $IR_{\mathbf{w}_{\mathrm{GSC}}}(k)$, long-term averages $DC_{\mathbf{B}}$ and $IR_{\mathbf{w}_{\mathrm{GSC}}}$, and estimates of the cancellation of the desired signal by the GSC, $DC_{\mathrm{f},\mathbf{w}_{\mathrm{GSC}}}(\omega)$, in the DTFT domain are obtained as described in Sect. 5.5.1.

during continuous double-talk between the desired speaker and interference. The GSC with a fixed blocking matrix, which suppresses only the direct signal path of the desired signal, distorts the desired signal mainly at low frequencies.

7.3.3 Comparison of the Proposed Adaptation Control with an 'Ideal' Adaptation Control

In this experiment, we compare the proposed adaptation control based on the spatial selectivity of a fixed beamformer with an 'ideal' adaptation control. Cancellation of the desired signal by the RGSC and interference rejection of the RGSC are studied. As 'ideal' adaptation control, we use a double-talk detector which is based on the true SINR at the sensors (see also Fig. 7.3): If the DFT-bin-wise SINR at the sensors is greater than 15 dB, the blocking matrix is adapted. If the DFT-bin-wise SINR is less than −15 dB, the interference canceller is adapted. Experiments showed that these thresholds give maximum interference rejection and minimum cancellation of the desired signal of the RGSC. The sensor array with $M = 4$ and with $M = 8$ according to Fig. B.1 is used. The desired (male) speaker is located at $\theta_\mathrm{d} = \pi/2$ at a distance of $d = 60$ cm from the array center. We perform the comparison for the multimedia room with $T_{60} = 250$ ms and with $T_{60} = 400$ ms and for the passenger cabin of the car. In the multimedia room, a (male) interferer speaker is located at $\theta_\mathrm{n} = 0$ at a distance of 1.2 m from the array center. In the passenger cabin of the car, the car noise according to App. B is present. The average SINR at the sensors for all three environments is 0 dB. The sensor signals are high-pass-filtered with cut-off frequency 200 Hz. The parameter setup as indicated in Sect. 7.2.3 is used for all three acoustic environments. Blocking matrix and interference canceller are adapted for continuous double-talk until the interference rejection of the RGSC increases by less than 0.2 dB over a data block of 100000 samples.

The results for $M = 4$ sensors are shown in Table 7.6. It can be seen that the cancellation of the desired signal by the blocking matrix, $DC_{\mathbf{B}_\mathrm{s}}$, is less

Table 7.6. Comparison of the real adaptation control with an 'ideal' adaptation control for different acoustic environments ($M = 4$, $10\log_{10} SINR_\mathrm{in} = 0$ dB); values in parentheses correspond to the 'ideal' adaptation control with knowledge of the SINR at the sensors

$M = 4$	competing speech ($T_{60} = 250$ ms)	competing speech ($T_{60} = 400$ ms)	car noise (car environment)
$DC_{\mathbf{B}_\mathrm{s}}$ (dB)	23.70 (23.74)	19.34 (19.35)	17.25 (17.35)
$IR_{\mathbf{w}_\mathrm{c}}$ (dB)	3.36	3.12	3.46
$IR_{\mathbf{w}_\mathrm{RGSC}}$ (dB)	12.72 (14.81)	10.22 (11.00)	7.75 (7.97)

than 0.1 dB smaller for the real adaptation control than for the 'ideal' one. For the multimedia room with $T_{60} = 250$ ms, $DC_{\mathbf{B}_s}$ is greater than for the multimedia room with $T_{60} = 400$ ms, since for $T_{60} = 400$ ms longer room impulse responses have to be modeled by the filters of the blocking matrix. However, even for the environment with $T_{60} = 400$ ms, $DC_{\mathbf{B}_s} = 19.34$ dB. $DC_{\mathbf{B}_s}$ is lower for the car environment (car noise) than for the multimedia room (competing speech). This can be explained by the fact that car noise has a wide frequency range (especially at low frequencies) where the car noise is continuously active. For these frequencies, the blocking matrix can only be adapted during double-talk, so that $DC_{\mathbf{B}_s}$ is reduced. (See Sects. 5.5.4, 5.6.) However, $DC_{\mathbf{B}_s} = 17.25$ dB, so that cancellation of the desired signal can efficiently be prevented.

The interference rejection of the RGSC, $IR_{\mathbf{w}_{\mathrm{RGSC}}}$ is maximally 2.05 dB less for the proposed adaptation mechanism than for the 'ideal' one. The largest difference is obtained with the moderately reverberant environment ($T_{60} = 250$ ms). For a long reverberation time ($T_{60} = 400$ ms) and for car noise in the passenger cabin of the car, the difference is only 0.78 dB and 0.22 dB, respectively. These unequal differences may be explained by the parameter setup, since the parameters are chosen such that the performance is optimized for all the considered scenarios.

Table 7.7. Comparison of the real adaptation control with an 'ideal' adaptation control for different acoustic environments ($M = 8$, $10\log_{10} SINR_{\mathrm{in}} = 0$ dB); values in parentheses correspond to the 'ideal' adaptation control with knowledge of the SINR at the sensors

$M = 8$	competing speech ($T_{60} = 250$ ms)	competing speech ($T_{60} = 400$ ms)	car noise (car environment)
$DC_{\mathbf{B}_s}$ (dB)	23.06 (23.09)	18.55 (18.61)	16.15 (16.21)
$IR_{\mathbf{w}_c}$ (dB)	4.67	3.90	3.64
$IR_{\mathbf{w}_{\mathrm{RGSC}}}$ (dB)	15.43 (18.07)	12.77 (14.33)	14.23 (17.69)

For $M = 8$, the results are given in Table 7.7. Compared to the sensor array with $M = 4$, we notice that $IR_{\mathbf{w}_{\mathrm{RGSC}}}$ is almost doubled for the car environment from $IR_{\mathbf{w}_{\mathrm{RGSC}}} = 7.75$ dB to $IR_{\mathbf{w}_{\mathrm{RGSC}}} = 14.23$ dB with the realistic adaptation control, while the improvement of $IR_{\mathbf{w}_{\mathrm{RGSC}}}$ is less for the multimedia room with competing speech. For the diffuse car noise, the spatial number of degrees of freedom is not sufficient for both arrays. However, the number of spatial degrees of freedom is doubled for $M = 8$, so that more spatial nulls can be placed for increasing the directivity of the beamformer. In the multimedia room, a single directional interferer is present so that the number of spatial degrees of freedom of the RGSC is sufficient for both sensor arrays. The temporal degrees of freedom, however, are not sufficient to completely cancel the interferer because of the small number of filter coefficients of the interference canceller ($N = 256$). As it was shown in Sect. 3.2.2 and illustrated in Sect. 5.5.3 for the RGSC, spatial degrees of freedom can be traded-off against temporal degrees of freedom, so that the interference rejection increases with increasing number of sensors. However, since complete interference rejection can only be obtained with filters of infinite length, the gain of $IR_{\mathbf{w}_{\mathrm{RGSC}}}$ by using $M = 8$ sensors instead of $M = 4$ sensors is limited.

The cancellation of the desired signal by the blocking matrix is slightly reduced for $M = 8$ relative to $M = 4$. There are several reasons for this observation: It was illustrated in Sect. 5.5.3 that systems with different poles have to be modeled by the blocking matrix for $M = 4$ and for $M = 8$, which may lead to a decrease or to an increase of $DC_{\mathbf{B}_\mathrm{s}}$. Furthermore, during adaptation of the blocking matrix there are always interference components present because of the adaptation for $SINR_{\mathrm{f,in}}(r, n) < \infty$ and because of errors of the adaptation control which may influence the adaptation behavior. Finally, the spectral characteristics of the desired source influences $DC_{\mathbf{B}_\mathrm{s}}$.

We conclude from this experiment that the proposed spatial double-talk detector assures high cancellation of the desired signal by the blocking matrix while efficiently suppressing the interference at the output of the RGSC even for a small number of sensors. The results show that the RGSC using the proposed adaptation control can be applied for a large variety of acoustic conditions without changing any parameters, so that our real-time system is highly portable. Finally, we have seen that the double-talk detector works down to very low frequencies ($f = 200$ Hz) relative to the array aperture (28 cm), where the simple uniformly weighted beamformer provides only very little directivity. (See also Figs. 4.7 and 4.8.)

7.3.4 Application of the RGSC as a Front-End for an Automatic Speech Recognizer (ASR)

For successfully using recognition systems in real-world scenarios, robustness is a crucial issue. As illustrated in Chap. 1, the robustness of an automatic speech recognizer is strongly influenced by the ability to cope with presence of interference and noise and with distortions by the frequency characteristic of the transmission channel.

While robustness can be achieved by various approaches [Jun00] that require changes in the ASR system, we concentrate in our work on the usage of the RGSC as a front-end for an ASR. The separation of the speech recognizer from the noise reduction does not require changes of the ASR so that all speech enhancement techniques can be combined with all speech recognizers by simple cascading. The pre-processing by the beamforming microphone array increases the SINR at the input of the speech recognizer, which increases the word accuracy in turn. (See Fig. 7.8a.) This, of course, supposes that the pre-processing does not introduce artifacts which affect the performance of the speech recognizer.

In this section, we study the usability of the RGSC for noise-robust speech recognition. We will show that the word accuracy can be considerably improved by the RGSC compared to a noisy sensor signal and compared to a simple uniformly weighted beamformer for a wide range of SINRs at the sensors. Adaptation of the ASR to the noisy conditions is not necessary so that the ASR can be trained with clean speech data. For evaluation, we define a speaker-independent connected digit recognizer based on the HTK (Hidden Markov Model Tool Kit) [YKO+00]. The setup is trained and tested with the TIDigits data base [Leo84]. The interference and the reverberation of the acoustic environment are artificially added to the clean test set of the data base.

First, the setup of the HTK recognizer is described. Second, we summarize the testing conditions. Finally, recognition results are given.

Description of the ASR System

The ASR is based on the HTK software version 3.0 [YKO+00]. The sampling rate is 8 kHz.[11] The ASR input signal is divided into blocks of length 25 ms using a frame shift of 10 ms (block overlap factor of 2.5). The input blocks are pre-emphasized with a factor of 0.97 and tapered by a Hamming window. The pre-processed input data blocks are transformed into the DFT domain and analyzed by a Mel-weighted filterbank with 23 frequency bands between 200 Hz and half of the sampling rate. From this Mel spectrum, 13 Mel-frequency cepstral coefficients (MFCCs) are calculated. Cepstral mean normalization is performed by subtracting the average of each cepstral parameter across each input speech file (corresponds to the data files of the TIDigits data base). The MFCCs are liftered using the HTK default cepstral liftering. A feature vector of length 26 is formed from the 13 MFCCs (including the coefficient of order 0) plus the corresponding delta coefficients. The digits are modeled as whole word Hidden Markov Models (HMMs) with 18 states per word including entry and exit states. The HMMs are simple left-to-right models without skips over states. Mixtures of 3 Gaussians per state with mean and with a diagonal covariance matrix are used. A voice activity detector is not used.

Description of the Testing Conditions

The TIDigits data base is taken as a basis. The part with the recordings of male and female US-American adults speaking sequences of up to 7 digits is considered. The original 20 kHz data is downsampled to 8 kHz. This data is considered as clean data. While for training the recognizer the entire training set (8624 utterances) is used, the test set is limited to a quarter of the original size (2176 utterances) in order to reduce the run time of the experiments. For

[11] We use $f_s = 8$ kHz for the speech recognition for minimizing the run time of the experiments.

reducing the number of utterances, we simply take only every fourth data file in alphabetical order.[12] The difference in word accuracy between the complete clean test set and the reduced clean test set is about 0.2%.

The training of the ASR is done by applying Baum-Welch reestimation on the clean training set of TIDigits. The word models are initialized with one Gaussian per state with the global means and variances. Then, the number of Gaussians is successively increased up to 3. After each incrementation, 7 iterations of Baum-Welch reestimation are performed.

The noisy test set is obtained by filtering the clean test data (=source signals) with the impulse responses between the desired source position and the microphone positions in the desired acoustic environment. Internal filter states are preserved between the test files in order to have a continuous stream of sensor data. Then, interference is artificially added with the desired SINR, yielding the noisy sensor signals. The interference data is 'continuously' added to the source data so that periods with silence of the interference signals to not appear if the lengths of the files do not match. If the total lengths of different interference files are less than that of the test set, interference files are repeated.

For pre-processing the sensor signals, the efficient real-time realization of the RGSC as described at the beginning of this chapter is used. Acoustic echoes are not considered, since on the one hand, it would be possible to define scenarios where the AEC is arbitrarily efficient. This, of course, is equal to a scenario without presence of acoustic echoes. On the other hand, situations can be easily defined where the AEC is completely inefficient such that the performance of the combined system corresponds to that of the RGSC alone.

Recognition Performance

The speaker-independent isolated word recognizer is applied in the car environment and in the multimedia room with $T_{60} = 250$ ms with a competing speech signal for the SINR varying between 30 dB and -10 dB in 5 dB steps. The arrays with $M = 4$ microphones and with $M = 8$ microphones as described in App. B with an aperture of 28 cm are used. Due to insufficient spatial selectivity at low frequencies, the sensor signals are highpass filtered with cut-off frequency 200 Hz. The desired source is located in broadside direction at a distance of 60 cm to the array center. The sensor array is steered to the fixed position of the desired source so that a simple uniformly weighted beamformer suffices as quiescent weight vector. For mismatch between the steering direction and the position of the desired source, a wideband Dolph-Chebyshev design as described in Sect. 4.3 might yield better results.

The word accuracy of the speech recognizer is evaluated for each of the sensor signals $x_m(k)$, $m = 0, 1, \ldots, M-1$, for the output signal of the fixed

[12] For the ASR experiments we used a Matlab implementation of the algorithm for better analysis, which runs, in contrast to our C^{++} realization not in real-time.

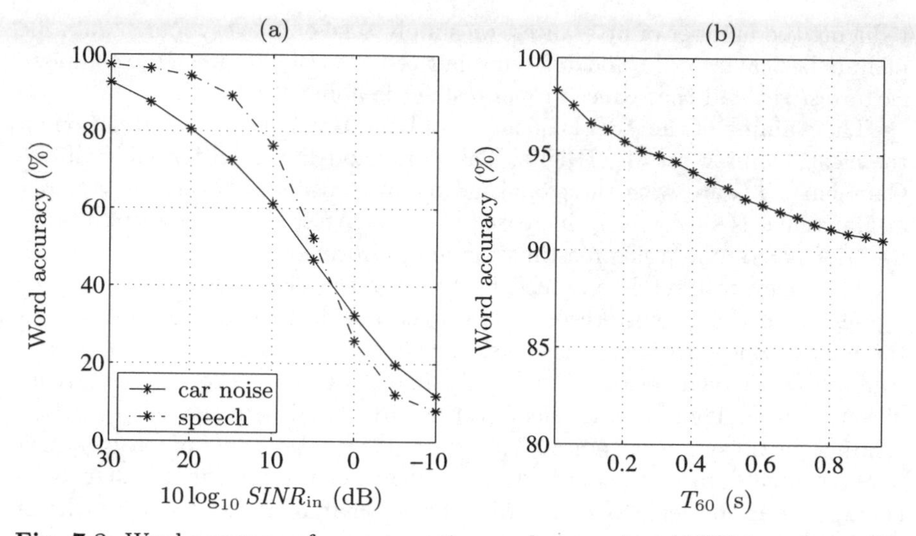

Fig. 7.8. Word accuracy of an automatic speech recognizer (ASR) in noisy conditions: **(a)** influence of additive car noise and of competing speech with varying signal-to-interference-plus-noise ratio (SINR), **(b)** influence of the reverberation time T_{60} of the acoustic environment; Connected digit recognition task based on the HTK toolkit [YKO+00] and on the TIDigits data base [Leo84] (see Chap. 7)

Table 7.8. Word accuracy (in %) in noisy environments as a function of the SINR

$10\log_{10} SINR_{\text{in}}$	car noise	speech
clean	98.29	98.29
30 dB	92.79	97.42
25 dB	87.54	96.43
20 dB	80.56	94.48
15 dB	87.54	89.20
10 dB	61.16	76.19
5 dB	46.70	52.27
0 dB	32.42	25.88
-10 dB	19.71	11.99
-20 dB	11.72	7.92

beamformer $y_{\mathbf{w}_c}(k)$ and for the output signal of the RGSC, $y_{\mathbf{w}_{\text{RGSC}}}(k)$. The word accuracy for the sensor signals $x_m(k)$ is averaged across the M sensor signals. The baseline word accuracy for clean speech is 98.29% (Fig. 7.8 and Tables 7.8, 7.9).

With decreasing $SINR_{\text{in}}$ (Fig. 7.8a) and with decreasing reverberation time T_{60} (Fig. 7.8b), the word accuracy decreases. In Fig. 7.8a, we see that car noise

Table 7.9. Word accuracy (in %) in a reverberant environment as a function of the reverberation time T_{60}

T_{60} (ms)	Word accuracy (%)
0	98.29
100	96.59
200	95.63
300	94.91
400	94.06
500	93.21
600	92.35
700	91.73
800	91.13
900	90.77

is more detrimental for the ASR than speech for $10 \log_{10} SINR_{\text{in}} > 3\,\text{dB}$, while it is the opposite for $10 \log_{10} SINR_{\text{in}} < 3\,\text{dB}$. This results from the fact that generally wideband slowly time-varying or stationary noise affects the word accuracy of an ASR system more than competing speech signals [Jun00]. For $10 \log_{10} SINR_{\text{in}} < 3\,\text{dB}$, the word accuracy for competing speech is less yet, since the competing speech phonetically overlaps with the desired speech so that for low SINRs the competing speech is recognized by the ASR as desired speech. In Fig. 7.8b, we notice the low dependency of the word accuracy on the reverberation time T_{60}. Even for $T_{60} = 0.9\,\text{s}$, the word accuracy is more than 90%. This can be explained by the rather simple recognition task which is robust against reverberation: First, the recognition task consists only of 11 digits. Therefore, the vector vectors have a high distance in the feature space. Second, we use HMMs which model entire words instead of smaller unities as, e.g., phonemes, which are less affected by reverberation, since the HMMs cover longer time intervals.

Car Environment

First, we consider the car environment as described in App. B.2. The files with the car noise at the sensors only contain 42 seconds of data, so that these files are concatenated until the length of the test set is obtained. The recognition results are depicted in Figs. 7.9a, b and in Tables 7.10a, b as a function of the SINR at the sensors for $M = 4$ and for $M = 8$, respectively. Figures 7.10a, b show the interference rejection of the fixed beamformer, $IR_{\mathbf{w}_c}$, of the RGSC, $IR_{\mathbf{w}_{\text{RGSC}}}$, and the cancellation of the desired signal by the blocking matrix, $DC_{\mathbf{B}_s}$ as a function of $10 \log_{10} SINR_{\text{in}}$, for $M = 4$ and for $M = 8$, respectively. The beamformer performance measures are averaged over the entire test set.

We see in Fig. 7.9 and in Table 7.10 that, generally, the word accuracy is greatly improved by using the RGSC instead of a single array microphone

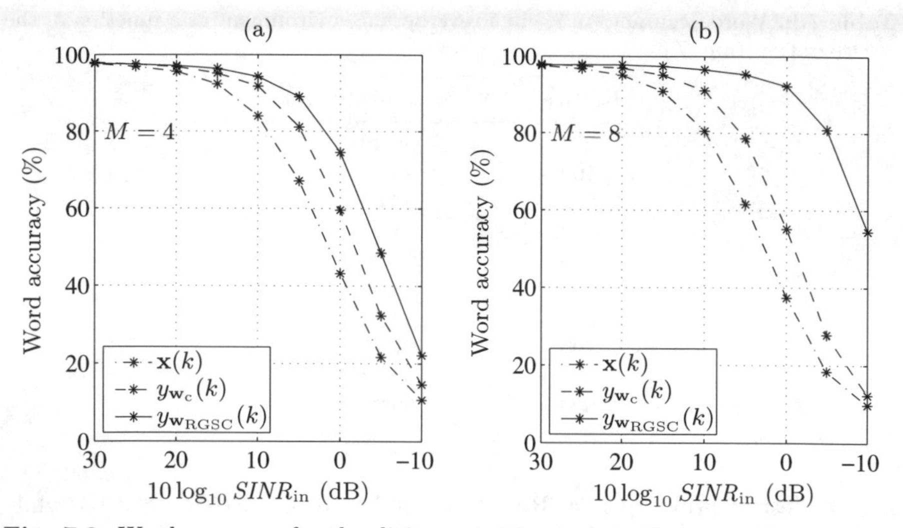

Fig. 7.9. Word accuracy for the digit recognition task in the car environment as a function of the average SINR at the sensors, $SINR_{\mathrm{in}}$; input signals of the ASR: microphone signals ($\mathbf{x}(k)$), output signal of a beamformer ($y_{\mathbf{w}_c}(k)$), and output signal of the RGSC ($y_{\mathbf{w}_{\mathrm{RGSC}}}(k)$)

Table 7.10. Word accuracy (%) in the car environment as a function of $SINR_{\mathrm{in}}$ **(a)** for $M = 4$ and **(b)** for $M = 8$

$10\log_{10} SINR_{\mathrm{in}}$ (dB)	(a) $\mathbf{x}(k)$	(a) $y_{\mathbf{w}_c}(k)$	(a) $y_{\mathbf{w}_{\mathrm{RGSC}}}(k)$	(b) $\mathbf{x}(k)$	(b) $y_{\mathbf{w}_c}(k)$	(b) $y_{\mathbf{w}_{\mathrm{RGSC}}}(k)$
30	97.64	97.95	97.70	97.52	97.97	97.65
25	96.87	97.49	97.47	96.68	97.44	97.77
20	95.58	96.84	97.12	95.02	96.89	97.77
15	92.35	95.17	96.38	90.84	95.12	97.38
10	83.94	91.73	94.25	80.48	90.98	96.50
5	67.12	81.10	89.02	61.50	78.59	95.30
0	43.15	59.55	74.53	37.36	55.14	92.10
-5	21.63	32.28	84.49	18.38	27.82	80.87
-10	10.64	14.62	22.11	9.79	12.30	54.41

or a fixed beamformer. For $M = 4$, the word accuracy for the RGSC is up to 124.13% ($10\log_{10} SINR_{\mathrm{in}} = -5$ dB) greater than for a single microphone and up to 51.23% greater than for the fixed beamformer ($10\log_{10} SINR_{\mathrm{in}} = -10$ dB). For $M = 8$, the improvement is even 455.42% ($10\log_{10} SINR_{\mathrm{in}} = -10$ dB) relative to a single microphone and 342.35% relative to the fixed beamformer ($10\log_{10} SINR_{\mathrm{in}} = -10$ dB). The word accuracy is greater than 90% for $10\log_{10} SINR_{\mathrm{in}} \geq 10$ dB ($M = 4$) and for $10\log_{10} SINR_{\mathrm{in}} \geq 0$ dB

($M = 8$), which shows the suitability of our approach for application in typical car environments where such SINRs are frequently encountered.

Figure 7.10 shows that the interference rejection of the RGSC is more than $IR_{\mathbf{w}_{\mathrm{RGSC}}} = 8\,\mathrm{dB}$ for $M = 4$ and more than $IR_{\mathbf{w}_{\mathrm{RGSC}}} = 17\,\mathrm{dB}$ for $M = 8$. Interpreting $IR_{\mathbf{w}_{\mathrm{RGSC}}}$ as an improvement of $SINR_{\mathrm{in}}$ in Fig. 7.9, it may be seen that $IR_{\mathbf{w}_{\mathrm{RGSC}}}$ translates nearly completely to a real increase of $SINR_{\mathrm{in}}$. This noticeable result reflects the high output signal quality of the RGSC and motivates that adaptive beamformers in general are well suited for application to ASRs.

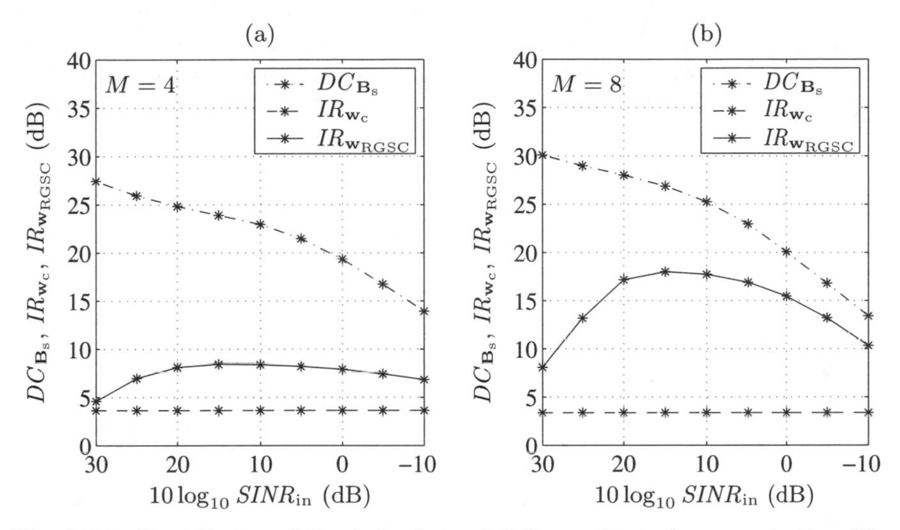

Fig. 7.10. Cancellation of the desired signal $DC_{\mathbf{B}_s}$ and interference rejection $IR_{\mathbf{w}_c}$ and $IR_{\mathbf{w}_{\mathrm{RGSC}}}$ over the average SINR at the sensors, $SINR_{\mathrm{in}}$, in the car environment

Figure 7.10 shows further that $IR_{\mathbf{w}_{\mathrm{RGSC}}}$ is maximum for $10\log_{10} SINR_{\mathrm{in}} = 15\,\mathrm{dB}$. The decrease of $IR_{\mathbf{w}_{\mathrm{RGSC}}}$ for $10\log_{10} SINR_{\mathrm{in}} < 15\,\mathrm{dB}$ results from less frequent adaptation of the interference canceller for high values of $SINR_{\mathrm{in}}$. The reduction of $IR_{\mathbf{w}_{\mathrm{RGSC}}}$ for $10\log_{10} SINR_{\mathrm{in}} > 15\,\mathrm{dB}$ is a consequence of the parameter choice, which is optimized for minimum distortion of the desired signal by the RGSC. The cancellation of the desired signal by the blocking matrix, $DC_{\mathbf{B}_s}$, decreases with decreasing $SINR_{\mathrm{in}}$. This effect results from the fact that the blocking matrix cannot be adapted for presence of only the desired signal due to the slow time-variance of car noise. (See also Sect. 5.5.4.) When real-world systems are developed with the RGSC, this effect has to be considered since it reduces the robustness of the RGSC against cancellation of the desired signal at the output of the RGSC. In contrast to Table 7.6 ($M = 4$) and Table 7.7 ($M = 8$), we notice that in Fig. 7.10 $DC_{\mathbf{B}_s}$ is greater for $M = 8$ than for $M = 4$. This can be explained by the different second-order statistics of the sensor data used in the experiments.

Multimedia Room with $T_{60} = 250$ ms

In the multimedia room ($T_{60} = 250$ ms), we tested the recognition performance for competing speech. As source signal of the interference, the data files of the test set of the TIDigits data base are arbitrarily permuted. The source signals are filtered by the room impulse responses between the position of the interference (endfire position, 1.2 m distance from the array center) and the sensor positions. Internal filter states are preserved between the files in order to have a continuous stream of sensor data. Then, the contribution of the desired signal and the contribution of the interference are added with the desired $SINR_{\text{in}}$, yielding the sensor signals. Due to the 'continuous' stream of data files for the interference and due to equal length of desired speech and interference, pauses of interference do (almost) not appear, and we have a continuous double-talk situation.

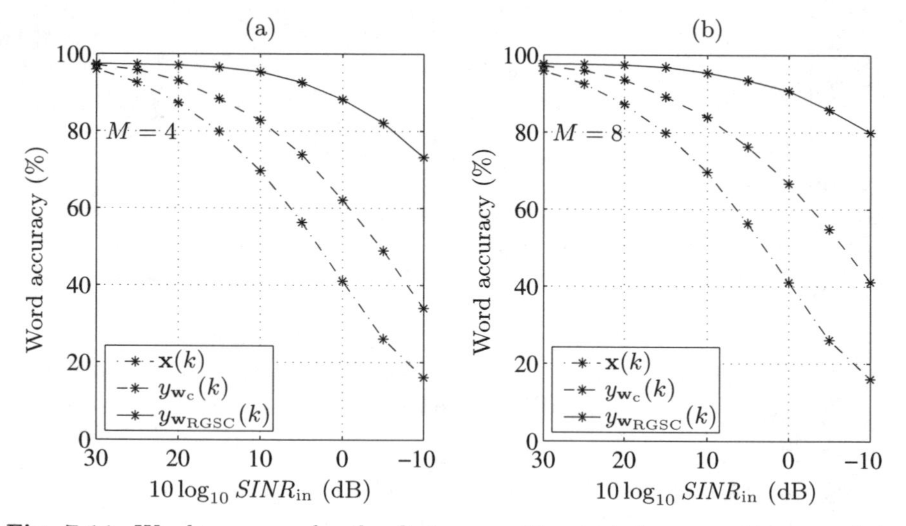

Fig. 7.11. Word accuracy for the digit recognition task for competing speech in the multimedia room ($T_{60} = 250$ ms) over the average SINR at the sensors, $SINR_{\text{in}}$; input signals of the ASR: microphone signals $\mathbf{x}(k)$, output signal $y_{\mathbf{w}_c}(k)$ of the fixed beamformer, and output signal $y_{\mathbf{w}_{\text{RGSC}}}(k)$ of the RGSC

The recognition results are illustrated in Fig. 7.11 and in Table 7.11 for $M \in \{4, 8\}$. Figure 7.12 shows the interference rejection $IR_{\mathbf{w}_{\text{RGSC}}}$ of the RGSC, the interference rejection $IR_{\mathbf{w}_c}$ of the fixed beamformer, and the cancellation $DC_{\mathbf{B}_s}$ of the desired signal by the blocking matrix. Compared to the car environment, we notice an even higher word accuracy of the ASR for the RGSC: The word accuracy is greater than 90% for $10 \log_{10} SINR_{\text{in}} \geq 5$ dB ($M = 4$) and for $10 \log_{10} SINR_{\text{in}} \geq 0$ dB ($M = 8$). The absolute word accuracy is thus greater for the multimedia room with competing speech than for

Table 7.11. Word accuracy (%) in the multimedia room ($T_{60} = 250$ ms) with competing speech as a function of $SINR_{\text{in}}$ **(a)** for $M = 4$ and **(b)** for $M = 8$

$10\log_{10} SINR_{\text{in}}$ [dB]	(a) $\mathbf{x}(k)$	(a) $y_{\mathbf{w}_c}(k)$	(a) $y_{\mathbf{w}_{\text{RGSC}}}(k)$	(b) $\mathbf{x}(k)$	(b) $y_{\mathbf{w}_c}(k)$	(b) $y_{\mathbf{w}_{\text{RGSC}}}(k)$
30	95.92	97.10	97.42	95.80	97.10	97.65
25	92.56	95.83	97.24	92.47	95.93	97.58
20	87.30	93.04	97.05	87.25	93.53	97.38
15	79.91	88.37	96.50	79.83	89.15	96.85
10	69.68	82.75	95.30	69.61	83.88	95.37
5	56.28	73.84	92.56	56.36	76.24	93.42
0	41.07	62.08	88.17	41.09	66.70	90.72
-5	26.13	48.97	82.09	26.13	54.85	85.81
-10	16.15	34.06	73.20	16.05	41.17	79.92

the car environment with car noise. This results from the fact that the RGSC is especially effective for non-stationary signals and from the fact that wide-band slowly time-varying or stationary noise signals affect the word accuracy of ASRs more than speech signals with varying frequency contents [Jun00]. The relative improvement to a single array microphone is up to 353.32% for $M = 4$ and up to 397.98% for $M = 8$ for $10\log_{10} SINR_{\text{in}} = -10$ dB. The relative improvement to the fixed beamformer is up to 114.91% for $M = 4$ and up to 94.12% for $M = 8$ for $10\log_{10} SINR_{\text{in}} = -10$ dB. As for the car environment, we see that the maximum relative improvement is obtained for low values of $SINR_{\text{in}}$.

In Fig. 7.12, it may be seen that $IR_{\mathbf{w}_{\text{RGSC}}} > 13$ dB for $M = 4$ and $IR_{\mathbf{w}_{\text{RGSC}}} > 16$ dB for $M = 8$ for a wide range of $SINR_{\text{in}}$ ($10\log_{10} SINR_{\text{in}} < 15$ dB). For the same range of $SINR_{\text{in}}$, the cancellation $DC_{\mathbf{B}_s}$ of the desired signal by the blocking matrix is greater than 17 dB, and does not decrease with decreasing $SINR_{\text{in}}$ in contrast to the car environment due to the non-stationarity of the interfering speech signal. $DC_{\mathbf{B}_s}$ rather increases with decreasing $SINR_{\text{in}}$. This effect can be explained by the parameter setup of the RGSC which is optimized for $10\log_{10} SINR_{\text{in}} = 0$ dB.

As for the car environment, $IR_{\mathbf{w}_{\text{RGSC}}}$ can be interpreted as an improvement of $SINR_{\text{in}}$ for the ASR. Consider, for example, $10\log_{10} SINR_{\text{in}} = 15$ dB for $M = 8$ (Fig. 7.11b), where the word accuracy is equal to 79.83%. The same word accuracy (79.92%) is obtained with the RGSC for $10\log_{10} SINR_{\text{in}} = -10$ dB. The improvement of $SINR_{\text{in}}$ is thus 25 dB for this example. However, $IR_{\mathbf{w}_{\text{RGSC}}}$ is only equal to 17 dB for $10\log_{10} SINR_{\text{in}} = -10$ dB (Fig. 7.12b), so that the improvement of $SINR_{\text{in}}$ by 17 dB yields a greater improvement of the word accuracy than expected. This can be explained by the non-stationarity of the competing speech signals and the fact that $SINR_{\text{in}}$ is obtained by averaging over the whole length of the test data set, which contains speech pauses.

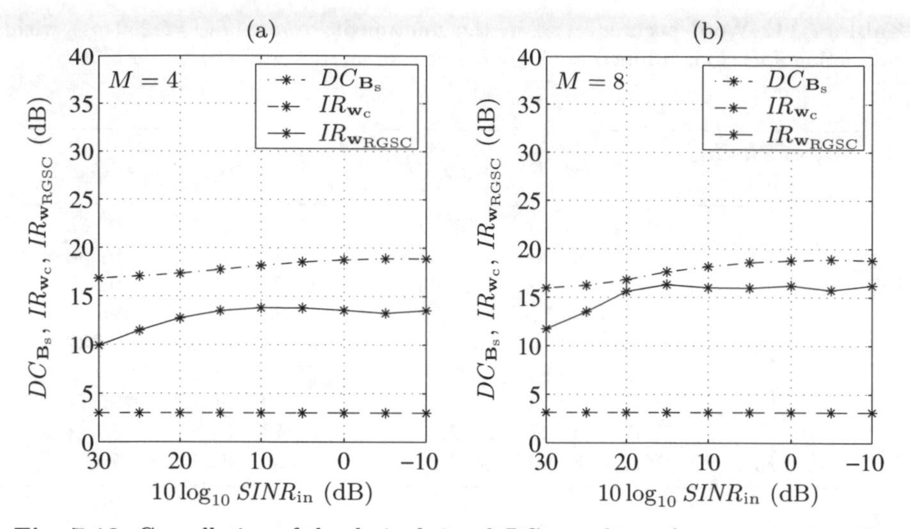

Fig. 7.12. Cancellation of the desired signal $DC_{\mathbf{B}_s}$ and interference rejection $IR_{\mathbf{w}_c}$ and $IR_{\mathbf{w}_{\mathrm{RGSC}}}$ over the average SINR at the sensors, $SINR_{\mathrm{in}}$, in the multimedia room ($T_{60} = 250$ ms) with competing speech

The interference rejection during activity of the interferer is thus greater than the average interference rejection $IR_{\mathbf{w}_{\mathrm{RGSC}}}$. This problem may be resolved by measuring the interference rejection of speech signals using averaging over the objective 'active' competing speech level [ITU93] instead of over the total length of the speech data.

In summary, the speech recognition experiments show that the proposed RGSC is well suited as a front-end for an automatic speech recognizer in such different acoustic conditions as passenger cabins of cars with presence of slowly time-varying car noise or reverberant rooms with presence of competing speech signals.

7.4 Discussion

In this chapter, we first introduced DFT-domain adaptive MIMO filtering as a powerful method for meeting the requirements of adaptive beamformers and multi-channel acoustic echo cancellers. Second, we applied this adaptive filtering method to the combinations of the RGSC and AECs. The decrosscorrelating property of multi-channel DFT-domain adaptive filters is used for maximum convergence speed. The direct access to distinct frequency bins of the DFT is used for the frequency-selective adaptation of the RGSC. Furthermore, the realization of the RGSC combined with acoustic echo cancellation in the DFT domain is highly computationally efficient compared to a realization in the time domain. Finally, we studied the performance of the proposed system experimentally. The experiments showed that the RGSC can be used with the same parameter setup in a wide variety of acoustic environments with highly varying SINRs and noise/coherence properties. The RGSC efficiently suppresses interference and noise while preserving a high quality of the desired signal. Especially, the output signal quality is sufficiently high to greatly improve speech recognition performance for hands-free scenarios even in very adverse conditions without retraining the speech recognizer for the characteristics of the acoustic front-end.

Audio examples which illustrate the performance of the real-time system can be found at www.wolfgangherbordt.de/micarraybook/.

8

Summary and Conclusions

Convenient human/machine interaction requires acoustic front-ends which allow seamless and hands-free audio communication. For suppressing interference, noise, and acoustic echoes of loudspeakers, interference/noise-reduction and acoustic echo cancellation should be integrated into the human/machine interface for maximum output signal quality and for optimum performance of speech recognizers.

Obviously, for suppressing interference and noise, spatio-temporal filtering is preferable over temporal filtering alone, because desired signal and interference often overlap in time and/or frequency but originate from different spatial coordinates. Spatio-temporal filtering by beamforming allows separation without distortion of the desired signal. In practical situations, beamforming microphone arrays have to cope with such diversified acoustic conditions as slowly time-varying and strongly time-varying wideband speech and audio signals, highly variable reverberation times, different noise/coherence conditions, and, finally, with restrictions in geometry, size, and number of sensors due to product design constraints. Adaptive optimum beamformers are desirable, since they optimally extract the desired signal from interference and noise for the observed sensor signals and for a given array geometry.

In this work, we studied adaptive optimum beamforming with a focus on the problems of real-world systems. In contrast to many traditional presentations, we explicitly dropped the narrowband and the stationarity assumptions and formulated optimum data-dependent beamformers using time-domain LS instead of frequency-domain MMSE criteria. We thus obtained a more general representation of various beamforming aspects which are relevant to our application. The relation to narrowband beamforming was illustrated.

The unified framework is based on optimum MIMO linear filtering using the LSE criterion in the discrete-time domain for optimization. In Chap. 3, we formally extended optimum MISO linear filtering to the MIMO case based on [BBK03]. This formalism was used throughout the entire work for describing the multi-channel systems. Particularly, we analyzed in Chap. 4 traditional LSE/MMSE beamformers and LCLSE/LCMV beamformers for their application in real-world systems for speech and audio signal processing and rigorously interpreted the beamforming methods in the vector space and in the eigenspace of the sensor signals. By doing so, it was possible to illustrate the influence of the spatial and temporal degrees of freedom on the properties of the LSE beamformer for non-stationary wideband signals, analogously to the interpretation of the MMSE beamformer in the DTFT domain according to [EFK67, SBM01].

The discussion of data-dependent optimum beamforming showed the optimality of LCLSE beamforming using the structure of the generalized sidelobe canceller [GJ82] for our scenario because of computational efficiency and because reference signals and, ideally, spatio-spectral correlation matrices w.r.t. the desired signal and w.r.t. interference-plus-noise are not required. However, LCLSE beamformers with traditional spatial constraints cannot be used for reverberant acoustic environments, since they do not account for reverberation w.r.t. the desired signal, but lead to cancellation of the desired signal at the output of the LCLSE beamformer instead. Therefore, we presented in Chap. 5 time-varying spatio-temporal constraints which allow to model reverberation and moving desired sources, and which take temporal characteristics of the source signals into account. By rigorously deriving a GSC using the spatio-temporal constraints, we obtained the robust GSC after [HSH99], which was not related to LCLSE/LCMV beamforming so far. Translating the spatio-temporal constraints to the special case of WSS signals in the DTFT domain, we found equivalence of the traditional spatial constraints with the proposed spatio-temporal constraints. Moreover, the DTFT-domain description showed formal equivalence of the GSC after [HSH99] and the GSC after [GBW01].

The blocking matrix and the interference canceller of the RGSC (as well as of other GSC realizations) can be interpreted as SIMO and MISO optimum linear filters. For the blocking matrix, these interpretations suggested other realizations. Then, they explained the dependency of the performance of the blocking matrix on the position of the desired source relative to the steering direction of the sensor array. Furthermore, they illustrated the behavior of the blocking matrix for presence of the desired signal and interference during adaptation. For the interference canceller, we found a relation between the interference rejection, the number of sensors, the number of interferers, and the filter lengths. These new insights cannot be obtained by a discussion in the DTFT domain.

The blocking matrix and the interference canceller should only be adapted during presence of the desired signal only and during presence of interference and noise only for maximum robustness against cancellation of the desired signal by the RGSC as suggested in, e.g., [Com90, HSH99]. However, a decision in the discrete time domain fails, since transient signal characteristics (and a transient blocking matrix) cannot be tracked during double-talk. For solving this problem, we proposed an adaptation in the eigenspace of the sensor signals, where (almost) simultaneous adaptation of the blocking matrix and of the interference canceller is possible.

Practical multi-channel audio acquisition systems for full-duplex applications require the combination of sensor array processing with acoustic echo cancellation such that optimum synergies are obtained. However, especially for systems with more than one reproduction channel and/or large numbers of sensors ($M \geq 4$), computational complexity has to be considered for real-world systems. Therefore, we studied in Chap. 6 various possibilities for combining beamforming and multi-channel acoustic echo cancellation based on [Kel97, HK00, Kel01] for the example of the RGSC. It was shown that for slowly time-varying echo paths optimum synergies – but with prohibitively high computational complexity for cost-sensitive systems – can be obtained with the AECs in the sensor channels ('AEC first'). The need for combined systems with less complexity lead to the RGSC with the AEC integrated into the fixed reference path (GSAEC). While providing for predominant acoustic echoes nearly the same performance as 'AEC first' with the equivalent of a single AEC for an arbitrary number of sensors, the spatial degrees of freedom for interference suppression is not increased relative to the RGSC alone. Therefore, we proposed the integration of the AEC into the interference canceller of the RGSC (GEIC). The number of spatial degrees of freedom of the RGSC for interference suppression corresponds to that of 'AEC first'. Additionally, this combination provides the advantage of simultaneous adaptation of the interference canceller and of the AEC if the proposed adaptation strategy is used. The AEC can be adapted during double-talk of the desired signal, interference-plus-noise, and acoustic echoes using the adaptation strategy in the eigenspace of the sensors signals. As a consequence, the high tracking capability of the interference canceller translates to the AEC, and a separate adaptation control is not required for the AEC. For strongly time-varying echo paths and for frequent double-talk, GEIC outperforms 'AEC first' due to the better tracking capability. However, the performance of GEIC after convergence of the AEC filters is reduced, since the AEC filters should not be longer than the filters of the interference canceller for optimum convergence, which reduces the number of temporal degrees of freedom of the AEC relative to 'AEC first'.

The block-based least-squares formulation of optimum data-dependent wideband beamformers and the short-time stationarity of speech and audio signals suggest to transform the RGSC combined with acoustic echo cancellation into the discrete Fourier transform domain for maximum computational efficiency for real-time systems. Therefore, DFT-domain realizations for the combinations of the RGSC with AECs are derived in Chap. 7 by rigorously applying multi-channel DFT-domain adaptive filtering after [BM00, BBK03] with an extension to constrained adaptive filtering in the DFT domain using penalty functions. Multi-channel DFT-domain adaptive filters meet well the requirements of both multi-channel acoustic echo cancellation and the RGSC: The de-crosscorrelating property is required by the AEC for maximum convergence speed in the presence of mutually strongly correlated loudspeaker signals. The direct access to distinct frequency bins of the DFT is exploited for the frequency-selective adaptation of the blocking matrix and of the interference canceller of the RGSC, which is required for solving the adaptation problems of the sidelobe cancelling path in an efficient way.

The frequency-selective adaptation of the RGSC requires a double-talk detector for classifying DFT bins into 'presence of the desired signal only', 'presence of interference-plus-noise only', and 'double-talk'. Therefore, we proposed in App. A a new technique which efficiently performs this classification for mixtures of both slowly time-varying and strongly time-varying desired signals and interference-plus-noise. This method was extended to estimate signal-to-interference-plus-noise ratios and spatial coherence functions for realizing LSE/MMSE beamformers for interference-plus-noise with strongly time-varying PSDs.

Experimental results confirm that the proposed system efficiently suppresses interference and noise while preserving a high output signal quality of the desired signal for such different acoustic conditions as passenger cabins of cars with presence of car noise and reverberant office environments with competing speech signals. The output signal quality is even sufficiently high to greatly improve speech recognition performance compared to far-field microphones or conventional fixed beamformers without retraining the speech recognizer for the characteristics of the acoustic front-end. As it is important for real-world systems, the same parameter setup of the RGSC can be used in a wide variety of acoustic environments with highly varying SINRs and noise/coherence properties.

Though the proposed acoustic human/machine front-end demonstrates the feasibility of high-quality multi-channel speech capture in adverse acoustic conditions using adaptive filtering, there are still particularly challenging problems to be addressed: One major problem is dereverberation, since beamformers only partially dereverberate the desired signal by increasing the power ratio between direct path and reverberation. While speech intelligibility [Kut91] and small-vocabulary automatic speech recognizers are less affected, large-vocabulary ASRs are very sensitive to reverberation [GOS96, Jun00].

A

Estimation of Signal-to-Interference-Plus-Noise Ratios (SINRs) Exploiting Non-stationarity

Our discussion about optimum data-dependent beamforming has shown that the second-order statistics of the sensor signals w.r.t. the desired signal and w.r.t. interference-plus-noise are required for realizing robust optimum data-dependent beamformers. For realizations in the DFT domain, especially, estimates of power spectral densities and of spatio-spectral correlation matrices w.r.t. the desired signal and w.r.t. interference-plus-noise are necessary. Consider, e.g., the optimum MMSE beamformer in the DTFT domain after (4.110) that requires the PSD and the spatio-spectral correlation matrix of interference-plus-noise. Such estimates can be obtained with, e.g., the minimum statistics after [Mar01a] using the generalization of [Bit02] for estimating spatio-spectral correlation matrices w.r.t. interference-plus-noise. However, these methods assume a slowly time-varying PSD of interference-plus-noise relative to a strongly time-varying PSD of the desired signal.

For mixtures of signals with strongly time-varying PSDs, these methods cannot be used, so that application of MMSE beamforming would be limited to situations with slowly time-varying wave fields of interference-plus-noise. However, as illustrated in Sect. 4.4.1, the optimum MMSE beamformer can be rewritten as a function of the frequency-dependent SINR at the sensors and as a function of the spatial coherence matrix w.r.t. interference-plus-noise. (See (4.115).) LCMV beamformers can be transformed in a similar way. For the realization of optimum data-dependent beamformers, thus, estimates of the frequency-dependent SINR at the sensors and estimates of the spatial coherence matrix w.r.t. interference-plus-noise are often sufficient for practical realizations.

For bridging this gap, we introduce in this chapter a new multi-channel method for estimating the DFT bin-wise SINR at the sensors for highly non-stationary PSDs and the spatial coherence functions of the desired signal and of interference-plus-noise between the sensors. Our approach thus relaxes the assumption of slowly time-varying PSDs of the interference relative to the PSD of the desired signal. The proposed algorithm requires a classifier for 'presence of desired signal only', 'presence of interference-plus-noise only', and 'double-talk between the desired signal and interference-plus-noise'. This classifier will be called *double-talk detector* in the following. It is used in Chap. 7 for controlling the adaptation of the RGSC. We assume that the position of the desired source is known within a given tolerance and that the spatial coherence functions are slowly time-varying relative to the PSDs. For mixtures of speech and/or music signals, this assumption is generally fulfilled. The described method was first published in [HTK03] based on [HSH99].

This chapter is organized as follows: In Sect. A.1, the biased SINR estimation is described. Section A.2 presents the method for correcting the bias. Section A.3 explains the resulting algorithm and gives experimental results.

A.1 Biased Estimation of the SINR Using Spatial Information

In this section, we describe the method for biased SINR estimation. Section A.1.1 introduces the principle of the proposed technique in the discrete time domain. The transition into the DTFT domain is performed in Sect. A.1.2. The properties of the biased SINR estimate are illustrated in Sect. A.1.3.

A.1.1 Principle

We assume presence of a desired signal and of interference-plus-noise so that the sensor signals are given by $\mathbf{x}(k) = \mathbf{d}(k) + \mathbf{n}(k)$ according to (4.28). A biased estimate of the SINR at the sensors is obtained as follows [HSH99]: The output signal $y(k) = \mathbf{w}^T\mathbf{x}(k)$ after (4.42) of a fixed beamformer $\mathbf{w}$, which is steered to the position of the desired source, can be interpreted as a spatially averaged estimate of the desired signal at the sensors. Accordingly, the output signals $\bar{y}_m(k)$ of *complementary fixed beamformers*

$$\bar{y}_m(k) = x_m(k-\kappa) - \mathbf{w}^T\mathbf{x}(k)\,, \tag{A.1}$$

$m = 0, 1, \ldots, M-1$, may be interpreted as estimates of interference-plus-noise, since the estimate $y(k) = \mathbf{w}^T\mathbf{x}(k)$ of the desired signal is subtracted from the sensor signals, where desired signal and interference-plus-noise is present. The delay κ synchronizes the output signal of the fixed beamformer with the sensor signals. The ratio

$$\upsilon(k) = \frac{\sigma_y^2(k)}{\sigma_{\bar{y}}^2(k)} \tag{A.2}$$

of the variance $\sigma_y^2(k)$ of the beamformer output $y(k)$,

$$\sigma_y^2(k) = \mathcal{E}\{\left(\mathbf{w}^T\mathbf{x}(k)\right)^2\} = \mathbf{w}^T\mathbf{R}_{\mathbf{xx}}(k)\mathbf{w}\,, \tag{A.3}$$

where $\mathbf{R}_{\mathbf{xx}} = \mathcal{E}\{\mathbf{x}(k)\mathbf{x}^T(k)\}$ according to (2.31), and the variance $\sigma_{\bar{y}}^2(k)$ of the output signals $\bar{y}_m(k)$ of the complementary fixed beamformers,

$$\begin{aligned}\sigma_{\bar{y}}^2(k) &= \frac{1}{M}\sum_{m=0}^{M-1}\mathcal{E}\{\bar{y}_m^2(k)\} = \frac{1}{M}\sum_{m=0}^{M-1}\mathcal{E}\left\{\left(x_m(k) - \mathbf{w}^T\mathbf{x}(k)\right)^2\right\}\\ &= \frac{1}{M}\sum_{m=0}^{M-1}\left[\mathcal{E}\{x_m^2(k)\} - \mathcal{E}\{x_m(k)\mathbf{x}^T(k)\}\mathbf{w} - \mathbf{w}^T\mathcal{E}\{\mathbf{x}(k)x_m(k)\}+\right.\\ &\quad\left.+\mathbf{w}^T\mathbf{R}_{\mathbf{xx}}(k)\mathbf{w}\right], \qquad (A.4)\end{aligned}$$

may thus be interpreted as an estimate of the SINR at the sensors, $SINR_{\text{in}}(k)$, after (4.64). If $\mathbf{w}$ completely suppresses interference-plus-noise and if $\mathbf{w}$ does not attenuate the desired signal, $\upsilon(k)$ is equal to $SINR_{\text{in}}(k)$. Otherwise, $\upsilon(k)$ only provides a biased estimate of $SINR_{\text{in}}(k)$.

A.1.2 Biased SINR Estimation in the DTFT Domain

For applying the SINR estimation to time-varying wideband signals in the frequency domain using a narrowband decomposition, we transform in this section the time-domain description into the DTFT domain. We assume $\kappa = 0$ for formal simplicity.

Assuming narrowband signals, we can write $\upsilon(k)$ after (A.2) in the DTFT domain as

$$\Upsilon(\omega) = \frac{S_{yy}(\omega)}{S_{\bar{y}\bar{y}}(\omega)}, \qquad (A.5)$$

where $S_{yy}(\omega)$ and $S_{\bar{y}\bar{y}}(\omega)$ are the PSD of the output signal of the fixed beamformer $y(k)$ and the average PSD of the output signals of the complementary fixed beamformers $\bar{y}_m(k)$, respectively, i.e.,

$$S_{yy}(\omega) = \mathbf{w}_{\text{f}}^H(\omega)\mathbf{S}_{\mathbf{xx}}(\omega)\mathbf{w}_{\text{f}}(\omega), \qquad (A.6)$$

$$S_{\bar{y}\bar{y}}(\omega) = \frac{1}{M}\left[\text{tr}\{\mathbf{S}_{\mathbf{xx}}(\omega)\} - \mathbf{1}_{M\times 1}^T\mathbf{S}_{\mathbf{xx}}(\omega)\mathbf{w}_{\text{f}}(\omega) - \mathbf{w}_{\text{f}}^H(\omega)\mathbf{S}_{\mathbf{xx}}(\omega)\mathbf{1}_{M\times 1}\right] \qquad (A.7)$$

where $\mathbf{S}_{\mathbf{xx}}(\omega)$ and $\mathbf{w}_{\text{f}}(\omega)$ are defined in (2.34) and in (4.46), respectively. For mutually uncorrelated desired signal and interference-plus-noise, the spatio-spectral correlation matrix $\mathbf{S}_{\mathbf{xx}}(\omega)$ may be separated according to (4.38) as $\mathbf{S}_{\mathbf{xx}}(\omega) = \mathbf{S}_{\mathbf{dd}}(\omega) + \mathbf{S}_{\mathbf{nn}}(\omega)$, and (A.6) and (A.7) may be written as

$$S_{yy}(\omega) = \mathbf{w}_{\text{f}}^H(\omega)\mathbf{S}_{\mathbf{dd}}(\omega)\mathbf{w}_{\text{f}}(\omega) + \mathbf{w}_{\text{f}}^H(\omega)\mathbf{S}_{\mathbf{nn}}(\omega)\mathbf{w}_{\text{f}}(\omega), \qquad (A.8)$$

$$\begin{aligned}S_{\bar{y}\bar{y}}(\omega) &= \frac{1}{M}\left[\text{tr}\{\mathbf{S}_{\mathbf{dd}}(\omega)\} - \mathbf{1}_{M\times 1}^T\mathbf{S}_{\mathbf{dd}}(\omega)\mathbf{w}_{\text{f}}(\omega) - \mathbf{w}_{\text{f}}^H(\omega)\mathbf{S}_{\mathbf{dd}}(\omega)\mathbf{1}_{M\times 1}+\right.\\ &\quad+M\mathbf{w}_{\text{f}}^H(\omega)\mathbf{S}_{\mathbf{dd}}(\omega)\mathbf{w}_{\text{f}}(\omega) + \text{tr}\{\mathbf{S}_{\mathbf{nn}}(\omega)\} - \mathbf{1}_{M\times 1}^T\mathbf{S}_{\mathbf{nn}}(\omega)\mathbf{w}_{\text{f}}(\omega)-\\ &\quad\left.-\mathbf{w}_{\text{f}}^H(\omega)\mathbf{S}_{\mathbf{nn}}(\omega)\mathbf{1}_{M\times 1} + M\mathbf{w}_{\text{f}}^H(\omega)\mathbf{S}_{\mathbf{nn}}(\omega)\mathbf{w}_{\text{f}}(\omega)\right], \qquad (A.9)\end{aligned}$$

respectively. We next assume spatially homogeneous wave fields. The spatio-spectral correlation matrix $\mathbf{S}_{\mathbf{xx}}(\omega)$ can then be separated as

$$\mathbf{S}_{\mathbf{xx}}(\omega) = \mathbf{S}_{\mathbf{dd}}(\omega) + \mathbf{S}_{\mathbf{nn}}(\omega) = S_{dd}(\omega)\mathbf{\Gamma}_{\mathbf{dd}}(\omega) + S_{nn}(\omega)\mathbf{\Gamma}_{\mathbf{nn}}(\omega)\,, \tag{A.10}$$

using (4.8) and (4.37). It follows for the PSDs $S_{yy}(\omega)$ and $S_{\bar{y}\bar{y}}(\omega)$ after (A.8) and (A.9) the correspondences

$$S_{yy}(\omega) = S_{dd}(\omega)F(\mathbf{\Gamma}_{\mathbf{dd}}(\omega)) + S_{nn}(\omega)F(\mathbf{\Gamma}_{\mathbf{nn}}(\omega))\,, \tag{A.11}$$
$$S_{\bar{y}\bar{y}}(\omega) = S_{dd}(\omega)\bar{F}(\mathbf{\Gamma}_{\mathbf{dd}}(\omega)) + S_{nn}(\omega)\bar{F}(\mathbf{\Gamma}_{\mathbf{nn}}(\omega))\,, \tag{A.12}$$

respectively, where

$$F(\mathbf{\Gamma}_{\mathbf{xx}}(\omega)) = \mathbf{w}_{\mathrm{f}}^H(\omega)\mathbf{\Gamma}_{\mathbf{xx}}(\omega)\mathbf{w}_{\mathrm{f}}(\omega)\,, \tag{A.13}$$
$$\bar{F}(\mathbf{\Gamma}_{\mathbf{xx}}(\omega)) = 1 - \frac{1}{M}\left[\mathbf{1}_{M\times 1}^T\mathbf{\Gamma}_{\mathbf{xx}}(\omega)\mathbf{w}_{\mathrm{f}}(\omega) + \mathbf{w}_{\mathrm{f}}^H(\omega)\mathbf{\Gamma}_{\mathbf{xx}}(\omega)\mathbf{1}_{M\times 1}\right] + + \mathbf{w}_{\mathrm{f}}^H(\omega)\mathbf{\Gamma}_{\mathbf{xx}}(\omega)\mathbf{w}_{\mathrm{f}}(\omega)\,. \tag{A.14}$$

The matrix $\mathbf{\Gamma}_{\mathbf{xx}}(\omega)$ is the spatial coherence matrix of the sensor signals according to (2.35), i.e., $\mathbf{S}_{\mathbf{xx}}(\omega) = S_{xx}(\omega)\mathbf{\Gamma}_{\mathbf{xx}}(\omega)$. The functions $F(\mathbf{\Gamma}_{\mathbf{xx}}(\omega))$ and $\bar{F}(\mathbf{\Gamma}_{\mathbf{xx}}(\omega))$ represent the PSDs $S_{yy}(\omega)$ and $S_{\bar{y}\bar{y}}(\omega)$ of the output signals of the fixed beamformer and of the complementary fixed beamformer normalized by the PSD of the sensor signals $S_{xx}(\omega)$, respectively.

The SINR at the sensors, $SINR_{\mathrm{f,in}}(\omega)$, for monochromatic signals in the DTFT domain is defined by (4.69). For homogeneous wave fields, we obtain with (4.8) and with (4.37)

$$SINR_{\mathrm{f,in}}(\omega) = \frac{\mathrm{tr}\,\{\mathbf{S}_{\mathbf{dd}}(\omega)\}}{\mathrm{tr}\,\{\mathbf{S}_{\mathbf{nn}}(\omega)\}} = \frac{S_{dd}(\omega)}{S_{nn}(\omega)}\,. \tag{A.15}$$

Replacing in (A.5) the PSDs $S_{yy}(\omega)$ and $S_{\bar{y}\bar{y}}(\omega)$ by (A.11) and by (A.12), respectively, and dividing the numerator and the denominator by $S_{nn}(\omega)$, we obtain with (A.15) the expression

$$\Upsilon(\omega) = \frac{SINR_{\mathrm{f,in}}(\omega)F(\mathbf{\Gamma}_{\mathbf{dd}}(\omega)) + F(\mathbf{\Gamma}_{\mathbf{nn}}(\omega))}{SINR_{\mathrm{f,in}}(\omega)\bar{F}(\mathbf{\Gamma}_{\mathbf{dd}}(\omega)) + \bar{F}(\mathbf{\Gamma}_{\mathbf{nn}}(\omega))}\,. \tag{A.16}$$

A.1.3 Illustration

Consider for illustration the special case of a uniformly weighted beamformer, i.e., $\mathbf{w}_{\mathrm{f}} = 1/M \cdot \mathbf{1}_{M\times 1}$. Equation (A.14) then simplifies to

$$\bar{F}(\mathbf{\Gamma}_{\mathbf{xx}}(\omega)) = 1 - F(\mathbf{\Gamma}_{\mathbf{xx}}(\omega))\,, \tag{A.17}$$

which justifies the term ‘complementary fixed beamformer’. Note that (A.17) holds for arbitrarily steered uniformly weighted fixed beamformers, if the steering delays are separated from the beamformer weights. We obtain with (A.11), (A.12), and (A.17) for $\Upsilon(\omega)$ the expression

$$\Upsilon(\omega) = \frac{SINR_{\mathrm{f,in}}(\omega) F(\mathbf{\Gamma}_{\mathbf{dd}}(\omega)) + F(\mathbf{\Gamma}_{\mathbf{nn}}(\omega))}{SINR_{\mathrm{f,in}}(\omega) \left[1 - F(\mathbf{\Gamma}_{\mathbf{dd}}(\omega))\right] + \left[1 - F(\mathbf{\Gamma}_{\mathbf{nn}}(\omega))\right]} . \tag{A.18}$$

For homogeneous wave fields, the array gain $A_{\mathrm{f}}(\omega)$ after (4.71) with can be simplified as

$$A_{\mathrm{f}}(\omega) = \frac{F(\mathbf{\Gamma}_{\mathbf{dd}}(\omega))}{F(\mathbf{\Gamma}_{\mathbf{nn}}(\omega))} . \tag{A.19}$$

Assuming $F(\mathbf{\Gamma}_{\mathbf{dd}}(\omega)) = 1$ (no steering errors), normalizing the right side of (A.18) by $F(\mathbf{\Gamma}_{\mathbf{nn}}(\omega))$, and introducing the array gain after (A.19) into (A.18), it follows:

$$\Upsilon(\omega) = \frac{SINR_{\mathrm{f,in}}(\omega) A_{\mathrm{f}}(\omega) + 1}{A_{\mathrm{f}}(\omega) - 1} . \tag{A.20}$$

We see that $\Upsilon(\omega)$ converges to the real SINR at the sensors for $A_{\mathrm{f}}(\omega) \to \infty$. For $SINR_{\mathrm{f,in}}(\omega) \to 0$, $\Upsilon(\omega)$ becomes independent of the sensor SINR, i.e., $\Upsilon(\omega) = (A_{\mathrm{f}}(\omega) - 1)^{-1}$, which means that the bias increases with decreasing $SINR_{\mathrm{f,in}}(\omega)$. For $A_{\mathrm{f}}(\omega) \to 1$, the ratio $\Upsilon(\omega)$ has a pole. Figure A.1 illustrates the influence of the array gain on the accuracy of the estimate of the SINR at the sensors over frequency for various values of $SINR_{\mathrm{f,in}}(\omega)$. We assume an array with $M = 8$ sensors with inter-sensor spacing $d = 4\,\mathrm{cm}$. The desired signal arrives from broadside. In Fig. A.1a, a directional plane-wave interferer from $\theta_{\mathrm{n}} = \pi/3$ is assumed. In Fig. A.1b diffuse interference is present. We clearly see the characteristics which were described above.

For this special case of a uniformly weighted beamformer, the SINR estimate is thus biased by a function of the array gain $A_{\mathrm{f}}(\omega)$. More generally, (A.16) shows that $\Upsilon(\omega)$ provides an estimate of $SINR_{\mathrm{f,in}}(\omega)$ which is biased by functions of $F(\mathbf{\Gamma}_*(\omega))$ and of $\bar{F}(\mathbf{\Gamma}_*(\omega))$, where $* \in \{\mathbf{dd}, \mathbf{nn}\}$.

A.2 Unbiased SINR Estimation Using Spatial Coherence Functions

In this section, the principle of the unbiased SINR estimation is described. In Sect. A.1, it was shown that $\Upsilon(\omega)$ provides an estimate of $SINR_{\mathrm{f,in}}(\omega)$ which is biased by functions of $F(\mathbf{\Gamma}_*(\omega))$ and of $\bar{F}(\mathbf{\Gamma}_*(\omega))$, where $* \in \{\mathbf{dd}, \mathbf{nn}\}$. Therefore, we solve (A.16) for $SINR_{\mathrm{f,in}}(\omega)$, and we obtain the expression

$$SINR_{\mathrm{f,in}}(\omega) = \frac{F(\mathbf{\Gamma}_{\mathbf{nn}}(\omega)) - \Upsilon(\omega) \bar{F}(\mathbf{\Gamma}_{\mathbf{nn}}(\omega))}{\Upsilon(\omega) \bar{F}(\mathbf{\Gamma}_{\mathbf{dd}}(\omega)) - F(\mathbf{\Gamma}_{\mathbf{dd}}(\omega))} . \tag{A.21}$$

We see that $SINR_{\mathrm{f,in}}(\omega)$ can be calculated from $\Upsilon(\omega)$ without bias, if $F(\mathbf{\Gamma}_*(\omega))$ and $\bar{F}(\mathbf{\Gamma}_*(\omega))$, where $* \in \{\mathbf{dd}, \mathbf{nn}\}$, are known. For calculating $F(\mathbf{\Gamma}_*(\omega))$ and $\bar{F}(\mathbf{\Gamma}_*(\omega))$, $* \in \{\mathbf{dd}, \mathbf{nn}\}$, we introduce the ratio $\Psi(\omega)$ between the PSD $S_{yy}(\omega)$

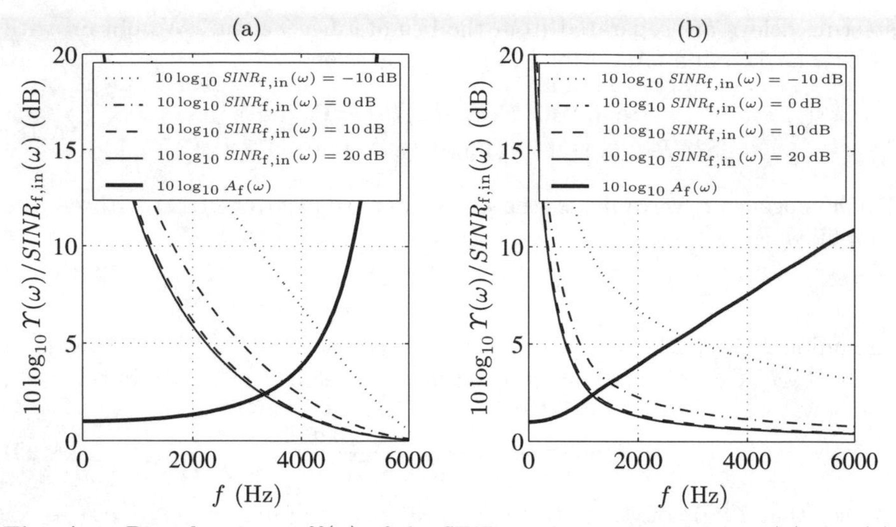

Fig. A.1. Biased estimate $\Upsilon(\omega)$ of the SINR at the sensors, $SINR_{\mathrm{f,in}}(\omega)$, for **(a)** directional interference from $\theta_{\mathrm{n}} = \pi/3$ and for **(b)** diffuse interference for a uniformly weighted beamformer with broadside steering without steering mismatch; $M = 8$, $d = 4\,\mathrm{cm}$

of the output signal of the fixed beamformer and the PSD $S_{xx}(\omega)$ of the sensor signals as

$$\Psi(\omega) = \frac{S_{yy}(\omega)}{S_{xx}(\omega)}\,, \tag{A.22}$$

where $S_{xx}(\omega)$ is given by

$$S_{xx}(\omega) = \frac{1}{M}\mathrm{tr}\{\mathbf{S}_{\mathbf{xx}}(\omega)\} = \frac{1}{M}\mathrm{tr}\{\mathbf{S}_{\mathbf{dd}}(\omega) + \mathbf{S}_{\mathbf{nn}}(\omega)\} = S_{dd}(\omega) + S_{nn}(\omega)\,. \tag{A.23}$$

Introducing (A.11) and (A.23) into (A.22), we see that $F(\mathbf{\Gamma}_{\mathbf{dd}}(\omega))$ and $F(\mathbf{\Gamma}_{\mathbf{nn}}(\omega))$ can be obtained by

$$F(\mathbf{\Gamma}_{\mathbf{dd}}(\omega)) = \Psi(\omega)\big|_{S_{nn}(\omega)=0}\,, \tag{A.24}$$

$$F(\mathbf{\Gamma}_{\mathbf{nn}}(\omega)) = \Psi(\omega)\big|_{S_{dd}(\omega)=0}\,, \tag{A.25}$$

respectively. The calculation of $\bar{F}(\mathbf{\Gamma}_{\mathbf{dd}}(\omega))$ and $\bar{F}(\mathbf{\Gamma}_{\mathbf{nn}}(\omega))$ from (A.12) is straightforward. With

$$\bar{\Psi}(\omega) = \frac{S_{\bar{y}\bar{y}}(\omega)}{S_{xx}(\omega)}\,, \tag{A.26}$$

$\bar{F}(\mathbf{\Gamma}_{\mathbf{dd}}(\omega))$ and $\bar{F}(\mathbf{\Gamma}_{\mathbf{nn}}(\omega))$ can be calculated as

$$\bar{F}(\mathbf{\Gamma}_{\mathbf{dd}}(\omega)) = \bar{\Psi}(\omega)\big|_{S_{nn}(\omega)=0}\,, \tag{A.27}$$

$$\bar{F}(\mathbf{\Gamma}_{\mathbf{nn}}(\omega)) = \bar{\Psi}(\omega)\big|_{S_{dd}(\omega)=0}\,, \tag{A.28}$$

respectively, where we substituted in (A.26) $S_{\bar{y}\bar{y}}(\omega)$ and $S_{xx}(\omega)$ by (A.12) and by (A.23), respectively.

In summary, we notice that an unbiased estimate of $SINR_{\mathrm{f,in}}(\omega)$ is given by (A.21) if $F(\mathbf{\Gamma}_*(\omega))$ and $\bar{F}(\mathbf{\Gamma}_*(\omega))$, $* \in \{\mathbf{dd}, \mathbf{nn}\}$, are estimated by (A.22)–(A.28). This requires inactivity of interference-plus-noise, i.e., $S_{nn}(\omega) = 0$ in (A.24) and in (A.27), and inactivity of the desired signal, i.e., $S_{dd}(\omega) = 0$ in (A.25) and in (A.28). For short-time stationary wideband speech and audio signals, this means that estimates of the PSDs $S_{xx}(\omega)$, $S_{yy}(\omega)$, and $S_{\bar{y}\bar{y}}(\omega)$ are required, and that periods of inactivity of interference-plus-noise and of the desired signal have to be detected.

A.3 Algorithm of the Unbiased SINR Estimation

In this section, the algorithm of the unbiased SINR estimation for wideband time-varying signals is introduced. In Sect. A.3.1, the method for estimating the PSDs $S_{xx}(\omega)$, $S_{yy}(\omega)$, and $S_{\bar{y}\bar{y}}(\omega)$ is specified. Section A.3.2 describes the double-talk detector. For practical realizations of the SINR estimation, robustness improvement is considered in Sect. A.3.3. Finally, Sect. A.3.4 summarizes the algorithm and gives experimental results.

A.3.1 Estimation of the PSDs in the DFT Domain

With regard to the application of the SINR estimation to short-time stationary wideband signals, we estimate PSDs by second-order periodograms in the DFT domain using data-overlapping and windowing in the time domain and exponential averaging in the DFT domain [Bri75]. The DFT length should be equal to $2N$ in accordance with Chap. 7. DFT bins are indexed by n, $n = 0, 1, \ldots, 2N-1$, where $\omega = \pi n/(NT_{\mathrm{s}})$. The PSD estimation is described for the example of the PSD $S_{x_m x_m}(\omega)$ of the m-th sensor signal. Let the second-order windowed periodogram $I_{x_m x_m}(r, n)$ be defined as

$$I_{x_m x_m}(r, n) = \left| \sum_{k=0}^{2N-1} w_k x_m(k + rR - N) \exp\left(-jn(\pi/N)k\right) \right|^2 , \qquad \text{(A.29)}$$

where $R = 2N/\alpha$. The factor $\alpha \geq 1$ is the block overlap factor. The windowing function w_k is normalized, i.e., $\sum_{k=0}^{2N-1} |w_k|^2 = 1$. w_k allows to control the frequency resolution and the sidelobe leakage [Har78]. The estimate $\hat{S}_{x_m x_m}(r, n)$ of $S_{x_m x_m}(\omega)$ at block time r for the n-th DFT bin is then obtained as

$$\hat{S}_{x_m x_m}(r, n) = (1 - \beta) \sum_{i=0}^{r} \beta^{r-i} I_{x_m x_m}(i, n) , \qquad \text{(A.30)}$$

which may be written as

$$\hat{S}_{x_m x_m}(r,n) = (1-\beta)\hat{S}_{x_m x_m}(r-1,n) + \beta I_{x_m x_m}(r,n), \tag{A.31}$$

where we extracted in (A.30) the term $\beta I_{x_m x_m}(r,n)$ from the sum and reintroduced $\hat{S}_{x_m x_m}(r-1,n)$. $0 < \beta < 1$ is the exponential forgetting factor. The normalization factor $(\beta-1)$ assures an asymptotically ($2N \to \infty, r \to \infty$) unbiased estimate of the PSD $S_{x_m x_m}(\omega)$ for $\omega T_s \neq 0 \pmod{2\pi}$ [Bri75]. The factor $(\beta-1)^{-1}$ can be interpreted as the effective memory for averaging the second-order windowed periodograms $I_{x_m x_m}(r,n)$ (for $r \to \infty$). The estimate $\hat{S}_{xx}(r,n)$ of $S_{xx}(\omega)$ is obtained according to (A.23) as

$$\hat{S}_{xx}(r,n) = \frac{1}{M} \sum_{m=0}^{M-1} \hat{S}_{x_m x_m}(r,n). \tag{A.32}$$

Estimates $\hat{S}_{yy}(r,n)$ and $\hat{S}_{\bar{y}\bar{y}}(r,n)$ of $S_{yy}(\omega)$ and of $S_{\bar{y}\bar{y}}(\omega)$ are obtained by replacing in (A.29) $x_m(k)$ by $y(k)$ and by $\bar{y}(k)$, respectively.

A.3.2 Double-Talk Detection

The functions $F(\mathbf{\Gamma}_*(\omega))$ and $\bar{F}(\mathbf{\Gamma}_*(\omega))$, $* \in \{\mathbf{dd}, \mathbf{nn}\}$, can only be estimated for presence of only desired signal and for presence of only interference-plus-noise, respectively. In order to optimally benefit of the diversity of the time-frequency representation of the PSDs, presence of desired signal only and presence of interference-plus-noise only should be detected over time and over frequency. For that, an activity detector is required, which classifies the sensor signals into 'presence of only desired signal', 'presence of only interference', and 'double-talk' over time and over frequency.

We propose a double-talk detector as follows: The output signals $y(k)$ and $\bar{y}(k)$ of the fixed beamformer and of the complementary fixed beamformer provide biased estimates of the desired signal and of the interference, respectively. These estimates are generally not sufficiently accurate for estimating the SINR at the sensors. However, the ratio $\hat{\Upsilon}(r,\omega)$,

$$\hat{\Upsilon}(r,n) = \frac{\hat{S}_{yy}(r,n)}{\hat{S}_{\bar{y}\bar{y}}(r,n)}, \tag{A.33}$$

which is obtained by replacing in (A.5) $S_{yy}(\omega)$ and $S_{\bar{y}\bar{y}}(\omega)$ by $\hat{S}_{yy}(r,n)$ and by $\hat{S}_{\bar{y}\bar{y}}(r,n)$, respectively, becomes maximum if only the desired signal is present and minimum for only interference activity for $A_f(\omega) \neq 1$ (see (A.20)). Application of thresholds to $\hat{\Upsilon}(r,n)$ thus allows the desired classification.

In Sect. A.1, it was shown that $\hat{\Upsilon}(r,n)$ is not only a function of the SINR at the sensors but that $\hat{\Upsilon}(r,n)$ also depends on the spatio-spectral coherence matrices of the desired signal and of interference-plus-noise, which strongly vary with the acoustic environment and with the wavelength. Especially for non-stationary conditions, it is therefore not practical (if not useless) to provide fixed thresholding to $\hat{\Upsilon}(r,n)$ for the double-talk detection.

A more robust approach leading to a more robust classification is dynamic thresholding: It allows to track the time variance of the maxima and of the minima of $\hat{\Upsilon}(r,n)$, and it does not require calibration of the dependency on frequency. For dynamically detecting the maxima and the minima of $\hat{\Upsilon}(r,n)$, the minimum statistics after [Mar01a] can be used: We search for the minima of $1/\hat{\Upsilon}(r-i,n)$ (maxima of $\hat{\Upsilon}(r-i,n)$) and for the minima of $\hat{\Upsilon}(r-i,n)$, $i = 0, 1, \ldots, D$, over a window of D subsequent blocks in all DFT bins. This gives the upper thresholds $\hat{\Upsilon}_{\mathrm{th,d}}(r,n)$ and $\hat{\Upsilon}_{\mathrm{th,n}}(r,n)$. For $1/\hat{\Upsilon}(r,n) \leq \hat{\Upsilon}_{\mathrm{th,d}}(r,n)$, activity of the desired signal only is assumed. For $\hat{\Upsilon}(r,n) \leq \hat{\Upsilon}_{\mathrm{th,n}}(r,n)$, activity of interference-plus-noise only is assumed.

A.3.3 Unbiased Estimation of the SINR

This double-talk detection can now be applied to the estimation of $F(\mathbf{\Gamma}_*(\omega))$ and $\bar{F}(\mathbf{\Gamma}_*(\omega))$, $* \in \{\mathbf{dd}, \mathbf{nn}\}$. Let the estimate $\hat{\Psi}(r,n)$ of $\Psi(\omega)$ after (A.22) be defined as

$$\hat{\Psi}(r,\omega) = \frac{\hat{S}_{yy}(r,n)}{\hat{S}_{xx}(r,n)}, \tag{A.34}$$

and let the estimate $\tilde{\Psi}(r,n)$ of $\bar{\Psi}(\omega)$ after (A.26) be given by

$$\tilde{\Psi}(r,n) = \frac{\hat{S}_{\bar{y}\bar{y}}(r,n)}{\hat{S}_{xx}(r,n)}. \tag{A.35}$$

Estimates $F(\hat{\mathbf{\Gamma}}_{\mathbf{dd}}(r,n))$ and $\bar{F}(\hat{\mathbf{\Gamma}}_{\mathbf{dd}}(r,n))$ of $F(\mathbf{\Gamma}_{\mathbf{dd}}(\omega))$ and of $\bar{F}(\mathbf{\Gamma}_{\mathbf{dd}}(\omega))$ can then be obtained according to (A.24) and (A.27) as

$$F(\hat{\mathbf{\Gamma}}_{\mathbf{dd}}(r,n)) = \hat{\Psi}(r,n)\Big|_{1/\hat{\Upsilon}(r,n) \leq \Upsilon_{0,\mathrm{d}}\hat{\Upsilon}_{\mathrm{th,d}}(r,n)}, \tag{A.36}$$

$$\bar{F}(\hat{\mathbf{\Gamma}}_{\mathbf{dd}}(r,n)) = \tilde{\Psi}(r,n)\Big|_{1/\hat{\Upsilon}(r,n) \leq \Upsilon_{0,\mathrm{d}}\hat{\Upsilon}_{\mathrm{th,d}}(r,n)}, \tag{A.37}$$

respectively. Estimates $F(\hat{\mathbf{\Gamma}}_{\mathbf{nn}}(r,n))$ and $\bar{F}(\hat{\mathbf{\Gamma}}_{\mathbf{nn}}(r,n))$ of $F(\mathbf{\Gamma}_{\mathbf{nn}}(\omega))$ and of $\bar{F}(\mathbf{\Gamma}_{\mathbf{nn}}(\omega))$ can be obtained according to (A.25) and (A.28) as

$$\bar{F}(\hat{\mathbf{\Gamma}}_{\mathbf{nn}}(r,n)) = \tilde{\Psi}(r,n)\Big|_{\hat{\Upsilon}(r,n) \leq \Upsilon_{0,\mathrm{n}}\hat{\Upsilon}_{\mathrm{th,n}}(r,n)}, \tag{A.38}$$

$$F(\hat{\mathbf{\Gamma}}_{\mathbf{nn}}(r,n)) = \hat{\Psi}(r,n)\Big|_{\hat{\Upsilon}(r,n) \leq \Upsilon_{0,\mathrm{n}}\hat{\Upsilon}_{\mathrm{th,n}}(r,n)}, \tag{A.39}$$

respectively. The multiplication of the dynamic thresholds $\hat{\Upsilon}_{\mathrm{th,d}}(r,n)$ and $\hat{\Upsilon}_{\mathrm{th,n}}(r,n)$ by factors $\Upsilon_{0,\mathrm{d}} > 0$ and $\Upsilon_{0,\mathrm{n}} > 0$, respectively, allows to vary the update rates.

Using (A.33) and (A.36)–(A.39) in (A.21), we can finally write the estimate $\widehat{SINR}_{\mathrm{f,in}}(r,n)$ of $SINR_{\mathrm{f,in}}(\omega)$ as follows:

$$\widehat{SINR}_{\mathrm{f,in}}(r,n) = \frac{F(\hat{\mathbf{\Gamma}}_{\mathbf{nn}}(r,n)) - \hat{\Upsilon}(r,n)\bar{F}(\hat{\mathbf{\Gamma}}_{\mathbf{nn}}(r,n))}{\hat{\Upsilon}(r,n)\bar{F}(\hat{\mathbf{\Gamma}}_{\mathbf{dd}}(r,n)) - F(\hat{\mathbf{\Gamma}}_{\mathbf{dd}}(r,n))}. \tag{A.40}$$

A.3.4 Robustness Improvement

In this section, we specify three methods for improving the robustness of the SINR estimation. First, a technique is presented for detecting and correcting inaccurate estimates of the spatial coherence functions. Second, we describe a measure for improving the tracking capability of small variations of the spatial coherence functions based on [Mar01a]. Third, a similar measure is defined in order to improve the accuracy of the double-talk detector.

Negative SINR Estimates

It cannot be assured that $F(\mathbf{\Gamma}_*(\omega))$, $\bar{F}(\mathbf{\Gamma}_*(\omega))$, where $* \in \{\mathbf{dd}, \mathbf{nn}\}$, and $\hat{\Upsilon}(r,n)$ are correctly estimated, since, generally, the number of data blocks for averaging is limited due to the non-stationarity of the sensor signals. This inaccuracy may lead to $\widehat{SINR}_{\mathrm{f,in}}(r,n) < 0$. From (A.40), we see that $\widehat{SINR}_{\mathrm{f,in}}(r,n) < 0$, if one of the following two conditions is fulfilled:

$$\hat{\Upsilon}(r,n) > \max\left\{\frac{F(\hat{\mathbf{\Gamma}}_{\mathbf{dd}}(r,n))}{\bar{F}(\hat{\mathbf{\Gamma}}_{\mathbf{dd}}(r,n))}, \frac{F(\hat{\mathbf{\Gamma}}_{\mathbf{nn}}(r,n))}{\bar{F}(\hat{\mathbf{\Gamma}}_{\mathbf{nn}}(r,n))}\right\} \tag{A.41}$$

$$\hat{\Upsilon}(r,n) < \min\left\{\frac{F(\hat{\mathbf{\Gamma}}_{\mathbf{dd}}(r,n))}{\bar{F}(\hat{\mathbf{\Gamma}}_{\mathbf{dd}}(r,n))}, \frac{F(\hat{\mathbf{\Gamma}}_{\mathbf{nn}}(r,n))}{\bar{F}(\hat{\mathbf{\Gamma}}_{\mathbf{nn}}(r,n))}\right\} . \tag{A.42}$$

If we assume that the interference rejection is greater than the cancellation of the desired signal, and if a uniformly weighted beamformer ($\mathbf{w}_{\mathrm{f}} = 1/M \cdot \mathbf{1}_{M\times 1}$) is used, we can simplify (A.41) and (A.42) using (A.17) as follows:

$$\hat{\Upsilon}(r,n) > \frac{F(\hat{\mathbf{\Gamma}}_{\mathbf{dd}}(r,n))}{1 - F(\hat{\mathbf{\Gamma}}_{\mathbf{dd}}(r,n))} \tag{A.43}$$

$$\hat{\Upsilon}(r,n) < \frac{F(\hat{\mathbf{\Gamma}}_{\mathbf{nn}}(r,n))}{1 - F(\hat{\mathbf{\Gamma}}_{\mathbf{nn}}(r,n))} . \tag{A.44}$$

Let the estimate $\hat{\Upsilon}(r,n)$ be more accurate than $F(\hat{\mathbf{\Gamma}}_{\mathbf{dd}}(r,n))$ and $F(\hat{\mathbf{\Gamma}}_{\mathbf{nn}}(r,n))$. This is reasonable since $\hat{\Upsilon}(r,n)$ is continuously estimated, while $F(\hat{\mathbf{\Gamma}}_{\mathbf{dd}}(r,n))$ and $F(\hat{\mathbf{\Gamma}}_{\mathbf{nn}}(r,n))$ cannot be estimated continuously, and while $F(\hat{\mathbf{\Gamma}}_{\mathbf{dd}}(r,n))$ and $F(\hat{\mathbf{\Gamma}}_{\mathbf{nn}}(r,n))$ depend on the accuracy of the double-talk detector. This allows to prevent $\widehat{SINR}_{\mathrm{f,in}}(r,n) < 0$ by replacing $F(\hat{\mathbf{\Gamma}}_{\mathbf{dd}}(r,n))$ and $F(\hat{\mathbf{\Gamma}}_{\mathbf{nn}}(r,n))$ by

$$F(\hat{\mathbf{\Gamma}}_{\mathbf{dd}}(r,n)) = \frac{\hat{\Upsilon}(r,n)}{\hat{\Upsilon}(r,n) + 1} , \tag{A.45}$$

$$F(\hat{\mathbf{\Gamma}}_{\mathbf{nn}}(r,n)) = \frac{\hat{\Upsilon}(r,n)}{\hat{\Upsilon}(r,n) + 1} , \tag{A.46}$$

if condition (A.43) and condition (A.44) are fulfilled, respectively. The right sides of (A.45) and (A.46) are obtained by solving (A.43) and (A.44) for $F(\hat{\mathbf{\Gamma}}_{\mathbf{dd}}(r,n))$ and for $F(\hat{\mathbf{\Gamma}}_{\mathbf{nn}}(r,n))$, respectively, and by setting the equality sign. During the update of $F(\hat{\mathbf{\Gamma}}_{\mathbf{dd}}(r,n))$ and $F(\hat{\mathbf{\Gamma}}_{\mathbf{nn}}(r,n))$ with (A.36), (A.39) or (A.45), (A.46), respectively, the SINR estimate becomes zero and infinity, respectively. Experiments showed that these limits can be efficiently prevented by using the right sides of (A.36), (A.39) and (A.45), (A.46) as updates for first-order recursive averages $F(\hat{\mathbf{\Gamma}}_{\mathbf{dd}}(r,n))$ and $F(\hat{\mathbf{\Gamma}}_{\mathbf{nn}}(r,n))$, respectively. The exponential forgetting factor is denoted by $0 < \beta_F < 1$.

Small Variations of the Estimates of the Spatial Coherence Functions

If the conditions

$$|\hat{\Psi}(r,n) - F(\hat{\mathbf{\Gamma}}_{\mathbf{dd}}(r-1,n))| \leq \Delta_F\,, \tag{A.47}$$

$$|\hat{\Psi}(r,n) - F(\hat{\mathbf{\Gamma}}_{\mathbf{nn}}(r,n)(r-1,n))| \leq \Delta_F\,, \tag{A.48}$$

where $\Delta_F \ll 1$, are fulfilled, it is very likely that only desired signal or only interference-plus-noise are active, respectively, though the double-talk detector might detect double-talk. In order to allow tracking of such small variations of $\hat{\Psi}(r,n)$, $F(\hat{\mathbf{\Gamma}}_{\mathbf{dd}}(r,n))$ and $F(\hat{\mathbf{\Gamma}}_{\mathbf{nn}}(r,n))$ are always updated if condition (A.47) and (A.48) are met, respectively.

Small Variations of the Thresholds of the Double-Talk Detector

A similar measure to that described in Sect. A.3.4 can be used for improving the accuracy of the double-talk detector: If the relative variation of $1/\hat{\Upsilon}(r,n)$ and of $\hat{\Upsilon}(r,n)$ relative to $\hat{\Upsilon}_{\mathrm{th,d}}(r-1,n)$ and relative to $\hat{\Upsilon}_{\mathrm{th,n}}(r-1,n)$ are less than given limits $\Delta\hat{\Upsilon}_{\mathrm{th,d}}$ and $\Delta\hat{\Upsilon}_{\mathrm{th,n}}$, respectively, i.e.,

$$|1/\hat{\Upsilon}(r,n) - \hat{\Upsilon}_{\mathrm{th,d}}(r-1,n)|/\hat{\Upsilon}_{\mathrm{th,d}}(r-1,n) \leq \Delta\hat{\Upsilon}_{\mathrm{th,d}}\,, \tag{A.49}$$

$$|\hat{\Upsilon}(r,n) - \hat{\Upsilon}_{\mathrm{th,n}}(r-1,n)|/\hat{\Upsilon}_{\mathrm{th,n}}(r-1,n) \leq \Delta\hat{\Upsilon}_{\mathrm{th,n}}\,, \tag{A.50}$$

the thresholds $\hat{\Upsilon}_{\mathrm{th,d}}(r,n) := 1/\hat{\Upsilon}(r,n)$ and $\hat{\Upsilon}_{\mathrm{th,n}}(r,n) := \hat{\Upsilon}(r,n)$ are updated immediately. In contrast to (A.47) and (A.48), the left sides of (A.49) and (A.50) are normalized by the thresholds $\hat{\Upsilon}_{\mathrm{th,d}}(r-1,n)$ and $\hat{\Upsilon}_{\mathrm{th,n}}(r-1,n)$, since their variations strongly depend on their magnitudes.

A.3.5 Summary of the Algorithm and Experimental Results

Table A.1 summarizes the algorithm for estimating the average SINR at the sensors for the case with a uniformly weighted beamformer, i.e., $\mathbf{w}_{\mathrm{f}}(\omega) = 1/M \cdot \mathbf{1}_{M\times 1}$.

Table A.1. r-th iteration for the n-th DFT bin of the spatial estimation of the SINR at the sensors for non-stationary wideband signals; Typical parameters for $f_s = 12\,\text{kHz}$: $N = 256$, $\alpha = 2$, $\beta = 0.59$, $D = 96$, $\beta_F = 0.3$, $\Delta\hat{\Upsilon}_{\text{th,d}} = 0.1$, $\Delta\hat{\Upsilon}_{\text{th,n}} = 0.3$, $\Upsilon_{0,\text{d}} = 1.0$, $\Upsilon_{0,\text{n}} = 2.0$

1 Compute $\hat{S}_{yy}(r,n)$, $\hat{S}_{\bar{y}\bar{y}}(r,n)$, and $\hat{S}_{xx}(r,n)$ using (A.29) and (A.31)

2 Compute $\hat{\Upsilon}(r,n) = \hat{S}_{yy}(r,n)/\hat{S}_{\bar{y}\bar{y}}(r,n)$ after (A.33)

3 Compute $\hat{\Psi}(r,n)$ after (A.34)

4 *Tracking of the minima of* $1/\hat{\Upsilon}(r,n)$
IF $|1/\hat{\Upsilon}(r,n) - \hat{\Upsilon}_{\text{th,d}}(r-1,n)| \leq \hat{\Upsilon}_{\text{th,d}}(r-1,n)\Delta\hat{\Upsilon}_{\text{th,d}}$, (A.49)
$\hat{\Upsilon}_{\text{th,d}}(r,n) = \hat{\Upsilon}(r,n)$
Replace all stored values of $\hat{\Upsilon}_{\text{d}}(r-i,n)$, $i = 1, 2, \ldots, D$, by $\hat{\Upsilon}_{\text{th,d}}(r,n)$
ELSE
Find $\hat{\Upsilon}_{\text{th,d}}(r,n)$, the minimum of $\hat{\Upsilon}_{\text{d}}(r-i,n)$, $i = 1, 2, \ldots, D$
$\hat{\Upsilon}_{\text{d}}(r,n) = \hat{\Upsilon}(r,n)$

5 IF $\hat{\Upsilon}(r,n) \leq \Upsilon_{0,\text{d}}\hat{\Upsilon}_{\text{th,d}}(r,n)$ OR $|\hat{\Psi}(r,n) - F(\hat{\mathbf{\Gamma}}_{\mathbf{dd}}(r-1,n))| \leq \Delta_F$, (A.47)
update $= \hat{\Psi}(r,n)$, (A.36)
ELSE *update* $= F(\hat{\mathbf{\Gamma}}_{\mathbf{dd}}(r-1,n))$
IF $\hat{\Upsilon}(r,n) > \textit{update}/(1-\textit{update})$, (A.43)
update $= \hat{\Upsilon}(r,n)/(\hat{\Upsilon}(r,n)+1)$, (A.45)

6 Compute $F(\hat{\mathbf{\Gamma}}_{\mathbf{dd}}(r,n)) = \beta_F F(\hat{\mathbf{\Gamma}}_{\mathbf{dd}}(r-1,n)) + (1-\beta_F)\textit{update}$

7 *Tracking of the minima of* $\hat{\Upsilon}(r,n)$
IF $|\hat{\Upsilon}(r,n) - \hat{\Upsilon}_{\text{th,n}}(r-1,n)| \leq \hat{\Upsilon}_{\text{th,n}}(r-1,n)\Delta\hat{\Upsilon}_{\text{th,n}}$, (A.50)
$\hat{\Upsilon}_{\text{th,n}}(r,n) = \hat{\Upsilon}(r,n)$
Replace all stored values of $\hat{\Upsilon}_{\text{n}}(r-i,n)$, $i = 1, 2, \ldots, D$, by $\hat{\Upsilon}_{\text{th,n}}(r,n)$
ELSE
Find $\hat{\Upsilon}_{\text{th,n}}(r,n)$, the minimum of $\hat{\Upsilon}_{\text{n}}(r-i,n)$, $i = 1, 2, \ldots, D$
$\hat{\Upsilon}_{\text{n}}(r,n) = \hat{\Upsilon}(r,n)$

8 IF $\hat{\Upsilon}(r,n) \leq \Upsilon_{0,\text{n}}\hat{\Upsilon}_{\text{th,n}}(r,n)$ OR $|\hat{\Psi}(r,n) - F(\hat{\mathbf{\Gamma}}_{\mathbf{nn}}(r-1,n))| \leq \Delta_F$, (A.48)
update $= \hat{\Psi}(r,n)$, (A.39)
ELSE *update* $= F(\hat{\mathbf{\Gamma}}_{\mathbf{nn}}(r-1,n))$
IF $\hat{\Upsilon}(r,n) < \textit{update}/(1-\textit{update})$, (A.44)
update $= \hat{\Upsilon}(r,n)/(\hat{\Upsilon}(r,n)+1)$, (A.46)

9 $F(\hat{\mathbf{\Gamma}}_{\mathbf{nn}}(r,n)) = \beta_F F(\hat{\mathbf{\Gamma}}_{\mathbf{nn}}(r-1,n)) + (1-\beta_F)\textit{update}$

10 Compute $\widehat{SINR}_{\text{f,in}}(r,n) = \frac{F(\hat{\mathbf{\Gamma}}_{\mathbf{nn}}(r,n)) - \hat{\Upsilon}(r,n)\left(1-F(\hat{\mathbf{\Gamma}}_{\mathbf{nn}}(r,n))\right)}{\hat{\Upsilon}(r,n)\left(1-F(\hat{\mathbf{\Gamma}}_{\mathbf{dd}}(r,n))\right) - F(\hat{\mathbf{\Gamma}}_{\mathbf{dd}}(r,n))}$ after (A.40)

11 Store $F(\hat{\mathbf{\Gamma}}_{\mathbf{dd}}(r,n))$, $F(\hat{\mathbf{\Gamma}}_{\mathbf{nn}}(r,n))$, $\hat{\Upsilon}_{\text{d}}(r,n)$, $\hat{\Upsilon}_{\text{th,d}}(r,n)$, $\hat{\Upsilon}_{\text{n}}(r,n)$, and $\hat{\Upsilon}_{\text{th,n}}(r,n)$

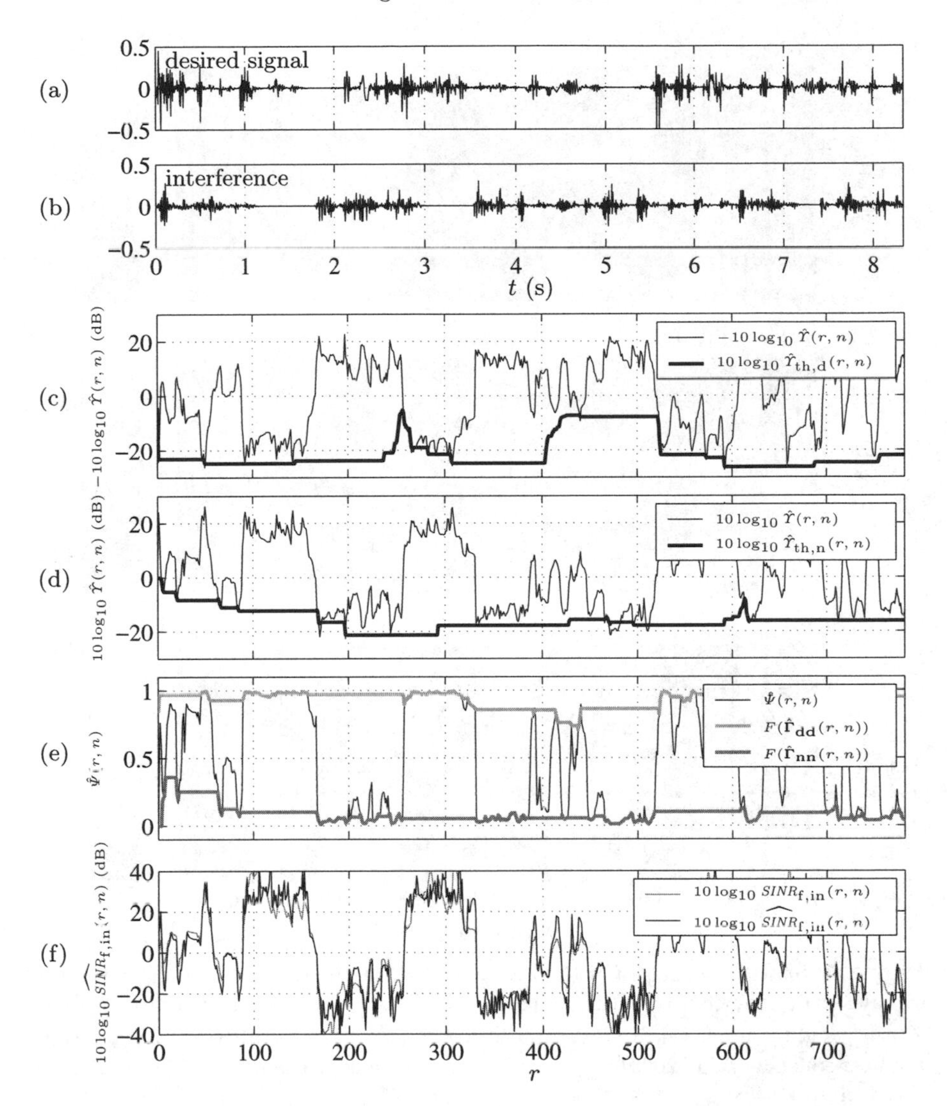

Fig. A.2. Illustration of the procedure of the SINR estimation: **(a)** desired signal and interference signal at the sensor $M/2$ as a function of time t, (c) activity detection of the desired signal, (d) activity detection of the interference, (e) estimation of the functions $F(\hat{\mathbf{\Gamma}}_{\mathbf{dd}}(r,n))$ and $F(\hat{\mathbf{\Gamma}}_{\mathbf{nn}}(r,n))$, (f) SINR estimate, $\widehat{SINR}_{\text{f,in}}(r,n)$, and reference SINR, $SINR_{\text{f,in}}(r,n)$; multimedia room with $T_{60} = 250\,\text{ms}$, $f \approx 1\,\text{kHz}$, $M = 8$, $d = 4\,\text{cm}$

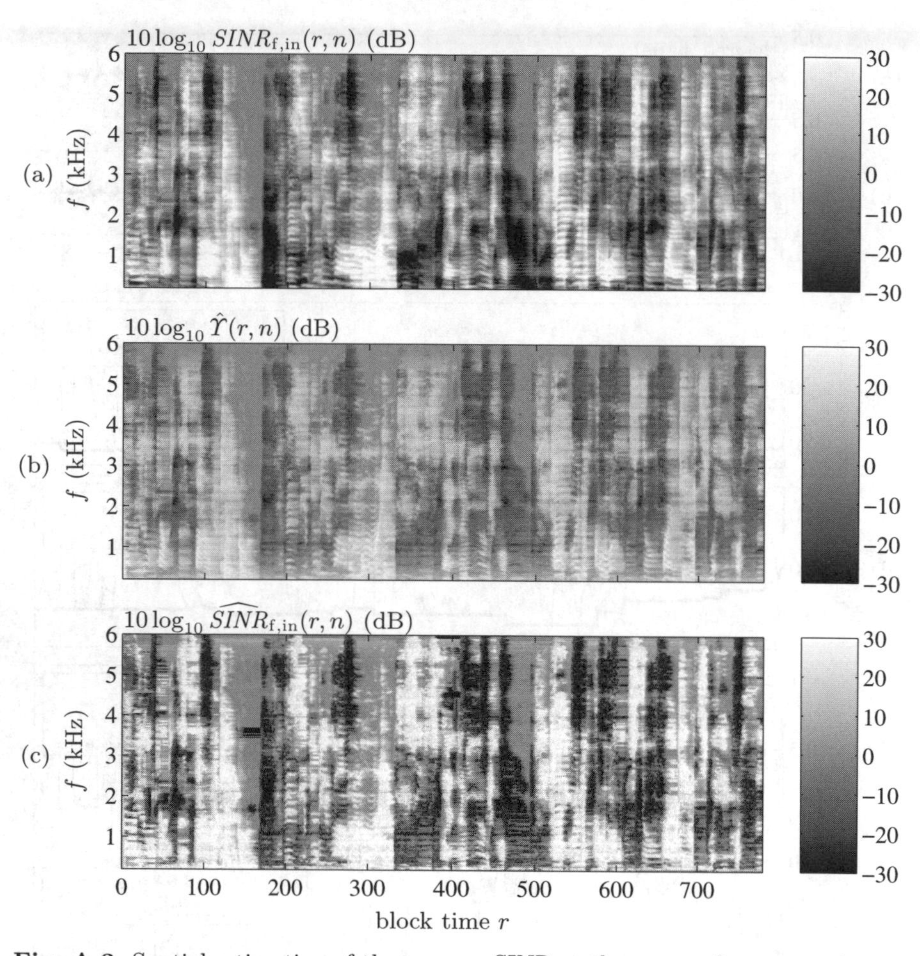

Fig. A.3. Spatial estimation of the average SINR at the sensors for non-stationary signals: **(a)** reference SINR at the sensors, $SINR_{\mathrm{f,in}}(r,n)$, **(b)** biased SINR estimate $\hat{\Upsilon}(r,n)$, and (c) unbiased SINR estimate, $\widehat{SINR}_{\mathrm{f,in}}(r,n)$, as a function of block time r and frequency f for $200\,\mathrm{Hz} \leq f \leq 6\,\mathrm{kHz}$; multimedia room with $T_{60} = 250\,\mathrm{ms}$, $f \approx 1\,\mathrm{kHz}$, $M = 8$, $d = 4\,\mathrm{cm}$

For illustrating the performance of the SINR estimation, we apply the algorithm to a mixture of two competing male speakers. We use the array with $M = 8$ sensors with inter-sensor spacing $d = 4\,\mathrm{cm}$ in the multimedia room with $T_{60} = 250\,\mathrm{ms}$ as described in App. B. The desired source is located in broadside direction ($\theta_{\mathrm{d}} = \pi/2$) at a distance of 60 cm to the array center. The interferer is placed in endfire direction ($\theta_{\mathrm{n}} = 0$) at a distance of 120 cm to the array center. The signals are chosen such that (almost) continuous double-talk is obtained in the discrete time domain. The source signals are high-pass filtered with cut-off frequency $f = 200\,\mathrm{Hz}$. The average SINR at the sensors is 0 dB. The parameters are chosen as defined in Table A.1.

The reference SINR at the sensors as a function of the block time r and of the DFT bin n,

$$SINR_{\mathrm{f,in}}(r,n) = \frac{\hat{S}_{dd}(r,n)}{\hat{S}_{nn}(r,n)}, \tag{A.51}$$

(see (A.15)) is obtained as follows: Estimates $\hat{S}_{dd}(r,n)$ and $\hat{S}_{nn}(r,n)$ of the PSDs $S_{dd}(\omega)$ and $S_{nn}(\omega)$ at block time r are calculated using recursively-averaged periodograms (A.31) by replacing in (A.29) $x_m(k)$ by $d_m(k)$ and by $n_m(k)$, respectively, and by averaging over the sensor signals according to (A.32).

The results are depicted in Fig. A.2 for $f \approx 1\,\mathrm{kHz}$. The signal of the desired source and the signal of the interferer at the $M/2$-th microphone are shown in Fig. A.2a and in Fig. A.2b as a function of time t, respectively. Figure A.2c illustrates the tracking of the minima of $1/\hat{\Upsilon}(r,n)$ as a function of the block time r, yielding the threshold $\hat{\Upsilon}_{\mathrm{th,d}}(r,n)$ for the activity detection of the desired signal (Step 4 in Table A.1). Figure A.2d shows the tracking of the minima of $\hat{\Upsilon}(r,n)$, which yields the threshold $\hat{\Upsilon}_{\mathrm{th,n}}(r,n)$ (Step 7 in Table A.1). The thresholds $\hat{\Upsilon}_{\mathrm{th,d}}(r,n)$ and $\hat{\Upsilon}_{\mathrm{th,n}}(r,n)$ are used in Step 5 and in Step 8 for classifying the frequency bins of the present signal frame. If activity of the desired signal only or activity of interference only is detected, $F(\hat{\mathbf{\Gamma}}_{\mathbf{dd}}(r,n))$ or $F(\hat{\mathbf{\Gamma}}_{\mathbf{nn}}(r,n))$ is estimated from $\hat{\Psi}(r,n)$, respectively (Fig. A.2e). The estimates $F(\hat{\mathbf{\Gamma}}_{\mathbf{dd}}(r,n))$ and $F(\hat{\mathbf{\Gamma}}_{\mathbf{nn}}(r,n))$ are finally used in Step 10 to estimate $\widehat{SINR}_{\mathrm{f,in}}(r,n)$, as illustrated in Fig. A.2f. It may be seen that the SINR estimate $\widehat{SINR}_{\mathrm{f,in}}(r,n)$ accurately follows $SINR_{\mathrm{f,in}}(r,n)$ for a wide range of $SINR_{\mathrm{f,in}}(r,n)$ while $\hat{\Upsilon}(r,n)$ in Fig. A.2d only provides a rough estimate of $SINR_{\mathrm{f,in}}(r,n)$. Finally, in Fig. A.3, the biased SINR estimate $\hat{\Upsilon}(r,n)$ (Fig. A.3b) and the bias-corrected SINR estimate $\widehat{SINR}_{\mathrm{f,in}}(r,n)$ (Fig. A.3c) are compared with the reference SINR $SINR_{\mathrm{f,in}}(r,n)$ (Fig. A.3a) as a function of the block time r and of the frequency f for $200\,\mathrm{Hz} \leq f \leq 6\,\mathrm{kHz}$. We notice that $\widehat{SINR}_{\mathrm{f,in}}(r,n)$ is an accurate estimate of $SINR_{\mathrm{f,in}}(r,n)$ for a wide range of frequencies while $\hat{\Upsilon}(r,n)$ is only a highly compressed and biased version of $SINR_{\mathrm{f,in}}(r,n)$.

B

Experimental Setups and Acoustic Environments

In this chapter, the experimental setup and the acoustic environments are described, which are used in this work for illustrating the properties of the proposed algorithms. For our experiments, we use a linear microphone array with a variable number of sensors $M \in \{4, 8, 12, 16\}$. The geometry is illustrated in Fig. B.1.

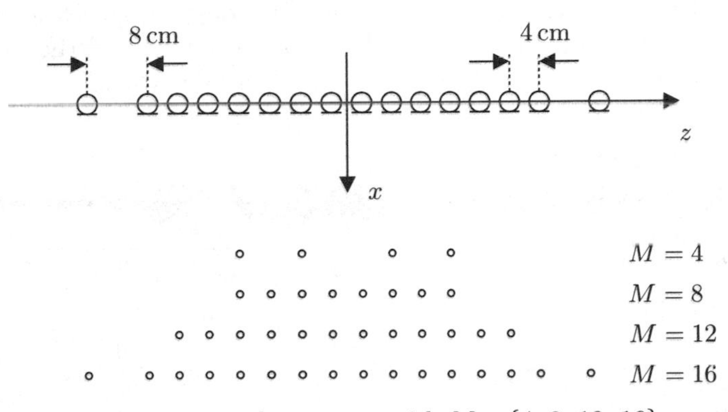

Fig. B.1. Linear microphone array with $M \in \{4, 8, 12, 16\}$ sensors

We cope with 3 acoustic environments with different reverberation times T_{60}: The first environment is the passenger cabin of a car (Sect. B.1) with a very low reverberation time of $T_{60} = 50$ ms with presence of typical car noise. The two other environments are the multimedia room of 'Multimedia Communications and Signal Processing' at the University Erlangen-Nuremberg with two reverberation times $T_{60} = 250, 400$ ms (Sect. B.2) with various interference signals.

B.1 Passenger Cabin of a Car

A fire emergency truck is equipped with an equidistant array with $M = 8$ uni-directional microphones of type Primo EM 118 with sensor spacing $d = 4\,\mathrm{cm}$ mounted on the sun-visor at the driver side. The sensor data is recorded using a Tascam MA-8 preamplifier, a Creamware A16 analog-digital converter, and a Laptop equipped with an RME Hammerfall DSP audio card. The sampling rate is 48 kHz with a resolution of 16 bits per sample [Aic02]. The fire emergency truck is driving in an urban area with variable speed ($\leq 80\,\mathrm{km/h}$) with closed windows. The car noise thus consists of motor noise, street noise, and some wind noise. The length of the recording which we are using in our experiments is about 42 seconds. For the experiments, the noise signal is downsampled to 8 kHz and to 12 kHz, respectively. An example of a sensor signal recorded in the emergency truck is illustrated in Fig. B.2a, the long-term averaged PSD of the noise signal is shown in Fig. B.2b. The long-term average of the PSD of the sensor signal is obtained using second-order windowed periodograms (A.29) averaged over the entire length of the noise data file using $N = 256$, $\alpha = 2$, and 'von Hann'-windowing [Har78] in the discrete time domain.

The desired speaker is located in broadside direction ($\theta_\mathrm{d} = \pi/2$) at a distance of 60 cm from the array center. The desired speech signal at the sensors is obtained by filtering room impulse responses with the clean source

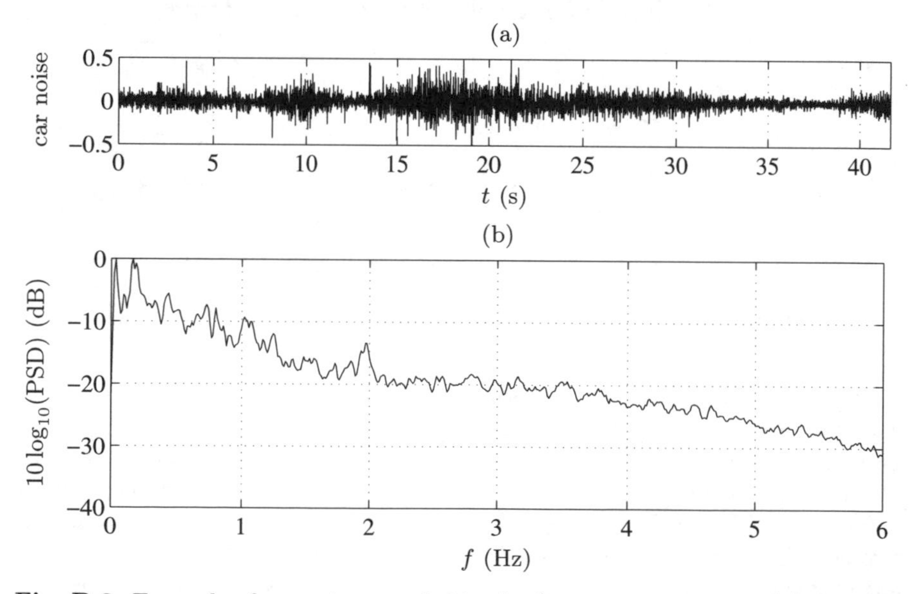

Fig. B.2. Example of car noise recorded in the fire emergency truck: **(a)** temporal signal, **(b)** long-term PSD

signals. The impulse responses between the source position and the sensors are simulated using the image method [AB79] with $T_{60} = 50\,\text{ms}$, since the noise data base [Aic02] does not provide these impulse responses. The mixture of desired speech and car noise at the sensors is obtained artificially by adding the desired speech and the noise. On the one hand, this provides the possibility to vary the SINR at the sensors arbitrarily. On the other hand, it allows to measure the beamformer performance in terms of, e.g., interference rejection or cancellation of the desired signal during continuous double-talk of desired speech and interference-plus-noise using a master/slave structure as shown in Fig. B.3 [Mar95, BSKD99, Bit02].

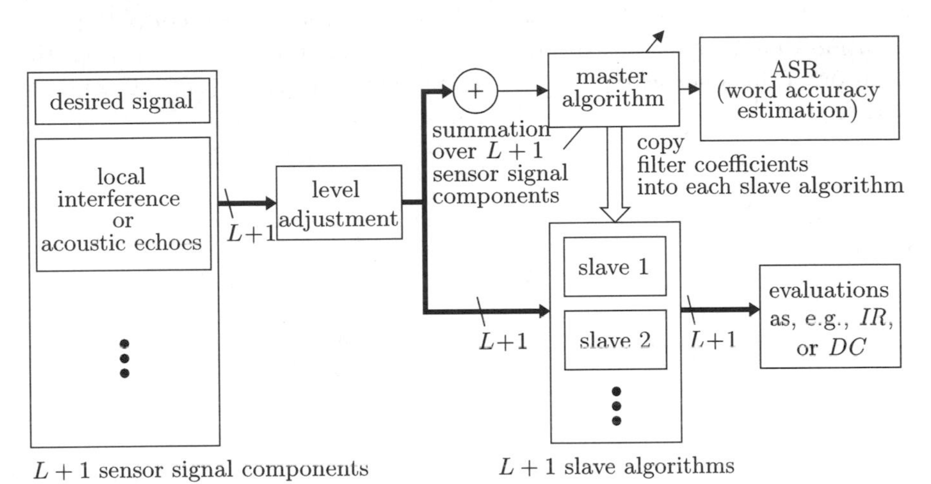

Fig. B.3. Evaluation of the proposed speech enhancement techniques

B.2 Multimedia Room

The multimedia room is a reverberant environment with modifiable reverberation time. The room with rigid walls and carpeted floor has the dimensions $5.9\,\text{m} \times 5\,\text{m} \times 3.1\,\text{m}$. The reverberation time can be changed by opening or closing sound absorbing curtains, which are mounted in front of the walls. Reverberation times are measured using [Sch65]. For closed curtains, the room has a reverberation time $T_{60} = 250\,\text{ms}$. For opened curtains, the reverberation time is $T_{60} = 400\,\text{ms}$. For this reverberant environment, we assume presence of point sources. Spatially extended sources, as, e.g., computer fans or air conditioning should not be present. Therefore, it is sufficient to use impulse response models for capturing the spatial characteristics of the room environment. This simplifies the recording of a data base since only room impulse responses between the source positions and the sensors need to be measured, which are convolved with desired source signals for obtaining the desired mixture of desired signal and interference at the sensors. The sensor array with $M = 16$ sensors with the geometry of Fig. B.3 is placed into the multimedia room in the middle of two opposite walls at 1.20 m from the third wall. The microphones are omni-directional sensors of type AKG CK92. For measuring the room impulse responses, maximum-length sequences (MLSs) are used [RV89, Rif92, Bur98, Mat99]. They are created in Matlab, and played back and recorded using the 24-channel audio card RME Hammerfall DSP and the 16-channel analog-digital/digital-analog converter Creamware A16. The loudspeaker is a Genelec 1029A, the microphone pre-amplifiers are two 8-channel Tascam MA-8. The sampling rate is 48 kHz with 16 bit resolution. The order of the MLS is 15. The MLS is repeated 20 times for averaging. For both reverberation times, we measured impulse responses for the distance between the position of the loudspeaker and the array center varying in $\{60, 120, 240\}$ cm and for the DOA varying in $\theta = 0, \pi/36, \ldots, \pi$.

C

Notations

C.1 Conventions

The following conventions are used: Lower case boldface denotes vectors, upper case boldface denotes matrices. The subscript $(\cdot)_{\mathrm{f}}$ describes a quantity in the DTFT domain if the same quantity exists in the discrete time domain and in the DTFT domain. Underlined quantities $\underline{(\cdot)}$ represent vectors or matrices in the DFT domain.

C.2 Abbreviations and Acronyms

AEC	acoustic echo canceller
ASR	automatic speech recognizer
BLUE	best linear unbiased estimator
BM	blocking matrix
CPSD	cross-power spectral density
D	dimensional
DFT	discrete Fourier transform
DOA	direction of arrival
DTFT	discrete-time Fourier transform
EIR	acoustic echo-to-local-interference (power) ratio
ERLE	echo return loss enhancement
FFT	fast Fourier transform
FIR	finite impulse response
GSAEC	GSC with AEC after the quiescent weight vector
GSC	generalized sidelobe canceller
GEIC	generalized echo/local-interference canceller
HMM	hidden Markov model
HTK	Hidden Markov Model Toolkit [YKO+00]
IC	interference canceller

IDTFT	inverse DTFT
IFFT	inverse fast Fourier transform
LCLSE	linearly-constrained least-squares error
LCMV	linearly-constrained minimum variance
LMS	least mean-squares
LS	least-squares
LSE	least-squares error
MFCC	Mel frequency cepstral coefficient
MIMO	multiple-input multiple-output
MINT	multiple-input/output inverse theorem
MISO	multiple-input single-output
MMSE	minimum mean-squared error
MSE	mean-squared error
MVDR	minimum-variance distortionless response
NRM	number of real multiplications
PSD	(auto-)power spectral density
RGSC	robust GSC
RLS	recursive least-squares
SIMO	single-input multiple-output
SINR	signal-to-interference-plus-noise (power) ratio
SISO	single-input single-output
SNR	signal-to-noise (power) ratio
SVD	singular value decomposition
WSS	wide-sense stationary

C.3 Mathematical Symbols

$(\cdot)^{-1}$	inverse of $(\cdot)$
$\mathrm{diag}\{\cdot\}$	diagonal matrix formed by the listed entries
$\dim\{\cdot\}$	dimension of the space which is spanned by the columns of a matrix
$\max\{\cdot\}$	maximum of the scalars given in the argument
mod	modulo operator
$\mathrm{rk}\{\cdot\}$	rank of a matrix
$\mathrm{span}\{\cdot\}$	space which is spanned by the columns of the given matrix
$\mathrm{tr}\{\cdot\}$	trace of a matrix, sum of the elements on the main diagonal of a matrix
$\lvert\cdot\rvert$	absolut value of the argument
$\lVert\cdot\rVert_2$	L_2-norm of a vector or of a matrix
$\lVert\cdot\rVert_F$	Frobenius norm of a matrix
$\lfloor\cdot\rfloor$	rounding toward the next smaller integer value
$(\cdot)^H$	Hermitian of $(\cdot)$
$(\cdot)^T$	transpose of a matrix or a vector
$(\cdot)^*$	conjugate complex of $(\cdot)$

$(\cdot)^{\perp}$	orthogonal complement of the space which is spanned by the columns of the given matrix
$(\cdot)^{+}$	pseudoinverse of a matrix
$[\cdot]_{m,n}$	the element in the m-th row and in the n-th column of a matrix
$\mathcal{E}\{\cdot\}$	expectation operator
$\otimes$	Kronecker product [Gra81]
∇^2	Laplacian operator
$\cup$	union set of
$\cap$	intersection with
$\subset$	subspace of
$\forall$	for all

Symbols

$\mathbf{0}_{m\times n}$	vector or matrix of size $m \times n$ with zeroes
$\mathbf{1}_{m\times n}$	vector or matrix of size $m \times n$ with ones
$\mathbf{1}_{m\times 1}^{(n)}$	vector of size $m \times 1$ with zeroes with the n-th element equal to one
α	temporal block overlap factor ($\alpha = K/R$)
β	exponential forgetting factor for first-order recursive averaging
$\beta_{\mathbf{a}}$	exponential forgetting factor for averaging $\hat{\underline{\mathbf{S}}}_{\mathbf{x}_\mathrm{a}\mathbf{x}_\mathrm{a}}(r')$
$\beta_{\mathbf{B}_\mathrm{b}}$	exponential forgetting factor for averaging the matrix $\hat{\underline{\mathbf{S}}}_{\mathbf{x}_{\mathbf{B}_\mathrm{b}}\mathbf{x}_{\mathbf{B}_\mathrm{b}}}(r)$
β_F	exponential forgetting factor for averaging $F(\hat{\mathbf{\Gamma}}_{\mathbf{dd}}(r,n))$ and of $F(\hat{\mathbf{\Gamma}}_{\mathbf{nn}}(r,n))$ over block time r
$\beta_{\mathbf{w}_\mathrm{a}}$	exponential forgetting factor for averaging $\hat{\underline{\mathbf{S}}}_{\mathbf{x}_{\mathbf{w}_\mathrm{a}}\mathbf{x}_{\mathbf{w}_\mathrm{a}}}(r)$
$\beta_{\mathbf{w}_\mathrm{c}}$	exponential forgetting factor for averaging $\underline{\mathbf{X}}_{\mathbf{B}_\mathrm{b}}^{H}(r)\underline{\mathbf{X}}_{\mathbf{B}_\mathrm{b}}(r)$
$\mathbf{\Gamma}_{\mathbf{dd}}(\omega)$	spatial coherence matrix of size $M \times M$ w.r.t. the desired signal at the sensors in the DTFT domain
$\hat{\mathbf{\Gamma}}_{\mathbf{dd}}(r,n)$	estimate of $\mathbf{\Gamma}_{\mathbf{dd}}(\omega)$ as a function of the block time r
$\mathbf{\Gamma}_{\mathbf{nn}}(\omega)$	spatial coherence matrix of size $M \times M$ w.r.t. interference-plus-noise at the sensors in the DTFT domain
$\mathbf{\Gamma}_{\mathbf{n}_\mathrm{c}\mathbf{n}_\mathrm{c}}(\omega)$	spatial coherence matrix of size $M \times M$ w.r.t. interference at the sensors in the DTFT domain
$\hat{\mathbf{\Gamma}}_{\mathbf{nn}}(r,n)$	estimate of $\mathbf{\Gamma}_{\mathbf{nn}}(\omega)$ as a function of the block time r in the DFT domain
$\mathbf{\Gamma}_{\mathrm{si}}(\omega)$	spatial coherence matrix of a spherically isotropic wave field for a uniform sensor array in the DTFT domain
$\mathbf{\Gamma}_{\mathbf{xx}}(\omega)$	spatial coherence matrix of size $M \times M$ of the sensor signals in the DTFT domain

$\gamma_{\mathrm{si},x_{m_1}x_{m_2}}(\omega)$	spatial coherence function of a diffuse wave field between two positions $\mathbf{p}_{m_1}$ and $\mathbf{p}_{m_2}$ in the DTFT domain
$\gamma_{x_{m_1}x_{m_2}}(\omega)$	spatial coherence function of a space-time random field $x(\mathbf{p},t)$ between two positions $\mathbf{p}_{m_1}$ and $\mathbf{p}_{m_2}$ in the DTFT domain
$\Delta\hat{\Psi}_{\mathrm{th,d}}(r,n)$	minimum of $1/\hat{\Psi}(r,n)$ over a data block of length D as a function of the block time r
$\Delta\hat{\Psi}_{\mathrm{th,n}}(r,n)$	minimum of $\hat{\Psi}(r,n)$ over a data block of length D as a function of the block time r
Δ_F	threshold for direct update of $F(\hat{\mathbf{\Gamma}}_{\mathbf{dd}}(r,n))$ and of $F(\hat{\mathbf{\Gamma}}_{\mathbf{nn}}(r,n))$
$\Delta W_{\mathrm{rel}}(k)$ (dB)	system error as a function of the discrete time k
$\delta(t)$	unit impulse in the continuous time domain
$\eta_x(\mathbf{p},t)$	mean value of the space-time random field $x(\mathbf{p},t)$ as a function of the position $\mathbf{p}$ and of time t
θ (rad)	elevation angle measured from the positive z-axis, incidence angle of a propagating wave onto a linear array
θ_{d} (rad)	elevation angle of the DOA of the desired signal
θ_{n} (rad)	elevation angle of the DOA of an interferer
κ	discrete time lag index, synchronization delay
$\kappa_{\mathbf{B_b}}$	synchronization delay between the sensor signals and the output signals of $\mathbf{B_b}(k)$
$\kappa_{\mathbf{w}_{\mathrm{a}}}$	synchronization delay between $y_{\mathbf{w}_{\mathrm{c}}}(k)$ and the output signal of $\mathbf{w}_{\mathrm{a,s}}(k)$
$\kappa_{\mathbf{w}_{\mathrm{c}}}$	synchronization delay between $y_{\mathbf{w}_{\mathrm{c}}}(k)$ and $x_m(k)$
$\mathbf{\Lambda}_{\mathbf{C}}(k)$	projection of $\mathbf{\Phi}_{\mathbf{xx}}^{-1}(k)$ into the space which is spanned by $\mathbf{C}(k)$
$\mathbf{\Lambda}_{\widetilde{\mathbf{C}}_{V'}}(k)$	projection of $\mathbf{\Sigma}_{\mathbf{X},V'}^{-1}(k)$ into the space which is spanned by $\widetilde{\mathbf{C}}_{V'}(k)$
$\mathbf{\Lambda}_{\mathbf{D}}(k)$	projection of $\mathbf{\Phi}_{\mathbf{nn}}^{-1}(k)$ into the space which is spanned by $\mathbf{D}(k)$
$\mathbf{\Lambda}_{\mathbf{W}_{\mathrm{c,s}}}(k)$	projection of $\mathbf{\Phi}_{\mathbf{dd},\mathrm{s}}(k)$ into the space which is spanned by $\mathbf{W}_{\mathrm{c,s}}(k)$
$\Lambda_{\mathbf{v}_{\mathrm{d}}}(\omega)$	projection of $\mathbf{S}_{\mathbf{nn}}^{-1}(\omega)$ into the space which is spanned by $\mathbf{v}_{\mathrm{d}}(\mathbf{k},\omega)$
$\Lambda'_{\mathbf{v}_{\mathrm{d}}}(\omega)$	projection of $\mathbf{\Gamma}_{\mathbf{nn}}^{-1}(\omega)$ into the space which is spanned by $\mathbf{v}_{\mathrm{d}}(\mathbf{k},\omega)$
λ (m)	wavelength of a monochromatic wave
λ_{u}	minimum wavelength without spatial aliasing with linear equally spaced sensor arrays
$\boldsymbol{\mu}(r)$	diagonal matrix of size $2MN \times 2MN$ with stepsizes $\mu(r,n)$, $n = 0, 1, \ldots, 2N-1$, on the main diagonal
$\boldsymbol{\mu}_{\mathbf{B_b}}(r)$	diagonal matrix of size $2MN \times 2MN$ with elements $\mu_{\mathbf{B_b}}(r,n)$, $n = 0, 1, \ldots, 2N-1$, on the main diagonal

$\boldsymbol{\mu}_{\mathbf{w}_\mathrm{a}}(r)$	diagonal matrix of size $2MN \times 2MN$ with elements $\mu_{\mathbf{w}_\mathrm{a}}(r,n)$, $n = 0, 1, \ldots, 2N-1$, on the main diagonal
$\mu(r,n)$	stepsize for the update of $\underline{\mathbf{W}}(r)$ as a function of the block-time r and of the DFT bin n
$\mu_{\mathbf{a}}$	stepsize for the update of $\underline{\mathbf{A}}^{(1)}(r')$
$\mu_{\mathbf{B_b}}(r,n)$	stepsize for the update of $\underline{\mathbf{B}}_{\mathbf{b}}(r)$ as a function of the block time r and of the DFT bin n
$\mu_{\mathbf{B_b},0}$	constant stepsize for the update of $\underline{\mathbf{B}}_{\mathbf{b}}(r)$
$\mu_{\mathbf{w}_\mathrm{a}}(r,n)$	stepsize for the update of $\underline{\mathbf{w}}_\mathrm{a}(r)$ as a function of the block time r and of the DFT bin n
$\mu_{\mathbf{w}_\mathrm{a},0}$	constant stepsize for the update of $\underline{\mathbf{w}}_\mathrm{a}(r)$
$\underline{\xi}(r)$	cost function of the MIMO optimum filter in the DFT domain as a function of the block time r
$\underline{\xi}_\mathrm{c}(r)$	constrained cost function of the MIMO optimum filter in the DFT domain as a function of the block time r
$\xi_{\mathrm{LCMV}}(k)$	LCMV cost function in the discrete time-domain
$\xi_{\mathrm{MMSE}}(k)$	MMSE cost function in the discrete time-domain
$\xi_{\mathrm{LC}}(r)$	LCLSE cost function as a function of block time r
$\xi_{\mathrm{LC,o}}(r)$	minimum of $\xi_{\mathrm{LC}}(r)$
$\xi_{\mathrm{LC}}^{(\mathrm{s})}(r)$	LCLSE cost function for spatio-temporal constraints as a function of block time r
$\xi_{\mathrm{LSE}}(r)$	LSE cost function as a function of the block time r
$\xi_{\mathrm{LSE,o}}(r)$	minimum of the cost function of the LSE beamformer as a function of block time r
ξ_{MMSE}	MMSE optimization criterion
$\boldsymbol{\pi}$ (m)	vector of size 3×1 with the position(s) of excitation
$\boldsymbol{\rho}$	spatial correlation matrix of size $M \times M$
$\rho_{x_{m_1} x_{m_2}}$	spatial correlation coefficient between two sensor positions
$\rho_x(\mathbf{p}_{m_1}, \mathbf{p}_{m_2}; t_1, t_2)$	correlation function coefficient between positions $\mathbf{p}_{m_1}$ and $\mathbf{p}_{m_2}$ and between time instants t_1 and t_2
$\boldsymbol{\Sigma}_{\mathbf{D}_\mathrm{s}}(k)$	diagonal matrix with the singular values of $\mathbf{U}_{\mathbf{X}_\mathrm{s}}(k)$ which correspond to $\mathbf{U}_{\mathbf{D}_\mathrm{s}}(k)$ on the main diagonal
$\boldsymbol{\Sigma}_{\mathbf{D}_\mathrm{s}+\mathbf{N}_\mathrm{s}}(k)$	diagonal matrix with the singular values of $\mathbf{U}_{\mathbf{X}_\mathrm{s}}(k)$ which correspond to $\mathbf{U}_{\mathbf{D}_\mathrm{s}+\mathbf{N}_\mathrm{s}}(k)$ on the main diagonal
$\boldsymbol{\Sigma}_{\mathbf{N}}(k)$	diagonal matrix of size $MN \times MN$ with the singular values of $\mathbf{N}(k)$ on the main diagonal
$\boldsymbol{\Sigma}_{\mathbf{N}_\mathrm{s}}(k)$	diagonal matrix of size $MN \times MN$ with the singular values of $\mathbf{U}_{\mathbf{X}_\mathrm{s}}(k)$ which correspond to $\mathbf{U}_{\mathbf{N}_\mathrm{s}}(k)$ on the main diagonal
$\boldsymbol{\Sigma}_{\mathbf{X}}(k)$	diagonal matrix of size $MN \times MN$ with the singular values of $\mathbf{X}(k)$ on the main diagonal
$\boldsymbol{\Sigma}_{\mathbf{X}_\mathrm{s}}(k)$	diagonal matrix of size $MN^2 \times MN^2$ with the singular values of $\mathbf{X}_\mathrm{s}(k)$ on the main diagonal

$\mathbf{\Sigma}_{\mathbf{X},V'}(k)$	diagonal matrix of size $V' \times V'$ with the V' largest singular values of $\mathbf{X}(k)$ on the main diagonal
σ	weighting term for penalty functions of constrained optimization
$\sigma^2_{\mathbf{E}_\mathrm{o}}(k)$	variance of the measurement error matrix $\mathbf{E}_\mathrm{o}(k)$
$\sigma_{\mathbf{X},v}(k)$	v-th singular value of $\mathbf{X}(k)$
$\sigma^2_d(k)$	variance of the spatially homogeneous desired signal at the sensors as a function of the discrete time k
$\sigma^2_{d_m}(k)$	variance of the desired signal at the m-th sensor as a function of the discrete time k
$\sigma^2_{\mathrm{dl}}(k)$	variance of the diagonal loading for $\mathbf{\Phi}_{\mathbf{xx},\mathrm{dl}}(k)$ as a function of the discrete time k
$\sigma^2_{n_\mathrm{w}}$	variance of the sensor noise
σ^2_x	variance of a spatially and temporally WSS stationary space-time random field $x(\mathbf{p},t)$
$\sigma^2_x(\mathbf{p},t)$	variance of a space-time random field $x(\mathbf{p},t)$ as a function of the position $\mathbf{p}$ and of time t
$\sigma^2_y(k)$	variance of $y(k)$ as a function of the discrete time k
$\sigma^2_{\bar{y}}(k)$	variance of $\bar{y}_m(k)$ averaged over m, $m = 0, 1, \ldots, M-1$, as a function of the discrete time k
τ (s)	continuous time lag or excitation time(s)
τ_m (s)	propagation delay at the m-th sensor for a plane wave with wavenumber vector $\mathbf{k}$ w.r.t. the spatial origin
τ_b (s)	propagation delay between two adjacent sensors for a plane wave which arrives from direction θ relative to a plane wave which arrives from the steering direction of the beamformer
τ'_d (s)	continuous time delay for causality of the beamsteering unit
$\tau_{\mathrm{d},m}$ (s)	propagation delay between position of the desired source and the position of the m-th sensor
$\mathbf{\Phi}_{\mathbf{dd}}(k)$	sample spatio-temporal correlation matrix of size $MN \times MN$ w.r.t. the desired signal at the sensors
$\mathbf{\Phi}_{\mathbf{dd},\mathrm{s}}(k)$	sample spatio-temporal correlation matrix of size $MN^2 \times MN^2$ w.r.t. the desired signal at the sensors
$\hat{\mathbf{\Phi}}_{\mathbf{dd},\mathrm{s}}(k)$	exponentially averaged $\tilde{\mathbf{\Phi}}_{\mathbf{dd},\mathrm{s}}(k)$
$\tilde{\mathbf{\Phi}}_{\mathbf{dd},\mathrm{s}}(k)$	spatio-temporal correlation matrix $\mathbf{\Phi}_{\mathbf{dd},\mathrm{s}}(k)$ in $\mathrm{span}\{\mathbf{U}_{\mathbf{D}_\mathrm{s}}(k)\}$
$\mathbf{\Phi}_{\mathbf{dd},\mathrm{ss}}(k)$	sample spatio-temporal correlation matrix of size $MN^3 \times MN^3$ w.r.t. the desired signal
$\mathbf{\Phi}_{\mathbf{nn}}(k)$	sample spatio-temporal correlation matrix of size $MN \times MN$ w.r.t. interference-plus-noise at the sensors
$\mathbf{\Phi}_{\mathbf{nn},\mathrm{s}}(k)$	sample spatio-temporal correlation matrix of size $MN^2 \times MN^2$ w.r.t. interference-plus-noise at the sensors

$\hat{\mathbf{\Phi}}_{\mathbf{nn},\mathrm{ss}}(k)$	exponentially averaged sample spatio-temporal correlation matrix of size $MN^3 \times MN^3$ w.r.t. interference-plus-noise in the eigenspace of interference-plus-noise only
$\mathbf{\Phi}_{\mathbf{n}_\mathrm{c}\mathbf{n}_\mathrm{c}}(k)$	sample spatio-temporal correlation matrix of size $MN \times MN$ w.r.t. interference at the sensors
$\mathbf{\Phi}_{\mathbf{n}_\mathrm{w}\mathbf{n}_\mathrm{w}}(k)$	sample spatio-temporal correlation matrix of size $MN \times MN$ w.r.t. sensor noise at the sensors
$\mathbf{\Phi}_{\mathbf{xx}}(k)$	sample spatio-temporal correlation matrix of size $MN \times MN$
$\mathbf{\Phi}_{\mathbf{xx},\mathrm{dl}}(k)$	$\mathbf{\Phi}_{\mathbf{xx}}(k)$ with diagonal loading
$\mathbf{\Phi}_{\mathbf{xx},\mathrm{s}}(k)$	sample spatio-temporal correlation matrix of size $MN^2 \times MN^2$ w.r.t. the sensor signals
$\mathbf{\Phi}_{\mathbf{xx},\mathrm{ss}}(k)$	sample spatio-temporal correlation matrix of size $MN^3 \times MN^3$ w.r.t. the sensor signals
ϕ (rad)	azimuthal angle measured counter-clockwise from the positive x-axis
$\underline{\mathbf{\Psi}}(r)$	matrix of size $2MN \times P$ with the penalty function for the update equation of $\underline{\mathbf{W}}(r)$
$\Psi(\omega)$	ratio of $S_{yy}(\omega)$ and $S_{xx}(\omega)$
$\hat{\Psi}(r,n)$	ratio of $\hat{S}_{yy}(r,n)$ and $\hat{S}_{xx}(r,n)$
$\bar{\Psi}(\omega)$	ratio of $S_{\bar{y}\bar{y}}(\omega)$ and $S_{xx}(\omega)$
$\tilde{\Psi}(r,n)$	ratio of $\hat{S}_{\bar{y}\bar{y}}(r,n)$ and $\hat{S}_{xx}(r,n)$
$\underline{\psi}(r)$	vector of size $2MN \times 1$ with the penalty function for the update equation of $\underline{\mathbf{w}}_\mathbf{a}(r)$
$\Upsilon(\omega)$	ratio of $S_{yy}(\omega)$ and $S_{\bar{y}\bar{y}}(\omega)$
$\hat{\Upsilon}(r,n)$	estimate of $\Upsilon(\omega)$ as a function of the block time r in the DFT domain
$\Upsilon_{0,\mathrm{d}}$	correction factor for $\hat{\Upsilon}_{\mathrm{th,d}}(r,n)$
$\Upsilon_{0,\mathrm{n}}$	correction factor for $\hat{\Upsilon}_{\mathrm{th,n}}(r,n)$
$\hat{\Upsilon}_{\mathrm{th,d}}(r,n)$	minima of $1/\hat{\Upsilon}(r,n)$ over a data block of length D
$\hat{\Upsilon}_{\mathrm{th,n}}(r,n)$	minima of $\hat{\Upsilon}(r,n)$ over a data block of length D
$\upsilon(k)$	ratio of $\sigma_y^2(k)$ and $\sigma_{\bar{y}}^2(k)$
ω (rad/s)	temporal radial frequency variable
ω_0 (rad/s)	temporal radial frequency
ω_u (rad/s)	maximum temporal frequency without spatial aliasing for linear equally spaced sensor arrays
$\mathbf{A}^{(1)}(k)$	stacked matrix of size $QN_\mathbf{a} \times M$ with vectors $\mathbf{a}_{q,m}^{(1)}(k)$, $q = 0, 1, \ldots, Q-1$, $m = 0, 1, \ldots, M-1$
$\underline{\mathbf{A}}^{(1)}(r')$	stacked matrix of size $2N_\mathbf{a}Q \times M$ with vectors $\underline{\mathbf{a}}_{q,m}^{(1)}(r')$, $q = 0, 1, \ldots, Q-1$, $m = 0, 1, \ldots, M-1$
$A(k)$	array gain as a function of the discrete time k
$A_\mathrm{f}(\omega)$	array gain for narrowband signals in the DTFT domain
$A_\mathrm{w}(k)$	white noise gain as a function of the discrete time k

$\mathbf{a}(\theta, \phi)$	unit vector in spherical coordinates that points into the direction of propagation of a plane wave
$\mathbf{a}_{q,m}^{(1)}(k)$	weight vector of size $N_{\mathbf{a}} \times 1$ of the AEC between the q-th loudspeaker signal $x_{\mathbf{a},q}(k)$ and the m-th sensor signal $x_m(k)$ as a function of discrete time k
$\underline{\mathbf{a}}_{q,m}^{(1)}(r')$	vector of size $2N_{\mathbf{a}} \times 1$ with the DFT of $\mathbf{a}_{q,m}^{(1)}(r'R')$
$\mathbf{a}^{(2)}(k)$	stacked vector of size $QN_{\mathbf{a}} \times 1$ of the AEC for 'beamformer first'
$\mathbf{a}^{(3)}(k)$	stacked vector of size $QN_{\mathbf{a}} \times 1$ of the AEC for GSAEC
$\mathbf{a}^{(4)}(k)$	stacked vector of size $QN_{\mathbf{a}} \times 1$ of the AEC for GEIC
$\underline{\mathbf{a}}^{(4)}(r)$	stacked vector of size $2N_{\mathbf{a}}Q \times 1$ with $\mathbf{a}^{(4)}(rR)$ in the DFT domain
$\mathbf{a}_{\mathrm{o}}^{(4)}(k)$	optimum of $\mathbf{a}^{(4)}(k)$ in the LSE sense
a	constant for general purpose
$\mathbf{B}(k)$	blocking matrix of size $MN \times BN$ as a function of the discrete time k
$\mathbf{B}_{\mathbf{b}}(k)$	stacked matrix of size $N \times M$ with vectors $\mathbf{b}_m(k)$, $m = 0, 1, \ldots, M-1$
$\mathbf{B}_{\mathbf{b},\mathrm{o}}(k)$	optimum of $\mathbf{B}_{\mathbf{b}}(k)$ in the LSE sense
$\underline{\mathbf{B}}_{\mathbf{b}}(r)$	matrix of size $2N \times M$ with $\mathbf{B}_{\mathbf{b}}(rR)$ in the DFT domain
$\mathbf{B}_{\mathrm{f}}(\omega)$	blocking matrix of size $M \times B$ in the DTFT domain
$\mathbf{B}_{\mathrm{f},\mathrm{o}}(\omega)$	optimum of $\mathbf{B}_{\mathrm{f}}(\omega)$ in the MMSE sense
$\mathbf{B}_{\mathrm{s}}(k)$	spatio-temporal blocking matrix of size $MN^2 \times M$ as a function of the discrete time k
$\mathbf{B}_{\mathrm{s},\mathrm{o}}(k)$	optimum of $\mathbf{B}_{\mathrm{s}}(k)$ in the LSE sense
$\mathbf{B}_{\mathrm{ss}}(k)$	block-diagonal matrix of size $MN^3 \times MN$ with matrices $\mathbf{B}_{\mathrm{s}}(k-n)$, $n = 0, 1, \ldots, N-1$, on the main diagonal
B	number of output channels of a blocking matrix
$B(\omega; \theta, \phi)$	beampattern in spherical coordinates
$B_m(\omega)$	coefficient vector $\mathbf{b}_m(k)$ in the DTFT domain
$\mathbf{b}_m(k)$	vector of size $N \times 1$ with elements $b_{n,m}(k)$, $n = 0, 1, \ldots, N-1$
$\mathbf{b}_{\mathrm{f}}(\omega)$	vector of size $M \times 1$ with elements $B_m^*(\omega)$, $m = 0, 1, \ldots, M-1$
$\mathbf{b}_{\mathrm{f},\mathrm{o}}(\omega)$	optimum of $\mathbf{b}_{\mathrm{f}}(\omega)$ in the MMSE sense
$b_{n,m}(k)$	n-th filter coefficient of the spatio-temporal blocking matrix for the m-th sensor channel as a function of the discrete time k
$\mathbf{C}(k)$	constraint matrix of size $MN \times CN$
$\widetilde{\mathbf{C}}(k)$	constraint matrix $\mathbf{C}(k)$ projected onto $\mathbf{\Sigma}_{\mathbf{X}}^{-1}(k)\mathbf{U}_{\mathbf{X}}(k)$
$\widetilde{\mathbf{C}}_{V'}(k)$	constraint matrix of size $V' \times CN$ projected onto $\mathbf{U}_{\mathbf{X},V'}(k)$
$\mathbf{C}_{\mathrm{f}}(\omega)$	constraint matrix of size $M \times C$ in the DTFT domain

$\mathbf{C}_{\mathrm{s}}(k)$	spatio-temporal constraint matrix of size $MN^2 \times CN^2$ as a function of the discrete time k
C	number of constraints for the LCLSE/LCMV beamformer
C'	number of penalty functions for $\underline{\xi}_{\mathrm{c}}(r)$
$C_x(\mathbf{p}_{m_1}, \mathbf{p}_{m_2}; t_1, t_2)$	covariance function of $x(\mathbf{p}, t)$ between two positions $\mathbf{p}_{m_1}$ and $\mathbf{p}_{m_2}$ and between two time instants t_1 and t_2
$\mathbf{c}(k)$	constraint vector of size $CN \times 1$
$\mathbf{c}_{\mathrm{f}}(\omega)$	constraint vector of size $C \times 1$ in the DTFT domain
$\mathbf{c}_{\mathrm{s}}(k)$	spatio-temporal constraint vector of size $CN^2 \times 1$ as a function of the discrete time k
c (m/s)	sound velocity
$c(\underline{\mathbf{w}}_{\mathrm{a}}(r))$	penalty function for the update of $\underline{\mathbf{w}}_{\mathrm{a}}(r)$
$c_i(\underline{\mathbf{W}}(r))$	i-th penalty function for $\underline{\xi}_{\mathrm{c}}(r)$ as a function of $\underline{\mathbf{W}}(r)$
$\mathbf{D}(k)$	stacked matrix of size $MN \times K$ with data vectors $\mathbf{d}(k+i)$, $i = 0, 1, \ldots, K-1$
$\widetilde{\mathbf{D}}(k)$	$\mathbf{D}(k)$ projected onto $\mathbf{\Sigma}_{\mathbf{N}}^{-1}(k)\mathbf{U}_{\mathbf{N}}(k)$
$\mathbf{D}_{\mathrm{s}}(k)$	stacked matrix of size $MN^2 \times K$ with matrices $\mathbf{D}(k-n)$ for $n = 0, 1, \ldots, N-1$
D	length of the temporal window for searching maxima and minima of $\hat{\Upsilon}(r, n)$
$D(\omega)$	directivity of a beamformer for monochromatic signals in the DTFT domain
$D_m(\omega)$	$d_m(k)$ in the DTFT domain
DC (dB)	cancellation of the desired signal for temporally WSS signals
$DC(k)$ (dB)	cancellation of the desired signal as a function of the discrete time k
$DC_{\mathbf{B}}(k)$ (dB)	cancellation of the desired signal by the blocking matrix $\mathbf{B}(k)$ as a function of the discrete time k
$DC_{\mathbf{B}_{\mathrm{s}}}$ (dB)	cancellation of the desired signal by the blocking matrix $\mathbf{B}_{\mathrm{s}}(k)$ for temporally WSS signals
$DC_{\mathbf{B}_{\mathrm{s}}}(k)$ (dB)	cancellation of the desired signal by the blocking matrix $\mathbf{B}_{\mathrm{s}}(k)$ as a function of the discrete time k
$DC_{\mathrm{f}}(\omega)$ (dB)	cancellation of the desired signal for narrowband signals in the DTFT domain
$DC_{\mathrm{f},\mathbf{B}_{\mathrm{s}}}(\omega)$ (dB)	cancellation of the desired signal by the blocking matrix $\mathbf{B}_{\mathrm{s}}(k)$ for narrowband signals in the DTFT domain
$DC_{\mathrm{f},\mathbf{w}_{\mathrm{GSC}}}(\omega)$ (dB)	cancellation of the desired signal by the GSC for narrowband signals in the DTFT domain
$DC_{\mathrm{f},\mathbf{w}_{\mathrm{RGSC}}}(\omega)$ (dB)	cancellation of the desired signal by the RGSC for narrowband signals in the DTFT domain
$DC_{\mathbf{w}_{\mathrm{RGSC}}}(k)$ (dB)	cancellation of the desired signal by the RGSC as a function of the discrete time k

$DI(\omega)$ (dB)	directivity index for monochromatic signals in the DTFT domain
$\mathbf{d}(k)$	stacked vector of size $MN \times 1$ with vectors $\mathbf{d}_m(k)$, $m = 0, 1, \ldots, M-1$
$\widetilde{\mathbf{d}}(k)$	vector $\mathbf{d}(k)$ projected onto $\mathbf{\Sigma}_{\mathbf{N}}^{-1}(k)\mathbf{U}_{\mathbf{N}}(k)$
$\mathbf{d}_m(k)$	vector of size $N \times 1$ with elements $d_m(k-n)$ $n = 0, 1, \ldots, N-1$
d (m)	spatial distance, sensor spacing
$d_m(k)$	desired signal at the m-th sensor as a function of the discrete time k
$\mathbf{E}(k)$	stacked matrix of size $K \times P$ with vectors of error signals $\mathbf{e}(k+i)$, $i = 0, 1, \ldots, K-1$
$\underline{\mathbf{E}}(r)$	stacked matrix of size $2N \times P$ with vectors $\mathbf{e}_p(rR)$, $p = 0, 1, \ldots, P-1$ in the DFT domain
$\mathbf{E}_{\mathbf{a}}(k)$	stacked matrix of size $N_{\mathbf{a}} \times M$ with vectors $\mathbf{e}_{\mathbf{a},m}(k)$, $m = 0, 1, \ldots, M-1$
$\underline{\mathbf{E}}_{\mathbf{a}}(r')$	matrix of size $2N_{\mathbf{a}} \times M$ with $\mathbf{E}_{\mathbf{a}}(r'R')$ in the DFT domain
$\mathbf{E}_{\mathbf{B}_{\mathrm{b}}}(k)$	matrix of size $K \times M$ with vectors $\mathbf{e}_{\mathbf{B}_{\mathrm{b}}}(k+i)$, $i = 0, 1, \ldots, K-1$
$\underline{\mathbf{E}}_{\mathbf{B}_{\mathrm{b}}}(r)$	matrix of size $2N \times M$ with $\mathbf{E}_{\mathbf{B}_{\mathrm{b}}}(rR)$ in the DFT domain
$\mathbf{E}_{\mathrm{o}}(k)$	matrix of size $K \times P$ with a measurement error
$\mathbf{e}(k)$	vector of size $P \times 1$ with elements $e_p(k)$, $p = 0, 1, \ldots, P-1$
$\mathbf{e}_{\mathbf{a},m}(k)$	vector of size $N_{\mathbf{a}} \times 1$ with elements $e_{\mathbf{a},m}(k+n)$, $n = 0, 1, \ldots, N_{\mathbf{a}}-1$
$\mathbf{e}_{\mathbf{B}_{\mathrm{b}}}(k)$	vector of size $M \times 1$ with elements $e_{\mathbf{B}_{\mathrm{b}},m}(k)$, $m = 0, 1, \ldots, M-1$
$\mathbf{e}_{\mathrm{f}}(k)$	vector of size $P \times 1$ with elements $E_p(\omega)$, $p = 0, 1, \ldots, P-1$
$\mathbf{e}_{\mathbf{w}_{\mathrm{a}}}(k)$	vector of size $N \times 1$ with elements $y_{\mathbf{w}_{\mathrm{RGSC}}}(k+n)$, $n = 0, 1, \ldots, N-1$
$\underline{\mathbf{e}}_{\mathbf{w}_{\mathrm{a}}}(r)$	vector of size $2N \times 1$ with the vector $\mathbf{e}_{\mathbf{w}_{\mathrm{a}}}(rR)$ in the DFT domain
$E_p(\omega)$	DTFT of $e_p(k)$
$EIR_{\mathrm{f,in}}(\omega)$	EIR at the sensors for narrowband signals in the DTFT domain
EIR_{in}	EIR at the sensors for temporally WSS signals
$EIR_{\mathrm{in}}(k)$	EIR at the sensors as a function of the discrete time k
$ERLE_{\mathrm{AEC}}$	ERLE of the AEC for temporally WSS signals
$ERLE_{\mathrm{AEC}}(k)$	ERLE of the AEC as a function of the discrete time k
$ERLE_{\mathrm{f,AEC}}(\omega)$	ERLE of the AEC for narrowband signals in the DTFT domain
$ERLE_{\mathrm{f},\mathbf{w}_{\mathrm{RGSC}}}(\omega)$	ERLE of the RGSC for narrowband signals in the DTFT domain

$ERLE_{\mathrm{f,tot}}(\omega)$	ERLE of the joint system of AEC and beamformer for narrowband signals in the DTFT domain
$ERLE_{\mathrm{tot}}$	ERLE of the joint system of AEC and beamformer for temporally WSS signals
$ERLE_{\mathrm{tot}}(k)$	ERLE of the joint system of AEC and RGSC as a function of the discrete time k
$ERLE_{\mathbf{w}_{\mathrm{RGSC}}}(k)$	ERLE of the RGSC as a function of the discrete time k
$e(k)$	error signal of an LSE beamformer as a function of the discrete time k
$e_p(k)$	error signal of the p-th output channel of a MIMO optimum filter as a function of the discrete time k
$e_{\mathbf{a},m}(k)$	error signal of the AEC in the m-th sensor channel as a function of the discrete time k
$e_{\mathbf{B}_\mathrm{b},m}(k)$	output signal of the m-th channel of the blocking matrix $\mathbf{B}_\mathrm{s}(k)$ as a function of the discrete time k
$\mathbf{F}_{2N\times 2N}$	DFT matrix of size $2N \times 2N$
$F(\mathbf{\Gamma}_{\mathbf{xx}}(\omega))$	$S_{yy}(\omega)$ normalized by $S_{xx}(\omega)$
$\bar{F}(\mathbf{\Gamma}_{\mathbf{xx}}(\omega))$	$S_{\bar{y}\bar{y}}(\omega)$ normalized by $S_{xx}(\omega)$
f (1/s)	temporal frequency variable
f_s (1/s)	sampling frequency
$\mathbf{G}(k)$	stacked matrix of size $MN \times P$ with vectors $\mathbf{g}_{m,p}(k)$, $m = 0, 1, \ldots, M-1$, $p = 0, 1, \ldots, P-1$
$\mathbf{G}'(k)$	stacked matrix of size $P'N \times P$ with vectors $\mathbf{g}'_{p',p}(k)$, $p' = 0, 1, \ldots, P'-1$, $p = 0, 1, \ldots, P-1$
$\tilde{\mathbf{G}}(k)$	stacked matrix of size $(N_\mathbf{g}+N-1)P' \times MN$ with matrices $\mathbf{G}_{p',m}(k)$, $p' = 0, 1, \ldots, P'-1$, $m = 0, 1, \ldots, M-1$
$\mathbf{G}_{p',m}(k)$	Toeplitz matrix of size $N_\mathbf{g} + N - 1 \times N$ formed by the vectors $\mathbf{g}_{p',m}(k)$,
$\mathbf{G}^{01}_{2N\times 2N}$	constraint matrix of size $2N \times 2N$
$\mathbf{G}^{10}_{2MN\times 2MN}$	block-diagonal matrix of size $2MN \times 2MN$ with matrices $\mathbf{G}^{10}_{2N\times 2N}$ on the main diagonal
$\mathbf{G}^{10}_{2N\times 2N}$	constraint matrix of size $2N \times 2N$
$G^{(3)}_{\mathbf{a}}(\omega)$	gain factor of the AEC in GSAEC in the DTFT domain
$G(\mathbf{k}, \omega)$	wavenumber-frequency response or beamformer response
$\mathbf{g}(k)$	vector which captures the impulse response of the combined system of $\mathbf{H}_\mathrm{d}(k)$ and $\mathbf{W}_{\mathrm{c,s}}(k)$
$\mathbf{g}_{m,p}(k)$	vector of size $N_\mathbf{g} \times 1$ with elements $g_{n,m,p}(k)$, $n = 0, 1, \ldots, N-1$
$\mathbf{g}'_{p',p}(k)$	vector of size $N_{\mathbf{g}'} \times 1$ with the impulse response between input channel p' and output channel p of a desired MIMO system $\mathbf{G}'(k)$
$g_{n,m,p}(k)$	n-th filter coefficient of the filter between the m-th input channel and the p-th output channel of a MIMO system $\mathbf{G}(k)$ as a function of the discrete time k

$\mathbf{H}_{\mathrm{d}}(k)$	stacked matrix of size $N_{\mathbf{h}_{\mathrm{d}}} \times M$ with vectors $\mathbf{h}_{\mathrm{d},m}(k)$, $m = 0, 1, \ldots, M-1$
$\mathbf{H}_{\mathrm{n}}(k)$	stacked matrix of size $LN_{\mathbf{h}_{\mathrm{n}}} \times M$ with vectors $\mathbf{h}_{\mathrm{n},m}(k)$, $m = 0, 1, \ldots, M-1$
$H(\mathbf{p}, \boldsymbol{\pi}; \omega)$	DTFT of the spatio-temporal impulse response $h(\mathbf{p}, \boldsymbol{\pi}; \tau)$
$\mathbf{h}_{\mathrm{d},m}(k)$	vector of size $N_{\mathbf{h}_{\mathrm{d}}} \times 1$ with elements $h(\mathbf{p}_m, \mathbf{p}_{\mathrm{d}}; kT_{\mathrm{s}}, \kappa T_{\mathrm{s}})$, $\kappa = 0, 1, \ldots, N_{\mathbf{h}_{\mathrm{d}}} - 1$
$\mathbf{h}_{\mathrm{f}}(\mathbf{p}_{\mathrm{s}}, \omega)$	vector of size $M \times 1$ with elements $H(\mathbf{p}_m, \mathbf{p}_{\mathrm{s}}; \omega)$, $m = 0, 1, \ldots, M-1$
$\mathbf{h}_{\mathrm{ls},q,m}(k)$	vector of size $N_{\mathbf{h}_{\mathrm{ls}}} \times 1$ with elements $h(\mathbf{p}_m, \mathbf{p}_{\mathrm{ls},q}; kT_{\mathrm{s}}, \kappa T_{\mathrm{s}})$, $\kappa = 0, 1, \ldots, N_{\mathbf{h}_{\mathrm{ls}}} - 1$
$\mathbf{h}_{\mathrm{n},l,m}(k)$	vector of size $N_{\mathbf{h}_{\mathrm{n}}} \times 1$ with elements $h(\mathbf{p}_m, \mathbf{p}_{\mathrm{n},l}; kT_{\mathrm{s}}, \kappa T_{\mathrm{s}})$, $\kappa = 0, 1, \ldots, N_{\mathbf{h}_{\mathrm{n}}} - 1$
$\mathbf{h}_{\mathrm{n},m}(k)$	stacked vector of size $LN_{\mathbf{h}_{\mathrm{n}}} \times 1$ with vectors $\mathbf{h}_{\mathrm{n},l,m}(k)$, $l = 0, 1, \ldots, L-1$, $m = 0, 1, \ldots, M-1$
$h(\mathbf{p}, \boldsymbol{\pi}; \tau)$	time-invariant spatio-temporal impulse response between the excitation position $\boldsymbol{\pi}$ and the observation position $\mathbf{p}$ for time lag τ
$h(\mathbf{p}, \boldsymbol{\pi}; t, \tau)$	spatio-temporal impulse response between the excitation position $\boldsymbol{\pi}$ and the observation position $\mathbf{p}$ for time lag τ as a function of the continuous time t
$\mathbf{I}_{m \times m}$	identity matrix of size $m \times m$
$I_{x_{m_1} x_{m_2}}(r, n)$	second-order periodogram between $x_{m_1}(k)$ and $x_{m_2}(k)$ in the DTFT domain using data-overlapping and windowing the time domain
IR (dB)	interference rejection for temporally WSS signals
$IR(k)$ (dB)	interference rejection as a function of the discrete time k
$IR_{\mathbf{B}_{\mathrm{s}}}(k)$ (dB)	interference rejection of the blocking matrix $\mathbf{B}_{\mathrm{s}}(k)$ as a function of the discrete time k
$IR_{\mathrm{f}}(\omega)$ (dB)	interference rejection for narrowband signals in the DTFT domain
$IR_{\mathrm{f},\mathbf{w}_{\mathrm{c}}}(\omega)$ (dB)	interference rejection of the quiescent weight vector $\mathbf{w}_{\mathrm{c}}$ for narrowband signals in the DTFT domain
$IR_{\mathrm{f},\mathbf{w}_{\mathrm{RGSC}}}(\omega)$ (dB)	interference rejection of the RGSC for narrowband signals in the DTFT domain
$IR_{\mathbf{w}_{\mathrm{c}}}$ (dB)	interference rejection of the quiescent weight vector $\mathbf{w}_{\mathrm{c}}$ for temporally WSS signals
$IR_{\mathbf{w}_{\mathrm{c}}}(k)$ (dB)	interference rejection of the quiescent weight vector $\mathbf{w}_{\mathrm{c}}(k)$ as a function of the discrete time k
$IR_{\mathbf{w}_{\mathrm{GSC}}}(k)$ (dB)	interference rejection of the GSC as a function of the discrete time k
$IR_{\mathbf{w}_{\mathrm{RGSC}}}$ (dB)	interference rejection of the RGSC for temporally WSS signals
$IR_{\mathbf{w}_{\mathrm{RGSC}}}(k)$ (dB)	interference rejection of the RGSC as a function of the discrete time k

i	index for general purpose
$\mathbf{J}^{(N)}_{MN^2 \times M}$	stacked matrix of size $MN^2 \times M$ with vectors $\mathbf{1}^{(n)}_{MN^2 \times 1}$, $n = 0, N, \ldots, MN - N$
$\mathbf{j}^{(M)}_{M^2 \times 1}$	stacked vector of size $M^2 \times 1$ with vectors $\mathbf{1}^{(m)}_{M \times 1}$, where $m = 0, 1, \ldots, M - 1$
j	square root of -1
K	length of a data block
$\mathbf{k}$ (rad/m)	wavenumber vector
k	discrete time index
L	number of interferers
l	index variable for interferers
M	number of sensors of an array or number of input channels of a MIMO system
m	index of a position of observation
$\mathbf{N}(k)$	stacked matrix of size $MN \times K$ with vectors $\mathbf{n}(k + i)$, $i = 0, 1, \ldots, K - 1$
$\widetilde{\mathbf{N}}(k)$	$\mathbf{N}(k)$ projected onto $\mathbf{\Sigma}^{-1}_{\mathbf{N}}(k)\mathbf{U}_{\mathbf{N}}(k)$
$\mathbf{N}_{\mathrm{c}}(k)$	stacked matrix of size $MN \times K$ with vectors $\mathbf{n}_{\mathrm{c}}(k + i)$, $i = 0, 1, \ldots, K - 1$
$\mathbf{N}_{\mathrm{s}}(k)$	stacked matrix of size $MN^2 \times K$ with matrices $\mathbf{N}(k - n)$ for $n = 0, 1, \ldots, N - 1$
$\mathbf{N}_{\mathrm{w}}(k)$	stacked matrix of size $MN \times K$ with vectors $\mathbf{n}_{\mathrm{c}}(k + i)$, $i = 0, 1, \ldots, K - 1$
N	filter length
$N_{\mathbf{a}}$	number of filter coefficients of $\mathbf{a}_{q,m}(k)$
$N_{\mathbf{g}}$	number of filter coefficients of $\mathbf{g}_{m,p}(k)$
$N_{\mathbf{g}'}$	number of filter coefficients of $\mathbf{g}'_{p',p}(k)$
$N_{\mathbf{h}_{\mathrm{d}}}$	number of filter coefficients of $\mathbf{h}_{\mathrm{d},m}(k)$
$N_{\mathbf{h}_{\mathrm{ls}}}$	number of filter coefficients of $\mathbf{h}_{\mathrm{ls},q,m}(k)$
$N_{\mathbf{h}_{\mathrm{n}}}$	number of filter coefficients of $\mathbf{h}_{\mathrm{n},l,m}(k)$
$\mathbf{n}(k)$	stacked vector of size $MN \times 1$ with vectors $\mathbf{n}_m(k)$, $m = 0, 1, \ldots, M - 1$
$\widetilde{\mathbf{n}}(k)$	vector $\mathbf{n}(k)$ projected onto $\mathbf{\Sigma}^{-1}_{\mathbf{N}}(k)\mathbf{U}_{\mathbf{N}}(k)$
$\mathbf{n}_m(k)$	vector of size $N \times 1$ with elements $n_m(k - n)$, $n = 0, 1, \ldots, N - 1$
$\mathbf{n}_{\mathrm{c}}(k)$	stacked vector of size $MN \times 1$ with vectors $\mathbf{n}_{\mathrm{c},m}(k)$, $m = 0, 1, \ldots, M - 1$
$\mathbf{n}_{\mathrm{c},m}(k)$	vector of size $N \times 1$ with elements $n_{\mathrm{c},m}(k - n)$, $n = 0, 1, \ldots, N - 1$
$\mathbf{n}_{\mathrm{w}}(k)$	stacked vector of size $MN \times 1$ with vectors $\mathbf{n}_{\mathrm{w},m}(k)$, $m = 0, 1, \ldots, M - 1$
$\mathbf{n}_{\mathrm{w},m}(k)$	vector of size $N \times 1$ with elements $n_{\mathrm{w},m}(k - n)$, $n = 0, 1, \ldots, N - 1$
n	index for filter coefficients or for DFT bins

$n_m(k)$	signal of interference-plus-noise at the m-th sensor as a function of the discrete time k
$n_{\mathrm{c},l}(\mathbf{p}_m, t)$	signal of the l-th interferer at the m-th sensor as a function of the continuous time t
$n_{\mathrm{c},l,m}(k)$	signal of the l-th interferer at the m-th sensor as a function of the discrete time k
$n_{\mathrm{c},m}(k)$	interference signal at the m-th sensor as a function of the discrete time k
$n_{\mathrm{ls},q,m}(k)$	acoustic echo of the q-th loudspeaker at the m-th sensor as a function of discrete time k
$n_{\mathrm{w},m}(k)$	sensor noise at the m-th sensor as a function of the discrete time k
P	number of output channels of a MIMO system
P'	number of input channels of the MIMO system
$P(\omega; \theta, \phi)$	power pattern in spherical coordinates
$\mathbf{p}$ (m)	position vector of size 3×1 with spatial coordinates
$\mathbf{p}_m$ (m)	position of the m-th sensor
$\mathbf{p}_{\mathrm{d}}$ (m)	position of the desired source
$\mathbf{p}_{\mathrm{ls},q}$	position of the q-th loudspeaker
$\mathbf{p}_{\mathrm{n},l}$	position of the l-th interferer
$\mathbf{p}_{\mathrm{s}}$ (m)	position of a point source
p	index of an output channel of a MIMO system
p'	index of an input channel of a MIMO system
$p_x(x(\mathbf{p}, t))$	probability density function of the space-time random field $x(\mathbf{p}, t)$
Q	number of loudspeakers
q	index of a loudspeaker
$\mathbf{R}_{\mathbf{dd}}(k)$	spatio-temporal correlation matrix of size $MN \times MN$ of the desired signal at the sensors as a function of the discrete time k
$\mathbf{R}_{\mathbf{nn}}(k)$	spatio-temporal correlation matrix of size $MN \times MN$ of interference-plus-noise at the sensors as a function of the discrete time k
$\mathbf{R}_{\mathbf{n}_\mathrm{c}\mathbf{n}_\mathrm{c}}(k)$	spatio-temporal correlation matrix of size $MN \times MN$ of interference at the sensors as a function of the discrete time k
$\mathbf{R}_{\mathbf{n}_\mathrm{w}\mathbf{n}_\mathrm{w}}$	spatio-temporal correlation matrix of the sensor noise at the sensors
$\mathbf{R}_{\mathbf{xx}}$	spatio-temporal correlation matrix of size $MN \times MN$ w.r.t. the temporally WSS sensor signals $\mathbf{x}(k)$
$\mathbf{R}_{\mathbf{xx}}(k)$	spatio-temporal correlation matrix of size $MN \times MN$ w.r.t. the sensor signals $\mathbf{x}(k)$ as a function of the discrete time k
$\mathbf{R}_{xx}$	temporal correlation matrix of size $N \times N$ of a temporally WSS random process

$\mathbf{R}_{x_{m_1}x_{m_2}}$	spatio-temporal correlation matrix between $x_{m_1}(k)$ and $x_{m_2}(k)$ for temporally WSS space-time random fields
$\mathbf{R}_{x_{m_1}x_{m_2}}(k)$	spatio-temporal correlation matrix between between $x_{m_1}(k)$ and $x_{m_2}(k)$ as a function of the discrete time k
R	block of data samples ($R = K/\alpha$)
R'	block of data samples ($R' = N_{\mathbf{a}}/\alpha_{\mathbf{a}}$)
$R_x(\boldsymbol{\pi};\tau)$	spatio-temporal correlation function for a spatially homogeneous and temporally WSS space-time random field $x(\mathbf{p},t)$ for a spatial lag vector $\boldsymbol{\pi}$ and time lag τ
$R_x(\boldsymbol{\pi};t_1,t_2)$	spatio-temporal correlation function for a spatially homogeneous space-time random field $x(\mathbf{p},t)$ for a spatial lag vector $\boldsymbol{\pi}$ between time instants t_1 and t_2
$R_x(\mathbf{p}_{m_1},\mathbf{p}_{m_2};\tau)$	spatio-temporal correlation function for a temporally WSS space-time random field $x(\mathbf{p},t)$ between the positions $\mathbf{p}_{m_1}$ and $\mathbf{p}_{m_2}$ for a time lag τ
$R_x(\mathbf{p}_{m_1},\mathbf{p}_{m_2};t_1,t_2)$	spatio-temporal correlation function for a space-time random field $x(\mathbf{p},t)$ between the positions $\mathbf{p}_{m_1}$ and $\mathbf{p}_{m_2}$ between time instants t_1 and t_2
$R_x^{(s)}(\boldsymbol{\pi})$	spatial correlation function for a spatially WSS space-time random field $x(\mathbf{p},t)$ for a spatial lag vector $\boldsymbol{\pi}$
$R_x^{(t)}(\tau)$	temporal correlation function for a temporally WSS space-time random field $x(\mathbf{p},t)$ for a time lag τ
r	block time index ($k = rR$)
r (m)	radius in a spherical coordinate system
r'	block time index ($k = r'R'$)
$\mathbf{S}_{\mathbf{dd}}(\omega)$	spatio-spectral correlation matrix of size $M \times M$ w.r.t. the desired signal at the sensors in the DTFT domain
$\mathbf{S}_{\mathbf{nn}}(\omega)$	spatio-spectral correlation matrix of size $M \times M$ w.r.t. interference-plus-noise at the sensors in the DTFT domain
$\mathbf{S}_{\mathbf{n}_c\mathbf{n}_c}(\omega)$	spatio-spectral correlation matrix of size $M \times M$ w.r.t. interference at the sensors in the DTFT domain
$\mathbf{S}_{\mathbf{n}_w\mathbf{n}_w}(\omega)$	spatio-spectral correlation matrix of size $M \times M$ w.r.t. the sensor noise at the sensors in the DTFT domain
$\mathbf{S}_{\mathbf{xx}}(\omega)$	spatio-spectral correlation matrix of size $M \times M$ in the DTFT domain
$\hat{\underline{\mathbf{S}}}_{\mathbf{xx}}(r)$	matrix of size $2MN \times 2MN$ with the (constrained) recursively averaged estimate of $\mathbf{S}_{\mathbf{xx}}(\omega)$ as a function of the block time r in the DFT domain
$\hat{\underline{\mathbf{S}}}'_{\mathbf{xx}}(r)$	matrix of size $2MN \times 2MN$ with the (unconstrained) recursively averaged estimate of $\mathbf{S}_{\mathbf{xx}}(\omega)$ as a function of the block time r in the DFT domain

$\hat{\underline{\mathbf{S}}}_{\mathbf{x_a x_a}}(r')$	matrix of size $2N_{\mathbf{a}}Q \times 2N_{\mathbf{a}}Q$ with the recursively averaged PSD of $\underline{\mathbf{X}}_{\mathbf{a}}(r')$
$\hat{\underline{\mathbf{S}}}_{\mathbf{x_{B_b} x_{B_b}}}(r)$	diagonal matrix of size $2N \times 2N$ with the recursively averaged PSD of $\underline{\mathbf{X}}_{\mathbf{B_b}}(r)$ on the main diagonal
$\hat{\underline{\mathbf{S}}}_{\mathbf{x_{w_a} x_{w_a}}}(r)$	matrix of size $2MN \times 2MN$ with the recursively averaged CPSD of $\underline{\mathbf{X}}_{\mathbf{w_a}}(r)$
$\mathbf{S}_{\mathbf{x}\mathbf{y}_{\mathrm{ref}}}(\omega)$	CPSD matrix of size $M \times P$ between the sensor signals and the reference signals of a MIMO system
$\hat{\underline{\mathbf{S}}}_{\mathbf{xy}}(r)$	matrix of size $2MN \times P$ with the recursively averaged estimate of the CPSD between $\underline{\mathbf{X}}(r)$ and $\underline{\mathbf{Y}}_{\mathrm{ref}}(r)$ in the DFT domain
$\mathbf{s}_{\mathrm{n}}(k)$	stacked vector of size $LN_{\mathbf{h}_{\mathrm{n}}} \times 1$ with vectors $\mathbf{s}_{\mathrm{n},l}(k)$, $l = 0, 1, \ldots, L-1$
$\mathbf{s}_{\mathrm{n},l}(k)$	vector of size $N_{\mathbf{h}_{\mathrm{n}}} \times 1$ with elements $s_{\mathbf{p}_{\mathrm{n},l}}((k-n)T_{\mathrm{s}})$, $n = 0, 1, \ldots, N_{\mathbf{h}_{\mathrm{n}}}$
$\mathbf{s}_{\mathbf{x}y_{\mathrm{ref}}}(\omega)$	CPSD vector of size $M \times 1$ between the sensor signals and the reference signal of an MMSE beamformer
S	amplitude of a monochromatic plane wave
$S_{dd}(\omega)$	PSD of the spatially homogeneous and temporally WSS desired signal in the DTFT domain
$\hat{S}_{dd}(r,n)$	recursively averaged estimate of the PSD $S_{dd}(\omega)$ in the DFT domain as a function of the block time r
$S_{nn}(\omega)$	PSD of spatially homogeneous and temporally WSS interference-plus-noise in the DTFT domain
$\hat{S}_{nn}(r,n)$	recursively averaged estimate of the PSD $S_{nn}(\omega)$ in the DFT domain as a function of the block time r
$S_{n_{\mathrm{c}}n_{\mathrm{c}}}(\omega)$	PSD of the spatially homogeneous and temporally WSS interference in the DTFT domain
$S_{n_{\mathrm{ls}}n_{\mathrm{ls}}}(\omega)$	PSD of spatially homogeneous and temporally WSS acoustic echoes in the DTFT domain
$S_{\mathbf{p}_{\mathrm{s}}}(\omega)$	DTFT $s_{\mathbf{p}_{\mathrm{s}}}(kT_{\mathrm{s}})$
$S_{\mathbf{p}_{\mathrm{d}}}(\omega)$	DTFT of $s_{\mathbf{p}_{\mathrm{d}}}(kT_{\mathrm{s}})$
$S_{ss}(\omega)$	PSD of $s_{\mathbf{p}_{\mathrm{s}}}(kT_{\mathrm{s}})$ in the DTFT domain
$\hat{S}_{xx}(r,n)$	average PSD of the sensor signals $x_m(k)$, $m = 0, 1, \ldots, M-1$, in the DFT domain as a function of the block time r
$S_{xx}(\omega)$	PSD of a spatially homogeneous and temporally WSS space-time random field in the DTFT domain
$S_{x_{m_1}x_{m_2}}(\omega)$	spatio-spectral correlation function between two sensor signals $x_{m_1}(k)$ and $x_{m_2}(k)$ in the DTFT domain
$\hat{S}_{x_{m_1}x_{m_2}}(r,n)$	recursively averaged estimate of the spatio-spectral correlation function $S_{x_{m_1}x_{m_2}}(\omega)$ in the DFT domain as a function of the block time r

$S_{yy}(\omega)$	PSD of $y(k)$ for spatially homogeneous and temporally WSS wave fields in the DTFT domain
$S_{yy}^{(3)}(\omega)$	PSD of $y^{(3)}(k)$ for spatially homogeneous and temporally WSS wave fields in the DTFT domain
$\hat{S}_{yy}(r,n)$	recursively averaged estimate of the PSD $S_{yy}(\omega)$ in the DFT domain as a function of the block time r
$S_{\bar{y}\bar{y}}(\omega)$	average PSD of the signals $\bar{y}_m(k)$, $m = 0, 1, \ldots, M-1$, in the DTFT domain
$\hat{S}_{\bar{y}\bar{y}}(r,n)$	recursively averaged estimate of the PSD $S_{\bar{y}\bar{y}}(\omega)$ in the DFT domain as a function of the block time r
$S_{y_{\mathbf{n}_c} y_{\mathbf{n}_c}}(\omega)$	PSD of spatially homogeneous and temporally WSS local interference at the output of GSAEC in the DTFT domain
$S_{y_{\mathbf{n}_{ls}} y_{\mathbf{n}_{ls}}}(\omega)$	PSD of spatially homogeneous and temporally WSS acoustic echoes at the output of GSAEC in the DTFT domain
$SINR_{\mathrm{f,in}}(\omega)$	SINR at the sensors for narrowband signals in the DTFT domain
$SINR_{\mathrm{f,out}}(\omega)$	SINR at the output of a beamformer for narrowband signals in the DTFT domain
$SINR_{\mathrm{in}}$	SINR at the sensors for for temporally WSS signals
$SINR_{\mathrm{in}}(k)$	SINR at the sensors as a function of the discrete time k
$SINR_{\mathrm{out}}(k)$	SINR at the beamformer output as a function of the discrete time k
$SINR_{\mathrm{f,in}}(r,n)$	recursively averaged estimate of the SINR at the sensors in the DFT domain as a function of the block time r
$\widehat{SINR}_{\mathrm{f,in}}(r,n)$	estimate of $SINR_{\mathrm{f,in}}(r,n)$
$s(\mathbf{p},t)$	propagating wave observed at position $\mathbf{p}$ at time t
$s_{\mathrm{n}}(\mathbf{p},t)$	interference signal as a function of the position $\mathbf{p}$ and of continuous time t
$s_{\mathrm{n},l}(\mathbf{p},t)$	signal of the l-th interferer as a function of the position $\mathbf{p}$ and of continuous time t
$s_{\mathbf{p}_{\mathrm{d}}}(t)$	signal of the desired source at position $\mathbf{p}_{\mathrm{d}}$ as a function of the continuous time t
$s_{\mathbf{p}_{\mathrm{ls},q}}(t)$	signal of the q-th loudspeaker located at position $\mathbf{p}_{\mathrm{ls},l}$ as a function of the continuous time t
$s_{\mathbf{p}_{\mathrm{n},l}}(t)$	signal of the l-th interferer, which is modeled as a point source at position $\mathbf{p}_{\mathrm{n},l}$ as a function of the continuous time t
$s_{\mathbf{p}_{\mathrm{s}}}(t)$	signal of a point source at position $\mathbf{p}_{\mathrm{s}}$ as a function of the continuous time t
$T(k)$	array sensitivity against uncorrelated random errors in the discrete time domain
$T_0(k)$	constrained upper limit of $T(k)$

T_{60} (s)	reverberation time
$T_\mathrm{f}(\omega)$	array sensitivity against uncorrelated random errors in the DTFT domain
T_s (s)	sampling period
t (s)	continuous observation time
$\mathbf{U}_{\mathbf{D}_\mathrm{s}}(k)$	matrix with eigenvectors which span a subspace of the eigenspace of $\mathbf{\Phi}_{\mathbf{dd},\mathrm{s}}(k)$ such that $\mathbf{U}_{\mathbf{D}_\mathrm{s}}(k)$ is orthogonal to $\mathbf{U}_{\mathbf{N}_\mathrm{s}}(k)$
$\mathbf{U}_{\mathbf{D}_\mathrm{s}+\mathbf{N}_\mathrm{s}}(k)$	matrix with columns of $\mathbf{U}_{\mathbf{D}_\mathrm{s}}(k)$ which are not contained in $\mathbf{U}_{\mathbf{D}_\mathrm{s}}(k)$ and $\mathbf{U}_{\mathbf{N}_\mathrm{s}}(k)$
$\mathbf{U}_{\mathbf{N}}(k)$	matrix of size $MN \times MN$ with left singular vectors of $\mathbf{N}(k)$
$\mathbf{U}_{\mathbf{N}_\mathrm{s}}(k)$	matrix with eigenvectors which span a subspace of the eigenspace of $\mathbf{\Phi}_{\mathbf{n},\mathrm{s}}(k)$ such that $\mathbf{U}_{\mathbf{D}_\mathrm{s}}(k)$ is orthogonal to $\mathbf{U}_{\mathbf{N}_\mathrm{s}}(k)$
$\mathbf{U}_{\mathbf{X}}(k)$	matrix of size $MN \times MN$ with left singular vectors of $\mathbf{X}(k)$
$\mathbf{U}_{\mathbf{X}_\mathrm{s}}(k)$	matrix of size $MN^2 \times MN^2$ with left singular vectors of $\mathbf{X}_\mathrm{s}(k)$
$\mathbf{U}_{\mathbf{X},V'}(k)$	matrix of size $MN \times V'$ with the left singular vectors of $\mathbf{X}(k)$ which correspond to the V' largest singular values of $\mathbf{X}(k)$
$\mathbf{V}_{\mathbf{X}}(k)$	matrix of size $K \times K$ with the right singular vectors of $\mathbf{X}(k)$
$\mathbf{V}_{\mathbf{N}}(k)$	matrix of size $K \times K$ with the right singular vectors of $\mathbf{N}(k)$
$\mathbf{v}(\mathbf{k},\omega)$	steering vector of size $M \times 1$
V	given number of singular values of $\mathbf{X}(k)$
V'	number of largest singular values of $\mathbf{X}(k)$
v	index of a singular value of $\mathbf{X}(k)$
$\mathbf{W}(k)$	stacked matrix of size $MN \times P$ with vectors $\mathbf{w}_{m,p}(k)$, $m = 0, 1, \ldots, M-1$, $p = 0, 1, \ldots, P-1$
$\underline{\mathbf{W}}(r)$	matrix of size $2MN \times P$ with $\mathbf{W}(rR)$ in the DFT domain
$\mathbf{W}^{01}_{2N\times 2N}$	windowing matrix of size $2N \times 2N$
$\mathbf{W}^{10}_{N\times 2N}$	windowing matrix of size $N \times 2N$
$\mathbf{W}^{10}_{2N\times N}$	windowing matrix of size $2N \times N$
$\mathbf{W}_\mathrm{f}(k)$	matrix of size $M \times P$ with elements $W^*_{m,p}(\omega)$, $m = 0, 1, \ldots, M-1$, $p = 0, 1, \ldots, P-1$
$\mathbf{W}_{\mathrm{c},\mathrm{s}}(k)$	block-diagonal matrix of size $MN^2 \times N$ with vectors $\mathbf{w}_\mathrm{c}(k-n)$, $n = 0, 1, \ldots, N-1$, on the main diagonal
$\mathbf{W}_\mathrm{LSE}(k)$	MIMO system $\mathbf{W}(k)$ which is optimum in the LSE sense
$W_m(\omega)$	DTFT of $\mathbf{w}_m$
$W_{m,p}(\omega)$	DTFT of $\mathbf{w}_{m,p}(k)$
$W_{\mathrm{c},m}(\omega)$	DTFT of $\mathbf{w}_{\mathrm{c},m}$

$\mathbf{w}$	stacked vector of size $MN \times 1$ with vectors $\mathbf{w}_m$, $m = 0, 1, \ldots, M-1$
$\mathbf{w}(k)$	stacked vector of size $MN \times 1$ with vectors $\mathbf{w}_m(k)$, $m = 0, 1, \ldots, M-1$
$\widetilde{\mathbf{w}}(k)$	weight vector $\mathbf{w}(k)$ in the eigenspace of $\mathbf{\Phi}_{\mathbf{xx}}^{-1}(k)$
$\mathbf{w}_m$	weight vector $\mathbf{w}_m(k)$ for fixed beamformers
$\mathbf{w}_m(k)$	vector of size $N \times 1$ with elements $w_{n,m}(k)$, $n = 0, 1, \ldots, N-1$
$\mathbf{w}_{m,p}(k)$	vector of size $N \times 1$ with elements $w_{n,m,p}(k)$, $n = 0, 1, \ldots, N-1$
$\mathbf{w}_{V',\mathrm{o}}(k)$	optimum weight vector of the LCLSE beamformer in the space which is spanned by the eigenvectors $\mathbf{U}_{\mathbf{X},V'}(k)$ of $\mathbf{\Phi}_{\mathbf{xx}}(k)$.
$\mathbf{w}_{\mathrm{a}}(k)$	stacked vector of size $MN \times 1$ with vectors $\mathbf{w}_{\mathrm{a},m}(k)$, $m = 0, 1, \ldots, M-1$
$\underline{\mathbf{w}}_{\mathrm{a}}(r)$	vector of size $2MN \times 1$ with the vector $\mathbf{w}_{\mathrm{a,s}}(rR)$ in the DFT domain
$\mathbf{w}_{\mathrm{a},m}(k)$	vector of size $BM \times 1$ with elements $w_{\mathrm{a},n,m}(k)$, $n = 0, 1, \ldots, N-1$, $m = 0, 1, \ldots, B-1$
$\mathbf{w}_{\mathrm{a,o}}(k)$	optimum of $\mathbf{w}_{\mathrm{a}}(k)$ in the LCLSE sense
$\mathbf{w}_{\mathrm{a,s}}(k)$	stacked vector of size $MN \times 1$ with elements $w_{\mathrm{a},n,m}(k)$ for the RGSC
$\mathbf{w}_{\mathrm{a,s}}^{(4)}(k)$	stacked vector of size $(MN + QN_{\mathbf{a}}) \times 1$ with $\mathbf{w}_{\mathrm{a,s}}(k)$ and $\mathbf{a}^{(4)}(k)$
$\mathbf{w}_{\mathrm{a,s,o}}(k)$	optimum of $\mathbf{w}_{\mathrm{a,s}}(k)$ in the LCLSE sense
$\mathbf{w}_{\mathrm{a,s,o}}^{(4)}(k)$	optimum of $\mathbf{w}_{\mathrm{a,s}}^{(4)}(k)$ in the LSE sense
$\mathbf{w}_{\mathrm{c}}$	stacked vector of size $MN \times 1$ with vectors $\mathbf{w}_{\mathrm{c},m}$, $m = 0, 1, \ldots, M-1$
$\underline{\mathbf{w}}_{\mathrm{c}}$	vector of size $2MN \times 1$ with the vector $\mathbf{w}_{\mathrm{c}}$ in the DFT domain
$\mathbf{w}_{\mathrm{c}}(k)$	stacked vector of size $MN \times 1$ with vectors $\mathbf{w}_{\mathrm{c},m}(k)$, $m = 0, 1, \ldots, M-1$
$\mathbf{w}_{\mathrm{c},m}$	weight vector $\mathbf{w}_{\mathrm{c},m}(k)$ for fixed beamformers
$\mathbf{w}_{\mathrm{c},m}(k)$	vector of size $MN \times 1$ with elements $w_{\mathrm{c},n,m}(k)$, $n = 0, 1, \ldots, N-1$
$\mathbf{w}_{\mathrm{c,s}}(k)$	stacked vector of size $MN^2 \times 1$ with vectors $\mathbf{w}_{\mathrm{c}}(k-n)$ for $n = 0, 1, \ldots, N-1$
$\mathbf{w}_{\mathrm{f}}(\omega)$	vector of size $M \times 1$ with elements $W_m^*(\omega)$, $m = 0, 1, \ldots, M-1$
$\mathbf{w}_{\mathrm{f,a}}(\omega)$	vector of size $M \times 1$ with $\mathbf{w}_{\mathrm{a},m}(k)$, $m = 0, 1, \ldots, M-1$, in the DTFT domain
$\mathbf{w}_{\mathrm{f,a,o}}(\omega)$	optimum of $\mathbf{w}_{\mathrm{f,a}}(\omega)$ in the LCMV sense
$\mathbf{w}_{\mathrm{f,a,o}}^{(3)}(\omega)$	optimum of $\mathbf{w}_{\mathrm{f,a}}(\omega)$ of the GSAEC in the LCMV sense
$\mathbf{w}_{\mathrm{f,c}}(\omega)$	vector of size $M \times 1$ with elements $W_{\mathrm{c}}(\omega)$

$\mathbf{w}_{\mathrm{f,LCMV}}(\omega)$	LCMV beamformer in the DTFT domain
$\mathbf{w}_{\mathrm{f,MMSE}}(\omega)$	MMSE beamformer in the DTFT domain
$\mathbf{w}_{\mathrm{f,MVDR}}(\omega)$	MVDR beamformer in the DTFT domain
$\mathbf{w}_{\mathrm{f,RGSC}}(\omega)$	weight vector of size $M \times 1$ of the RGSC in the DTFT domain
$\mathbf{w}_{\mathrm{LC}}(k)$	LCLSE beamformer as a function of the discrete time k
$\widetilde{\mathbf{w}}_{\mathrm{LC}}(k)$	LCLSE beamformer in the eigenspace of $\mathbf{\Phi}_{\mathbf{xx}}^{-1}(k)$
$\mathbf{w}_{\mathrm{LC}}^{(\mathrm{s})}(k)$	LCLSE beamformer with spatio-temporal constraints
$\mathbf{w}_{\mathrm{LSE}}(k)$	LSE beamformer as a function of the discrete time k
$\widetilde{\mathbf{w}}_{\mathrm{LSE}}(k)$	LSE beamformer in the eigenspace of $\mathbf{\Phi}_{\mathbf{nn}}^{-1}(k)$
$\mathbf{w}_{\mathrm{RGSC}}(k)$	weight vector of size $MN^3 \times 1$ of the RGSC as a function of the discrete time k
$\mathbf{w}_{\mathrm{s}}(k)$	stacked vector of size $MN^2 \times 1$ with vectors $\mathbf{w}(k-n)$, $n = 0, 1, \ldots, N-1$
w_k	windowing function in the discrete time-domain
$w_{n,m}$	n-th filter weight in the m-th sensor channel of a fixed beamformer
$w_{n,m}(k)$	n-th filter weight in the m-th sensor channel of a beamformer as a function of the discrete time k
$\widetilde{w}_{n,m}(k)$	elements of the weight vector $\widetilde{\mathbf{w}}(k)$
$w_{n,m,p}(k)$	n-th filter weight of the filter between the m-th input channel and the p-th output channel of a MIMO system as a function of the discrete time k
$w_{\mathrm{a},n,m}(k)$	n-th filter weight in the m-th channel of the interference canceller as a function of the discrete time k
$w_{\mathrm{c},n,m}(k)$	n-th filter weight in the m-th sensor channel of the quiescent weight vector as a function of the discrete time k
$\mathbf{X}(k)$	data matrix of size $MN \times K$ of the sensor signals
$\underline{\mathbf{X}}(r)$	stacked matrix of size $2N \times 2MN$ with matrices $\underline{\mathbf{X}}_m(r)$, $m = 0, 1, \ldots, M-1$
$\underline{\mathbf{X}}_m(r)$	diagonal matrix of size $2N \times 2N$ with the DFT of $x_m(rR+n)$, $n = -N, -N+1, \ldots, N-1$, on the main diagonal
$\mathbf{X}_{\mathbf{a}}(k)$	stacked matrix of size $QM \times K$ with vectors $\mathbf{x}_{\mathbf{a}}(k+i)$ for $i = 0, 1, \ldots, K-1$
$\underline{\mathbf{X}}_{\mathbf{a}}(r')$	stacked matrix of size $2N_{\mathbf{a}} \times 2MN_{\mathbf{a}}$ with matrices $\underline{\mathbf{X}}_{\mathbf{a},q}(r')$, $q = 0, 1, \ldots, Q-1$
$\underline{\mathbf{X}}_{\mathbf{a},q}(r')$	diagonal matrix of size $2N_{\mathbf{a}} \times 2N_{\mathbf{a}}$ with the DFT of $x_{\mathbf{a},q}(r'R'+n)$, $n = -N_{\mathbf{a}}, -N_{\mathbf{a}}+1, \ldots, N_{\mathbf{a}}-1$, on the main diagonal
$\underline{\mathbf{X}}_{\mathbf{B}_{\mathrm{b}}}(r)$	diagonal matrix of size $2N \times 2N$ with the DFT of $\mathbf{x}_{\mathbf{B}_{\mathrm{b}}}(rR)$ on the main diagonal
$\mathbf{X}_{\mathrm{s}}(k)$	stacked matrix of size $MN^2 \times K$ with matrices $\mathbf{X}(k-n)$ for $n = 0, 1, \ldots, N-1$

$\mathbf{X}_{\mathrm{ss}}(k)$	stacked matrix of size $MN^3 \times K$ with matrices $\mathbf{X}_{\mathrm{s}}(k-n)$ for $n = 0, 1, \ldots, N-1$
$\underline{\mathbf{X}}_{\mathbf{w}_{\mathrm{a}}}(r)$	stacked matrix of size $2N \times 2MN$ with matrices $\underline{\mathbf{X}}_{\mathbf{w}_{\mathrm{a}},m}(r)$, $m = 0, 1, \ldots, M-1$
$\underline{\mathbf{X}}_{\mathbf{w}_{\mathrm{a}},m}(r)$	diagonal matrix of size $2N \times 2N$ with the DFT of $\mathbf{x}_{\mathbf{w}_{\mathrm{a}},m}(rR)$ on the main diagonal
$\underline{\mathbf{X}}_{\mathbf{w}_{\mathrm{c}}}(r)$	stacked matrix of size $2N \times 2MN$ with matrices $\underline{\mathbf{X}}_{\mathbf{w}_{\mathrm{c}},m}(r)$, $m = 0, 1, \ldots, M-1$
$\underline{\mathbf{X}}_{\mathbf{w}_{\mathrm{c}},m}(r)$	diagonal matrix of size $2N \times 2N$ with the DFT of $\mathbf{x}_{\mathbf{w}_{\mathrm{c}},m}(rR)$ on the main diagonal
$X_m(\omega)$	DTFT of $x_m(k)$
$\mathbf{x}(k)$	stacked vector of size $MN \times 1$ with vectors $\mathbf{x}_m(k)$, $m = 0, 1, \ldots, M-1$
$\widetilde{\mathbf{x}}(k)$	data vector $\mathbf{x}(k)$ projected onto $\mathbf{\Sigma}_{\mathbf{X}}^{-1}(k)\mathbf{U}_{\mathbf{X}}(k)$ or onto $\mathbf{\Sigma}_{\mathbf{N}}^{-1}(k)\mathbf{U}_{\mathbf{N}}(k)$
$\mathbf{x}_m(k)$	vector of size $N \times 1$ with elements $x_m(k-n)$, $n = 0, 1, \ldots, N-1$
$\mathbf{x}_{\mathbf{a}}(k)$	stacked vector of size $QN_{\mathbf{a}} \times 1$ with vectors $\mathbf{x}_{\mathbf{a},q}(k)$, $q = 0, 1, \ldots, Q-1$
$\mathbf{x}_{\mathbf{a},q}(k)$	vector of size $N_{\mathbf{a}} \times 1$ with elements $x_{\mathbf{a},q}(k-n)$, $n = 0, 1, \ldots, N_{\mathbf{a}}-1$
$\mathbf{x}_{\mathbf{B}_{\mathrm{b}}}(k)$	vector of size $2N \times 1$ with elements $y_{\mathbf{w}_{\mathrm{c}}}(k+n)$, $n = -N, -N+1, \ldots, N-1$
$\mathbf{x}_{\mathrm{f}}(k)$	vector of size $M \times 1$ with elements $X_m(\omega)$, $m = 0, 1, \ldots, M-1$
$\mathbf{x}_{\mathrm{s}}(k)$	stacked vector of size $MN^2 \times 1$ with vectors $\mathbf{x}(k-n)$ for $n = 0, 1, \ldots, N-1$
$\mathbf{x}_{\mathrm{ss}}(k)$	stacked vector of size $MN^3 \times 1$ with vectors $\mathbf{x}_{\mathrm{s}}(k-n)$ for $n = 0, 1, \ldots, N-1$
$\mathbf{x}_{\mathbf{w}_{\mathrm{a}},m}(k)$	vector of size $2N \times 1$ with elements $e_{\mathbf{B}_{\mathrm{b}},m}(k+n)$, $n = -N, -N+1, \ldots, N-1$
$\mathbf{x}_{\mathbf{w}_{\mathrm{c}},m}(k)$	vector of size $2N \times 1$ with elements $x_m(k+n)$, $n = -N, -N+1, \ldots, N-1$
x (m)	x-coordinate in a Cartesian coordinate system
$x(\mathbf{p}, t)$	space-time observation signal (deterministic signal or random field)
$x_m(k)$	m-th sensor signal as a function of the discrete time k
$x_{\mathbf{a},q}(k)$	q-th loudspeaker signal as a function of the discrete time k
$\mathbf{Y}(k)$	stacked matrix of size $K \times P$ with vectors $\mathbf{y}(k+i)$, $i = 0, 1, \ldots, K-1$
$\mathbf{Y}_{\mathbf{a}}(k)$	stacked matrix of size $N_{\mathbf{a}} \times M$ with vectors $\mathbf{y}_{\mathbf{a},m}(k)$, $m = 0, 1, \ldots, M-1$
$\mathbf{Y}_{\mathbf{B}_{\mathrm{b}}}(k)$	stacked matrix of size $N \times M$ with vectors $\mathbf{y}_{\mathbf{B}_{\mathrm{b}},m}(k)$, $m = 0, 1, \ldots, M-1$

$\mathbf{Y}_{\mathrm{ref}}(k)$	reference matrix of size $K \times P$ of an optimum MIMO filter
$\underline{\mathbf{Y}}_{\mathrm{ref}}(r)$	stacked matrix of size $2N \times P$ with vectors $\mathbf{y}_{\mathrm{ref},p}(rR)$, $p = 0, 1, \ldots, P-1$ in the DFT domain
$Y(\omega)$	DTFT of $y(k)$
$Y^{(3)}(\omega)$	DTFT of $y^{(3)}(k)$
$Y_{\mathrm{ref},p}(\omega)$	DTFT of $y_{\mathrm{ref},p}(k)$
$\mathbf{y}(k)$	vector of size $P \times 1$ with elements $y_p(k)$, $p = 0, 1, \ldots, P-1$
$\mathbf{y}_{\mathbf{a},m}(k)$	vector of size $N \times 1$ with elements $x_m(k+n)$, $n = 0, 1, \ldots, N_{\mathbf{a}}-1$
$\mathbf{y}_{\mathbf{B}_{\mathrm{b}},m}(k)$	vector of size $N \times 1$ with elements $x_m(k - \kappa_{\mathbf{B}_{\mathrm{b}}} + n)$, $n = 0, 1, \ldots, N-1$
$\mathbf{y}_{\mathrm{f,ref}}(\omega)$	vector of size $P \times 1$ with elements $Y_{\mathrm{ref},p}(\omega)$, $p = 0, 1, \ldots, P-1$
$\mathbf{y}_{\mathrm{r}}(k)$	vector of size $K \times 1$ with elements $y_{\mathrm{ref}}(k+i)$, $i = 0, 1, \ldots, K-1$
$\mathbf{y}_{\mathrm{ref}}(k)$	vector of size $P \times 1$ with elements $y_{\mathrm{ref},p}(k)$, $p = 0, 1, \ldots, P-1$
$\mathbf{y}_{\mathrm{ref},p}(k)$	vector of size $N \times 1$ with elements $y_{\mathrm{ref},p}(k+n)$, $n = 0, 1, \ldots, N-1$
$\mathbf{y}_{\mathbf{w}_{\mathrm{a}}}(k)$	vector of size $N \times 1$ with elements $y_{\mathbf{w}_{\mathrm{c}}}(k - \kappa_{\mathbf{w}_{\mathrm{a}}} + n)$, $n = 0, 1, \ldots, N-1$
$\mathbf{y}_{\mathbf{w}_{\mathrm{c}}}(k)$	vector of size $N \times 1$ with elements $y_{\mathbf{w}_{\mathrm{c}}}(k+n)$, $n = 0, 1, \ldots, N-1$
y (m)	y-coordinate in a Cartesian coordinate system
$y(k)$	output signal of a beamformer as a function of the discrete time k
$y^{(3)}(k)$	output signal of GSAEC as a function of the discrete time k
$y^{(4)}(k)$	output signal of GEIC as a function of the discrete time k
$\bar{y}_m(k)$	output signal of a complementary fixed beamformer as a function of the discrete time k with the m-th sensor signal as reference
$y_p(k)$	output signal of the p-th output channel of a MIMO system as a function of the discrete time k
$y_{\mathrm{ref}}(k)$	reference signal for the LSE beamformer as a function of the discrete time k
$y_{\mathrm{ref},p}(k)$	reference signal for the p-th output channel of an optimum MIMO filter as a function of the discrete time k
$y_{\mathbf{w}_{\mathrm{c}}}(k)$	output signal of $\mathbf{w}_{\mathrm{c}}(k)$ as a function of discrete time k
$y_{\mathbf{w}_{\mathrm{RGSC}}}(k)$	output signal of $\mathbf{w}_{\mathrm{RGSC}}(k)$ as a function of discrete time k
z (m)	z-coordinate in a Cartesian coordinate system

References

[AB79] J.B. Allen and D.A. Berkley. Image method for efficiently simulating small-room acoustics. *Journal of the Acoustical Society of America*, 65(4):954–950, April 1979.

[Abh99] T.D. Abhayapala. *Modal Analysis and Synthesis of Broadband Nearfield Beamforming Arrays*. PhD thesis, Research School of Information Sciences and Engineering, Australian National University, Canberra, Australia, 1999.

[AC76] S.P. Applebaum and D.J. Chapman. Adaptive arrays with main beam constraints. *IEEE Trans. on Antennas and Propagation*, 24(9):650–662, September 1976.

[AC93] J. An and B. Champagne. Adaptive beamforming via two-dimensional cosine transform. *Proc. IEEE Pacific Rim Conf. on Communications, Computers, and Signal Processing*, 1:13–16, May 1993.

[AG97] S. Affes and Y. Grenier. A signal subspace tracking algorithm for microphone array processing of speech. *IEEE Trans. on Speech and Audio Processing*, 5(5):425–437, September 1997.

[AHBK03] R. Aichner, W. Herbordt, H. Buchner, and W. Kellermann. Least-squares error beamforming using minimum statistics and multichannel frequency-domain adaptive filtering. *Proc. Int. Workshop on Acoustic Echo and Noise Control*, pages 223–226, September 2003.

[AHYN00] F. Asano, S. Hayamizu, T. Yamada, and S. Nakamura. Speech enhancement based on the subspace method. *IEEE Trans. on Speech and Audio Processing*, 8(5):497–507, September 2000.

[Aic02] R. Aichner. Documentation of multichannel speech recordings (ANITA project funded by the European Commission under contract IST-2001-34327). Technical report, Multimedia Communications and Signal Processing, University Erlangen-Nuremberg, Erlangen, Germany, October 2002.

[AKW99] T.D. Abhayapala, R.A. Kennedy, and R.C. Williamson. Spatial aliasing for near-field sensor arrays. *IEE Electronic Letters*, 35(10):764–765, May 1999.

[Ali98] M. Ali. Stereophonic acoustic echo cancellation system using time-varying all-pass filtering for signal decorrelation. *Proc. IEEE Int. Conf. on Acoustics, Speech, and Signal Processing*, 6:3689–3692, May 1998.

[BBK03] H. Buchner, J. Benesty, and W. Kellermann. Multichannel frequency-domain adaptive filtering with application to multichannel acoustic echo cancellation. In J. Benesty and Y. Huang, editors, *Adaptive Signal Processing: Applications to Real-World Problems.* Springer, Berlin, 2003.

[BD91] J. Benesty and P. Duhamel. Fast constant modulus adaptive algorithm. *IEE Proceedings-F*, 138:379–387, August 1991.

[BD92] J. Benesty and P. Duhamel. A fast exact least mean square adaptive algorithm. *IEEE Trans. on Signal Processing*, 40(12):2904–2920, December 1992.

[BDH+99] C. Breining, P. Dreiseitel, E. Hänsler, A. Mader, B. Nitsch, H. Puder, T. Schertler, G. Schmidt, and J. Tilp. Acoustic echo control - an application of very-high-order adaptive filters. *IEEE Signal Processing Magazine*, 16(4):42–69, July 1999.

[Bel72] R. Bellmann. *Introduction to matrix analysis.* McGraw-Hill, New York, NY, 1972.

[Bel01] M.G. Bellanger. *Adaptive Digital Filters.* Marcel Dekker, New York, 2001.

[BG86] K.M. Buckley and L.J. Griffiths. An adaptive generalized sidelobe canceller with derivative constraints. *IEEE Trans. on Antennas and Propagation*, 34(3):311–319, March 1986.

[BH03] J. Benesty and Y. Huang, editors. *Adaptive Signal Processing: Applications to Real-World Problems.* Springer, Berlin, 2003.

[Bit02] J. Bitzer. *Mehrkanalige Geräuschunterdrückungssysteme – eine vergleichende Analyse.* PhD thesis, Universität Bremen, Bremen, Germany, May 2002.

[BK01] H. Buchner and W. Kellermann. Acoustic echo cancellation for two or more reproduction channels. *Proc. Workshop on Acoustic Echo and Noise Control*, pages 99–102, September 2001.

[BK02] H. Buchner and W. Kellermann. Improved Kalman gain computation for multi-channel frequency-domain adaptive filtering and application to acoustic echo cancellation. *Proc. IEEE Int. Conf. on Acoustics, Speech, and Signal Processing*, 2:1909–1912, May 2002.

[BM00] J. Benesty and D.R. Morgan. Multi-channel frequency-domain adaptive filtering. In S.L. Gay and J. Benesty, editors, *Acoustic Signal Processing for Telecommunication*, chapter 7, pages 121–133. Kluwer Academic Publishers, Boston, MA, 2000.

[BMS98] J. Benesty, D.R. Morgan, and M.M. Sondhi. A better understanding and an improved solution to the specific problems of stereophonic acoustic echo cancellation. *IEEE Trans. on Speech and Audio Processing*, 6(2):156–165, 1998.

[BP71] J.S. Bendat and A.G. Piersol. *Random Data: Analysis and Measurement Procedures.* John Wiley & Sons, Inc., New York, 1971.

[Bri75] D.R. Brillinger. *Time Series-Data Analysis and Theory.* Holt, Rinehart, and Winston, Inc., New York, 1975.

[BS01] J. Bitzer and K.U. Simmer. Superdirective microphone arrays. In M.S. Brandstein and D.B. Ward, editors, *Microphone Arrays: Signal Processing Techniques and Applications*, chapter 2, pages 19–38. Springer, Berlin, 2001.

[BSKD99] J. Bitzer, K.U. Simmer, and K.-D.Kammeyer. Multi-microphone noise reduction by post-filter and superdirective beamformer. *Proc. IEEE Int. Workshop on Acoustics, Echo, and Noise Control*, pages 100–103, September 1999.

[Buc86] K.M. Buckley. Broad-band beamforming and the generalized sidelobe canceller. *IEEE Trans. on Acoustics, Speech, and Signal Processing*, 34(5):1322–1323, October 1986.

[Buc87] K.M. Buckley. Spatial/spectral filtering with linearly constrained minimum variance beamformers. *IEEE Trans. on Acoustics, Speech, and Signal Processing*, 35(3):249–266, March 1987.

[Bur98] P.M.S. Burt. Measuring acoustic responses with maximum-length sequences. *Proc. IEEE Int. Telecommunications Symposium*, 1:284–289, August 1998.

[BW01] M.S. Brandstein and D.B. Ward, editors. *Microphone Arrays: Signal Processing Techniques and Applications.* Springer, Berlin, 2001.

[Car87] G.C. Carter. Coherence and time delay estimation. *Proceedings of the IEEE*, 75(2):236–255, February 1987.

[CB02] I. Cohen and B. Berdugo. Microphone array post-filtering for nonstationary noise suppression. *Proc. IEEE Int. Conf. on Acoustics, Speech, and Signal Processing*, 1:901–904, May 2002.

[CG80] A. Cantoni and L.C. Godara. Resolving the direction of sources in a correlated field incident on an array. *Journal of the Acoustical Society of America*, 67(4):1247–1255, April 1980.

[Cho95] T. Chou. Frequency-independent beamformer with low response error. *Proc. IEEE Int. Conf. on Acoustics, Speech, and Signal Processing*, 5:2995–2998, 1995.

[CK84] T.K. Citron and T. Kailath. An improved eigenvector beamformer. *Proc. IEEE Int. Conf. on Acoustics, Speech, and Signal Processing*, 3:33.3.1–33.3.4, 1984.

[Coh02] I. Cohen. Optimal speech enhancement under signal presence uncertainty using log-spectral amplitude estimator. *IEEE Signal Processing Letters*, 9(4):113–116, April 2002.

[Com80] R.T. Compton, Jr. Pointing accuracy and dynamic range in a steered beam adaptive array. *IEEE Trans. on Aerospace and Electronic Systems*, 16(3):280–287, May 1980.

[Com82] R.T. Compton, Jr. The effect of random steering vector errors in the Applebaum adaptive array. *IEEE Trans. on Aerospace and Electronic Systems*, 18(5):392–400, September 1982.

[Com88] R.T. Compton, Jr. *Adaptive Antennas: Concepts and Performance.* Prentice-Hall, Englewood Cliffs, NJ, 1988.

[Com90] D. Van Compernolle. Switching adaptive filters for enhancing noisy and reverberant speech from microphone array recordings. *Proc. IEEE Int. Conf. on Acoustics, Speech, and Signal Processing*, 2:833–836, 1990.

[Cox73a] H. Cox. Resolving power and sensitivity to mismatch of optimum array processors. *Journal of the Acoustical Society of America*, 54(3):771–785, 1973.

[Cox73b] H. Cox. Spatial correlation in arbitrary noise fields with application to ambient sea noise. *Journal of the Acoustical Society of America*, 54(5):1289–1301, November 1973.

[Cro98] M.J. Crocker. *Handbook of Acoustics.* John Wiley & Sons, Inc., New York, 1998.

[CY92] L. Chang and C.C. Yeh. Performance of DMI and eigenspace-based beamformers. *IEEE Trans. on Antennas and Propagation*, 40(11):1336–1347, November 1992.

[CZK86] H. Cox, R.M. Zeskind, and T. Kooij. Practical supergain. *IEEE Trans. on Acoustics, Speech, and Signal Processing*, 34(3):393–398, June 1986.

[CZK87] H. Cox, R.M. Zeskind, and T. Kooij. Robust adaptive beamforming. *IEEE Trans. on Acoustics, Speech, and Signal Processing*, 35(10):1365–1376, October 1987.

[DBC91] M. Dendrinos, S. Bakamidis, and G. Carayannis. Speech enhancement from noise. *Speech Communication*, 10(2):45–57, February 1991.

[DG01] B. Debail and A. Gilloire. Microphone array design with improved acoustic echo rejection. *Int. Workshop on Acoustic Echo and Noise Control*, pages 55–58, September 2001.

[DM01] S. Doclo and M. Moonen. GSVD-based optimal filtering for multi-microphone speech enhancement. In M.S. Brandstein and D.B. Ward, editors, *Microphone Arrays: Signal Processing Techniques and Applications*, chapter 6, pages 111–132. Springer, Berlin, 2001.

[DM03] S. Doclo and M. Moonen. Design of broadband beamformers robust against microphone position errors. *Int. Workshop on Acoustic Echo and Noise Control*, pages 267–270, September 2003.

[Doc03] S. Doclo. *Multi-microphone noise reduction and dereverberation techniques for speech applications.* PhD thesis, Katholieke Universiteit Leuven, Leuven, Belgium, 2003.

[Dol46] C.L. Dolph. A current distribution for broadside arrays which optimize the relationship between beam width and side-lobe level. *Proc. I.R.E. and Waves and Electrons*, 34(6):335–348, June 1946.

[Duv83] K. Duvall. *Signal Cancellation in Adaptive Arrays: Phenomena and a Remedy.* PhD thesis, Department of Electrical Engineering, Stanford University, 1983.

[EC83] M.H. Er and A. Cantoni. Derivative constraints for broad-band element space antenna array processors. *IEEE Trans. on Acoustics, Speech, and Signal Processing*, 31(6):1378–1393, December 1983.

[EC85] M.H. Er and A. Cantoni. An alternative formulation for an optimum beamformer with robustness capability. *IEE Proceedings-F*, 132:447–460, October 1985.

[Eck53] C. Eckart. The theory of noise in continuous media. *Journal of the Acoustical Society of America*, 25(2):195–199, March 1953.

[EDG03] G. Elko, E. Diethorn, and Tomas Gänsler. Room impulse response variations due to thermal fluctuations and its impact on acoustic echo cancellation. *Proc. Int. Workshop on Acoustic Echo and Noise Control*, pages 67–70, September 2003.

[EFK67] D.J. Edelblute, J.M. Fisk, and G.L. Kinneson. Criteria for optimum-signal-detection theory for arrays. *Journal of the Acoustical Society of America*, 41(1):199–205, January 1967.

[Elk96] G.W. Elko. Microphone array systems for hands-free telecommunication. *Speech Communication*, 20:229–240, 1996.

[Elk00] G.W. Elko. Superdirectional microphone arrays. In S.L. Gay and J. Benesty, editors, *Theory and Applications of Acoustic Signal Processing for Telecommunications*, chapter 10, pages 181–237. Kluwer Academic Publishers, London, 2000.

[EM85] Y. Ephraim and D. Malah. Speech enhancement using a minimum mean-square error log-spectral amplitude estimator. *IEEE Trans. on Acoustics, Speech, and Signal Processing*, 33(2):443–445, April 1985.

[ET95] Y. Ephraim and H. L. Van Trees. A signal subspace approach for speech enhancement. *IEEE Trans. on Speech and Audio Processing*, 3(2):251–266, July 1995.

[Fla85a] L.J. Flanagan. Beamwidth and usable bandwidth of delay-steered microphone arrays. *AT&T Technical Journal*, 64(4):983–995, April 1985.

[Fla85b] L.J. Flanagan. Computer-steered microphone arrays for sound transduction in large rooms. *Journal of the Acoustical Society of America*, 78(5):1508–1518, November 1985.

[Fle65] W.H. Fleming. *Functions of Several Variables.* Addison-Wesley Publishing Company, Inc., Reading, MA, 1965.

[Fle87] R. Fletcher. *Practical methods of optimization.* Wiley, New York, NY, 1987.

[FM01] D.A. Florencio and H.S. Malvar. Multichannel filtering for optimum noise reduction in microphone arrays. *Proc. IEEE Int. Conf. on Acoustics, Speech, and Signal Processing*, 1:197–200, May 2001.

[FMB89] G. Faucon, S.T. Mezalek, and R. Le Bouquin. Study and comparison of three structures for enhancement of noisy speech. *Proc. IEEE Int. Conf. on Acoustics, Speech, and Signal Processing*, 1:385–388, 1989.

[Fri88] B. Friedlander. A signal subspace method for adaptive interference cancellation. *IEEE Trans. on Acoustics, Speech, and Signal Processing*, 36(12):1835–1845, December 1988.

[Fro72] O.L. Frost. An algorithm for linearly constrained adaptive processing. *Proceedings of the IEEE*, 60(8):926–935, August 1972.

[Fur01] K. Furuya. Noise reduction and dereverberation using correlation matrix based on the multiple-input/output inverse-filtering theorem (MINT). *Int. Workshop on Hands-Free Speech Communication*, pages 59–62, April 2001.

[Gab86] W.F. Gabriel. Using spectral estimation techniques in adaptive processing antenna systems. *IEEE Trans. on Antennas and Propagation*, 34(3):291–300, March 1986.

[Gar92] W.A. Gardner. A unifying view of coherence in signal processing. *Signal Processing*, 29(2):113–140, November 1992.

[GB87] L.J. Griffiths and K.M. Buckley. Quiescent pattern control in linearly constrained adaptive arrays. *IEEE Trans. on Acoustics, Speech, and Signal Processing*, 35(7):917–926, July 1987.

[GB00] S.L. Gay and J. Benesty, editors. *Acoustic Signal Processing for Telecommunications.* Kluwer Academic Publishers, Boston, 2000.

[GB02] T. Gänsler and J. Benesty. New insight into the stereophonic acoustic echo cancellation problem and an adaptive nonlinearity solution. *IEEE Trans. on Speech and Audio Processing*, 10(5):257–267, July 2002.

[GBW01] S. Gannot, D. Burshtein, and E. Weinstein. Signal enhancement using beamforming and nonstationarity with applications to speech. *IEEE Trans. on Signal Processing*, 49(8):1614–1626, August 2001.

[GE93] M.M. Goodwin and G.W. Elko. Constant beamwidth beamforming. *Proc. IEEE Int. Conf. on Acoustics, Speech, and Signal Processing*, 1:169–172, 1993.

[GE98] T. Gänsler and P. Eneroth. Influence of audio coding on stereophonic acoustic echo cancellation. *Proc. IEEE Int. Conf. on Acoustics, Speech, and Signal Processing*, 6:3649–3652, May 1998.

[GJ82] L.J. Griffiths and C.W. Jim. An alternative approach to linearly constrained adaptive beamforming. *IEEE Trans. on Antennas and Propagation*, 30(1):27–34, January 1982.

[GJV98] S. Gustafsson, P. Jax, and P. Vary. A novel psychoacoustically motivated audio enhancement algorithm preserving background noise characteristics. *Proc. IEEE Int. Conf. in Acoustics, Speech, and Signal Processing*, 1:397–400, May 1998.

[GM55] E.N. Gilbert and S.P. Morgan. Optimum design of directive antenna arrays subject to random variables. *Bell Systems Technical Journal*, 34:637–663, May 1955.

[God86] L.C. Godara. Error analysis of the optimal antenna array processors. *IEEE Trans. on Aerospace and Electronic Systems*, 22(4):395–409, July 1986.

[Goo63] N.R. Goodman. Statistical analysis based on an certain multivariate complex Gaussian distribution. *Annales des Mathématiques Statistiques*, 34:152–177, March 1963.

[Goo68] J.W. Goodman. *Introduction to Fourier Optics*. McGraw-Hill Book Company, San Francisco, CA, 1968.

[GOS96] D. Giuliani, M. Omologo, and P. Svaizer. Experiments of speech recognition in noisy and reverberant environments. *Proc. Int. Conf. on Spoken Language Processing*, 3:1329–1332, October 1996.

[GP77] G.C. Goodwin and R.L. Payne. *Dynamic System Identification: Experimental Design and Data Analysis*. Academic Press, New York, 1977.

[Gra72] R.M. Gray. On the asymptotic eigenvalue distribution of Toeplitz matrices. *IEEE Trans. on Information Theory*, 18(6):725–730, November 1972.

[Gra81] A. Graham. *Kronecker Products and Matrix Calculus with Applications*. Wiley, New York, NY, 1981.

[GT98] A. Gilloire and V. Turbin. Using auditory properties to improve the behavior of stereophonic acoustic echo cancellers. *Proc. IEEE Int. Conf. on Acoustics, Speech, and Signal Processing*, 6:3681–3684, May 1998.

[GvL89] G.H. Golub and C.F. van Loan. *Matrix Computations*. John Hopkins University Press, Baltimore, MD, 2-nd edition, 1989.

[GZ01] J.E. Greenberg and P.M. Zurek. Microphone-array hearing aids. In M.S. Brandstein and D.B. Ward, editors, *Microphone Arrays: Signal Processing Techniques and Applications*, chapter 11, pages 229–253. Springer, Berlin, 2001.

[Hai91] A.M. Haimovich. An eigenanalysis interference canceller. *IEEE Trans. on Signal Processing*, 39(1):76–84, January 1991.

[Hai96] A. Haimovich. The eigencanceler: Adaptive radar by eigenanalysis methods. *IEEE Trans. on Aerospace and Electronic Systems*, 32(2):532–542, April 1996.

[Har78] F.J. Harris. On the use of windows for harmonic analysis with the discrete fourier transform. *Proceedings of the IEEE*, 66(1):51–83, January 1978.

[Hay96] S. Haykin. *Adaptive Filter Theory.* Prentice Hall, New Jersey, 3rd edition, 1996.

[HB03] Y. Huang and J. Benesty. Adaptive multichannel time delay estimation based on blind system identification for acoustic source localization. In J. Benesty and Y. Huang, editors, *Adaptive Signal Processing: Applications to Real-World Problems*, chapter 8, pages 227–247. Springer, Berlin, 2003.

[HBK03a] W. Herbordt, H. Buchner, and W. Kellermann. An acoustic human-machine front-end for multimedia applications. *EURASIP Journal on Applied Signal Processing*, 2003(1):1–11, January 2003.

[HBK+03b] W. Herbordt, H. Buchner, W. Kellermann, R. Rabenstein, S. Spors, and H. Teutsch. Full-duplex multichannel communication: real-time implementations in a general framework. *Proc. IEEE Int. Conf. on Multimedia and Expo*, 3:49–52, July 2003.

[HK00] W. Herbordt and W. Kellermann. Acoustic echo cancellation embedded into the generalized sidelobe canceller. *Proc. EURASIP European Signal Processing Conference*, 3:1843–1846, September 2000.

[HK01a] W. Herbordt and W. Kellermann. Computationally efficient frequency-domain combination of acoustic echo cancellation and robust adaptive beamforming. *Proc. EUROSPEECH, Aalborg*, 2:1001–1004, September 2001.

[HK01b] W. Herbordt and W. Kellermann. Computationally efficient frequency-domain robust generalized sidelobe canceller. *Proc. Int. Workshop on Acoustic Echo and Noise Control*, pages 51–55, 2001.

[HK01c] W. Herbordt and W. Kellermann. Efficient frequency-domain realization of robust generalized sidelobe cancellers. *IEEE Workshop on Multimedia Signal Processing*, pages 377–382, October 2001.

[HK01d] W. Herbordt and W. Kellermann. Limits for generalized sidelobe cancellers with embedded acoustic echo cancellation. *Proc. IEEE Int. Conf. on Acoustics, Speech, and Signal Processing*, 5:3241–3244, May 2001.

[HK02a] W. Herbordt and W. Kellermann. Analysis of blocking matrices for generalized sidelobe cancellers for non-stationary broadband signals. *Proc. IEEE Int. Conf. on Acoustics, Speech, and Signal Processing, Student Forum*, May 2002.

[HK02b] W. Herbordt and W. Kellermann. Frequency-domain integration of acoustic echo cancellation and a generalized sidelobe canceller with improved robustness. *European Transactions on Telecommunications*, 13(2):123–132, March 2002.

[HK03] W. Herbordt and W. Kellermann. Adaptive beamforming for audio signal acquisition. In J. Benesty and Y. Huang, editors, *Adaptive Signal Processing: Applications to Real-World Problems*, pages 155–194. Springer, Berlin, 2003.

[HL01] Y. Hu and P.C. Loizou. A subspace approach for enhancing speech corrupted by colored noise. *Proc. IEEE Int. Conf. on Acoustics, Speech, and Signal Processing*, 1:573–576, May 2001.

[Hod76] W.S. Hodgkiss. Covariance between fourier coefficients representing the time waveforms observed from an array of sensors. *Journal of the Acoustical Society of America*, 59(3):582–590, March 1976.

[Hol87] S. Holm. FFT pruning applied to time domain interpolation and peak localization. *IEEE Trans. on Acoustics, Speech, and Signal Processing*, 35(12):1776–1778, December 1987.

[HS03] E. Hänsler and G. Schmidt. Single-channel acoustic echo cancellation. In J. Benesty and Y. Huang, editors, *Adaptive Signal Processing: Applications to Real-World Problems*. Springer, Berlin, 2003.

[HSH99] O. Hoshuyama, A. Sugiyama, and A. Hirano. A robust adaptive beamformer for microphone arrays with a blocking matrix using constrained adaptive filters. *IEEE Trans. on Signal Processing*, 47(10):2677–2684, October 1999.

[HT83] E.K.L. Hung and R.M. Turner. A fast beamforming algorithm for large arrays. *IEEE Trans. on Aerospace and Electronic Systems*, 19(4):598–607, July 1983.

[HTK03] W. Herbordt, T. Trini, and W. Kellermann. Robust spatial estimation of the signal-to-interference ratio for non-stationary mixtures. *Proc. Int. Workshop on Acoustic Echo and Noise Control*, pages 247–250, September 2003.

[Hud81] J.E. Hudson. *Adaptive array principles.* Peter Peregrinus Ltd., London, GB, 1981.

[HYBK02] W. Herbordt, J. Ying, H. Buchner, and W. Kellermann. A real-time acoustic human-machine front-end for multimedia applications integrating robust adaptive beamforming and stereophonic acoustic echo cancellation. *Proc. Int. Conf. on Spoken Language Processing*, 2:773–776, September 2002.

[ITU93] International Telecommunication Union ITU. Objective measurement of active speech level. *ITU-T Recommendation*, P.56, March 1993.

[Jab86a] N.K. Jablon. Adaptive beamforming with the generalized sidelobe canceller in the presence of array imperfections. *IEEE Trans. on Antennas and Propagation*, 34(8):996–1012, August 1986.

[Jab86b] N.K. Jablon. Steady state analysis of the generalized sidelobe canceller by adaptive noise cancelling techniques. *IEEE Trans. on Antennas and Propagation*, 34(3):330–337, March 1986.

[Jac62] M.J. Jacobson. Space-time correlation in spherical and circular noise fields. *Journal of the Acoustical Society of America*, 34(7):971–978, July 1962.

[JC01a] F. Jabloun and B. Champagne. A multi-microphone signal subspace approach for speech enhancement. *Proc. IEEE Int. Conf. on Acoustics, Speech, and Signal Processing*, 1:205–208, May 2001.

[JC01b] F. Jabloun and B. Champagne. On the use of masking properties of the human ear in the signal subspace speech enhancement approach. *Proc. Int. Workshop on Acoustic Echo and Noise Control*, pages 199–202, September 2001.

[JC02] F. Jabloun and B. Champagne. A perceptual signal subspace approach for speech enhancement in colored noise. *Proc. IEEE Int. Conf. on Acoustics, Speech, and Signal Processing*, 1:569–572, May 2002.

[JD93] D.H. Johnson and D.E. Dudgeon. *Array Signal Processing.* Prentice Hall, New Jersey, 1993.

[JHHS95] S.H. Jensen, P.C. Hansen, S.D. Hansen, and J.A. Sorensen. Reduction of broad-band noise in speech by truncated QSVD. *IEEE Trans. on Speech and Audio Processing*, 3(6):439–448, November 1995.

[Jim77] C.W. Jim. A comparison of two LMS constrained optimal array structures. *Proceedings of the IEEE*, 65(12):1730–1731, December 1977.

[JS98] Y. Joncour and A. Sugyiama. A stereo echo canceler with pre-processing for correct echo path identification. *Proc. IEEE Int. Conf. on Acoustics, Speech, and Signal Processing*, 6:3677–3680, May 1998.

[Jun00] J.-C. Junqua. *Robust Speech Recognition in Embedded Systems and PC Applications.* Kluwer Academic Publishers, Boston, MA, 2000.

[KAW98] R.A. Kennedy, T. Abhayapala, and D.B. Ward. Broadband nearfield beamforming using a radial beampattern transformation. *IEEE Trans. on Signal Processing*, 46(8):2147–2155, August 1998.

[KCY92] K.M. Kim, I. W. Cha, and D. H. Youn. On the performance of the generalized sidelobe canceller in coherent situations. *IEEE Trans. on Antennas and Propagation*, 40(4):465–468, April 1992.

[Kel97] W. Kellermann. Strategies for combining acoustic echo cancellation and adaptive beamforming microphone arrays. *Proc. IEEE Int. Conf. on Acoustics, Speech, and Signal Processing*, 1:219–222, April 1997.

[Kel01] W. Kellermann. Acoustic echo cancellation for beamforming microphone arrays. In M.S. Brandstein and D.B. Ward, editors, *Microphone Arrays: Signal Processing Techniques and Applications*, chapter 13, pages 281–306. Springer, Berlin, 2001.

[KHB02] W. Kellermann, W. Herbordt, and H. Buchner. Signalverarbeitung für akustische Mensch-Maschine-Schnittstellen. *Tagungsband zur 13. Konf. elektronische Sprachsignalverarbeitung (ESSV)*, pages 49–57, September 2002.

[KK02a] M. Klein and P. Kabal. Signal subspace speech enhancement with perceptual post-filtering. *Proc. IEEE Int. Conf. on Acoustics, Speech, and Signal Processing*, 1:537–540, May 2002.

[KK02b] F. Kuech and W. Kellermann. Nonlinear line echo cancellation using a simplified second-order Volterra filter. *Proc. IEEE Int. Conf. on Acoustics, Speech and Signal Processing*, 2:1117–1120, May 2002.

[KK03] F. Kuech and W. Kellermann. Proportionate NLMS algorithm for second-order Volterra filters and its application to nonlinear echo cancellation. *Int. Workshop on Acoustic Echo and Noise Control*, pages 75–78, September 2003.

[KU89] J.W. Kim and C.K. Un. Noise subspace approach for interference cancellation. *IEE Electronic Letters*, 25(11):712–713, May 1989.

[Kut91] H. Kuttruff. *Room Acoustics.* Spon Press, London, 4th edition, 1991.

[KWA99] R.A. Kennedy, D.B. Ward, and T. Abhayapala. Near-field beamforming using radial reciprocity. *IEEE Trans. on Signal Processing*, 47(1):33–40, January 1999.

[Leo84] R.G. Leonard. A database for speaker independent digit recognition. *Proc. IEEE Int. Conf. on Acoustics, Speech, and Signal Processing*, 3:42.11.1–42.11.4, March 1984.

[LH74] C.L. Lawson and R.J. Hanson. *Solving Least Squares Problems.* Prentice Hall, Englewood Cliffs, NJ, 1974.

[LH93] M. Lu and Z. He. Adaptive beam forming using split-polarity transformation for coherent signal and interference. *IEEE Trans. on Antennas and Propagation*, 41(3):314–324, March 1993.

[LHA02] L. Lin, W.H. Holmes, and E. Ambikairajah. Speech denoising using perceptual modification of Wiener filtering. *IEE Electronic Letters*, 38(23):1486–1487, November 2002.

[Lim83] J.S. Lim. *Speech Enhancement.* Prentice Hall, New Jersey, 1983.

[LL97] C.-C. Lee and J.-H. Lee. Eigenspace-based adaptive array beamforming with robust capabilities. *IEEE Trans. on Antennas and Propagation*, 45(12):1711–1716, December 1997.

[LL00] J.-H. Lee and C.-C. Lee. Analysis of the performance and sensitivity of an eigenspace-based interference canceller. *IEEE Trans. on Antennas and Propagation*, 48(5):826–835, May 2000.

[LO79] J.S. Lim and A.V. Oppenheim. Enhancement and bandwidth compression of noisy speech. *Proceedings of the IEEE*, 67(12):1586–1604, December 1979.

[LVKL96] T.I. Laakso, V. Välimäki, M. Karjalainen, and U.K. Laine. Splitting the unit delay - tools for fractional filter design. *IEEE Signal Processing Magazine*, 13(1):30–60, January 1996.

[MAG95] E. Moulines, O. Ait Amrane, and Y. Grenier. The generalized multidelay adaptive filter: structure and convergence analysis. *IEEE Trans. on Signal Processing*, 43(1):14–28, January 1995.

[Mar71] J.D. Markel. FFT pruning. *IEEE Trans. on Audio and Electroacoustics*, 19:305–311, December 1971.

[Mar95] R. Martin. *Freisprecheinrichtungen mit mehrkanaliger Echokompensation und Störgeräuschreduktion.* PhD thesis, Aachener Institut für Nachrichtengeräte und Datenkommunikation, 1995.

[Mar01a] R. Martin. Noise power spectral density estimation based on optimal smoothing and minimum statistics. *IEEE Trans. on Speech and Audio Processing*, 9(5):504–512, July 2001.

[Mar01b] R. Martin. Small microphone arrays with postfilters for noise and acoustic echo cancellation. In M.S. Brandstein and D.B. Ward, editors, *Microphone Arrays: Signal Processing Techniques and Applications*, pages 255–279. Springer, Berlin, 2001.

[Mat99] I. Mateljan. Signal selection for the room acoustics measurement. *Proc. Int. Workshop on Applications of Signal Processing to Audio and Acoustics*, pages 199–202, October 1999.

[MB02] I. McCowan and H. Bourlard. Microphone array post-filter for diffuse noise field. *Proc. IEEE Int. Conf. on Acoustics, Speech, and Signal Processing*, 1:905–908, May 2002.

[McD71] R.N. McDonough. Degraded performance of nonlinear array processors in the presence of data modeling errors. *Journal of the Acoustical Society of America*, 51(4):1186–1193, April 1971.

[MG97] T.R. Messerschmitt and R.A. Gramann. Evaluation of the dominant mode rejection beamformer using reduced integration times. *IEEE Journal of Oceanic Engineering*, 22(2):385–392, April 1997.

[MIK00] D.G. Manolakis, V.K. Ingle, and S.M. Kogon. *Statistical and Adaptive Signal Processing.* McGraw-Hill, Singapore, 2000.

[Mil74] K.S. Miller. *Complex Stochastic Processes: An Introduction to Theory and Applications.* Addison-Wesley, Reading, MA, 1974.

[MK88] M. Miyoshi and Y. Kaneda. Inverse filtering of room acoustics. *IEEE Trans. on Acoustics, Speech, and Signal Processing*, 36(2):145–153, February 1988.

[MM80] R. Monzingo and T. Miller. *Introduction to adaptive arrays.* Wiley and Sons, New York, NY, 1980.

[MP00] U. Mittal and N. Phamdo. Signal/Noise KLT based approach for enhancing speech degraded by colored noise. *IEEE Trans. on Speech and Audio Processing*, 8(2):159–167, March 2000.

[MPS00] A. Mader, H. Puder, and G.U. Schmidt. Step-size controls for acoustic echo cancellation filters - an overview. *Signal Processing*, 80(9):1697–1719, September 2000.

[MS01] I.A. McCowan and S. Sridharan. Microphone array sub-band speech recognition. *Proc. IEEE Int. Conf. on Acoustics, Speech, and Signal Processing*, 1:185–188, May 2001.

[MV93] R. Martin and P. Vary. Combined acoustic echo cancellation, dereverberation, and noise-reduction: a two microphone approach. *Proc. Int. Workshop on Acoustic Echo Control*, pages 125–132, September 1993.

[NA79] S.T. Neely and J.T. Allen. Invertibility of a room impulse response. *Journal of the Acoustical Society of America*, 66(1):165–169, July 1979.

[NCB93] S. Nordholm, I. Claesson, and B. Bengtsson. Adaptive array noise suppression of handsfree speaker input in cars. *IEEE Trans. on Vehicular Technology*, 42:514–518, November 1993.

[NCE92] S. Nordholm, I. Claesson, and P. Eriksson. The broad-band Wiener solution for Griffiths-Jim beamformers. *IEEE Trans. on Signal Processing*, 40(2):474–478, February 1992.

[NFB01] W.H. Neo and B. Farhang-Boroujeny. Robust microphone arrays using subband adaptive filters. *Proc. IEEE Int. Conf. on Acoustics, Speech, and Signal Processing*, 6:3721–3724, 2001.

[NFB02] W.H. Neo and B. Farhang-Boroujeny. Robust microphone arrays using subband adaptive filters. *IEE Proc. on Vision, Image and Signal Processing*, 147(1):17–25, February 2002.

[OMS01] M. Omologo, M. Matassoni, and P. Svaizer. Speech recognition with microphone arrays. In M.S. Brandstein and D.B. Ward, editors, *Microphone Arrays: Signal Processing Techniques and Applications*, chapter 15, pages 331–353. Springer, Berlin, 2001.

[OS75] A.V. Oppenheim and R.W. Schafer. *Digital Signal Processing.* Prentice Hall, New Jersey, 1975.

[Ows80] N.L. Owsley. An overview of optimum-adaptive control in sonar array processing. In K.S. Narendra and R.V. Monopoli, editors, *Applications of Adaptive Control*, pages 131–164. Academic Press, New York, NY, 1980.

[Ows85] N.L. Owsley. Sonar array processing. In S. Haykin, editor, *Array Signal Processing*, chapter 3, pages 115–193. Prentice Hall, Englewood Cliffs, New Jersey, 1985.

[PB87] T.W. Parks and C. S. Burrus. *Digital Filter Design.* John Wiley & Sons, Inc., New York, 1987.

[PK89] S.U. Pillai and B.H. Kwon. Forward/Backward spatial smoothing techniques for coherent signal identification. *IEEE Trans. on Acoustics, Speech, and Signal Processing*, 37(1):8–15, January 1989.

[PM94] J. Prado and E. Moulines. Frequency-domain adaptive filtering with applications to acoustic echo cancellation. *Annales des Telecommunications*, 49:414–428, 1994.

[PM96] J.G. Proakis and D.G. Manolakis. *Digital Signal Processing: Principles, Algorithms, and Applications.* Prentice Hall, Upper Saddle River, NJ, 1996.

[PP02] A. Papoulis and S.U. Pillai. *Probability, Random Variables, and Stochastic Processes.* McGraw-Hill, New York, 4th edition, 2002.

[PWC84] L.A. Poole, G.E. Warnaka, and R.C. Cutter. The implementation of digital filters using a modified Widrow-Hoff algorithm for the adaptive cancellation of acoustic noise. *Proc. IEEE Int. Conf. on Acoustics, Speech, and Signal Processing*, 3:21.7.1–4, March 1984.

[PYC88] S.-C. Pei, C.-C. Yeh, and S.-C. Chiu. Modified spatial smoothing for coherent jammer suppression without signal cancellation. *IEEE Trans. on Acoustics, Speech, and Signal Processing*, 36(3):412–414, March 1988.

[QV95a] F. Qian and B.D. Van Veen. Partially adaptive beamforming for correlated interference rejection. *IEEE Trans. on Signal Processing*, 43(2):506–515, February 1995.

[QV95b] F. Qian and D.B. Van Veen. Quadratically constrained adaptive beamforming for coherent signals and interference. *IEEE Trans. on Signal Processing*, 43(8):1890–1900, August 1995.

[Rai00] D.R. Raichel. *The Science and Applications of Acoustics.* Springer, Berlin, 2000.

[RBB03] J. Rosca, R. Balan, and C. Beaugeant. Multi-channel psychoacoustically motivated speech enhancement. *Proc. IEEE Int. Conf. on Acoustics, Speech, and Signal Processing*, 1:84–87, April 2003.

[RE00] B. Rafaely and S.J. Elliot. A computationally efficient frequency-domain LMS algorithm with constraints on the adaptive filter. *IEEE Trans. on Signal Processing*, 48(6):1649–1655, June 2000.

[RG01] A. Rezayee and S. Gazor. An adaptive KLT approach for speech enhancement. *IEEE Trans. on Speech and Signal Processing*, 9(2):87–95, February 2001.

[Rif92] D.D. Rife. Modulation transfer function measurement with maximum-length sequences. *Journal of the Audio Engineering Society*, 40(10):779–790, October 1992.

[RMB74] I.S. Reed, J.D. Mallett, and L.E. Brennan. Rapid convergence rate in adaptive arrays. *IEEE Trans. on Aerospace and Electronic Systems*, 10(6):853–863, November 1974.

[RPK87] V.U. Reddy, A. Pauraj, and T. Kailath. Performance analysis of the optimum beamformer in the presence of correlated sources and its behavior under spatial smoothing. *IEEE Trans. on Acoustics, Speech, and Signal Processing*, 35(7):927–936, July 1987.

[RS78] L.R. Rabiner and R.W. Schafer. *Digital processing of speech signals.* Prentice Hall, Englewood Cliffs, NJ, 1978.

[RV89] D.D. Rife and J. Vanderkooy. Transfer function measurement with maximum-length sequences. *Journal of the Audio Engineering Society*, 37(6):419–444, June 1989.

[Sab93] W.C. Sabine. *Collected Papers on Acoustics.* Peninsula Pub, Los Altos, CA, reprint edition, 1993.

[SBB93a] W. Soede, A.J. Berkhout, and F.A. Binsen. Development of a directional hearing instrument based on array technology. *Journal of the Acoustical Society of America*, 94(2):785–798, August 1993.

[SBB93b] W. Soede, F.A. Binsen, and A.J. Berkhout. Assessment of a directional microphone array for hearing impaired listeners. *Journal of the Acoustical Society of America*, 94(2):799–808, August 1993.

[SBM01] K.U. Simmer, J. Bitzer, and C. Marro. Post-filtering techniques. In M.S. Brandstein and D.B. Ward, editors, *Microphone Arrays: Signal Processing Techniques and Applications*, chapter 3, pages 39–60. Springer, Berlin, 2001.

[Sch65] M.R. Schroeder. New method for measuring reverberation time. *Journal of the Acoustical Society of America*, 37(3):409–412, March 1965.

[Sch99] Daniël Schobben. *Efficient adaptive multi-channel concepts in acoustics: blind signal separation and echo cancellation.* PhD thesis, Technische Universiteit Eindhoven, Eindhoven, Netherlands, 1999.

[SE96] S.L. Sim and M.H. Er. An effective quiescent pattern control strategy for GSC structure. *IEEE Signal Processing Letters*, 3(8):236–238, August 1996.

[Shy92] J.J. Shynk. Frequency-domain and multirate adaptive filtering. *IEEE Signal Processing Magazine*, pages 14–37, January 1992.

[SK85] T.-J. Shan and T. Kailath. Adaptive beamforming for coherent signals and interference. *IEEE Trans. on Acoustics, Speech, and Signal Processing*, 33(6):527–534, June 1985.

[SK91] M.M Sondhi and W. Kellermann. Echo cancellation for speech signals. In S. Furui and M.M. Sondhi, editors, *Advances in Speech Signal Processing*. Marcel Dekker, 1991.

[SM94] I. Scott and B. Mulgrew. A sparse approach in partially adaptive linearly constrained arrays. *Proc. IEEE Int. Conf. on Acoustics, Speech, and Signal Processing*, 4:541–544, 1994.

[SMH95] M.M. Sondhi, D.R. Morgan, and J.L. Hall. Stereophonic acoustic echo cancellation - an overview of the fundamental problem. *IEEE Signal Processing Letters*, 2:148–151, August 1995.

[SP90] J.-S. Soo and K.K. Pang. Multidelay block frequency domain adaptive filter. *IEEE Trans. on Acoustics, Speech, and Signal Processing*, 38(2):373–376, February 1990.

[SR93] R.W. Stadler and W.M. Rabinowitz. On the potential of fixed arrays for hearing aids. *Journal of the Acoustical Society of America*, 94(3):1332–1342, September 1993.

[SSW86] Y.-L. Su, T.-J. Shan, and B. Widrow. Parallel spatial processing: A cure for signal cancellation in adaptive arrays. *IEEE Trans. on Antennas and Propagation*, 34(3):347–355, March 1986.

[Ste73] G.W. Stewart. *Introduction to Matrix Computations.* Academic Press, San Diego, 1973.

[Ste83] A.K. Steele. Comparison of directional and derivative constraints for beamformers subject to multiple linear constraints. *IEE Proceedings-F/H*, 130:41–45, 1983.

[Ste00] A. Stenger. *Kompensation akustischer Echos unter Einfluß von nichtlinearen Audiokomponenten.* PhD thesis, University Erlangen-Nuremberg, Erlangen, Germany, 2000.

[Syd94] C. Sydow. Broadband beamforming for a microphone array. *Journal of the Acoustical Society of America*, 96(2):845–849, August 1994.

[TCH99] N. Tangsangiumvisai, J.A. Chambers, and J.L. Hall. Higher-order time-varying allpass filters for signal decorrelation in stereophonic acoustic echo cancellation. *IEE Electronic Letters*, 2(8):148–151, August 1999.

[TCL95] I. Thng, A. Cantoni, and Y.H. Leung. Constraints for maximally flat optimum broadband antenna arrays. *IEEE Trans. on Signal Processing*, 43(6):1334–1347, June 1995.

[TE01] H. Teutsch and G.W. Elko. An adaptive close-talking microphone array. *IEEE Workshop on Applications of Signal Processing to Audio and Acoustics*, pages 163–166, October 2001.

[TG92] C.-Y. Tseng and L.J. Griffiths. A simple algorithm to achieve desired pattern for arbitrary arrays. *IEEE Trans. on Signal Processing*, 40(11):2737–2746, November 1992.

[TK87] K. Takao and N. Kikuma. Adaptive array utilizing an adaptive spatial averaging technique for multipath environments. *IEEE Trans. on Antennas and Propagation*, 35(12):1389–1396, December 1987.

[TKY86] K. Takao, N. Kikuma, and T. Yano. Toeplitzization of correlation matrix in multipath environment. *Proc. IEEE Int. Conf. on Acoustics, Speech, and Signal Processing*, 3:1873–1876, April 1986.

[TMK97] D.E. Tsoukalas, J.N. Mourjopoulos, and G. Kokkinakis. Speech enhancement based on audible noise suppression. *IEEE Trans. on Speech and Audio Processing*, 5(6):479–514, November 1997.

[Tre02] H.L. Van Trees. *Optimum Array Processing, Part IV of Detection, Estimation, and Modulation Theory.* John Wiley & Sons, Inc., New York, 2002.

[Tse92] C.-Y. Tseng. Minimum variance beamforming with phase-independent derivative constraints. *IEEE Trans. on Antennas and Propagation*, 40(3):285–294, March 1992.

[US56] M. Uzsoky and L. Solymar. Theory of super-directive linear arrays. *Acta Physica, Acad. Sci. Hung.*, 6:195–204, 1956.

[VB88] B.D. Van Veen and K.M. Buckley. Beamforming: A versatile approach to spatial filtering. *IEEE ASSP Magazine*, 5(2):4–24, April 1988.

[Vee88] B.D. Van Veen. Eigenstructure based partially adaptive array design. *IEEE Trans. on Antennas and Propagation*, 36(3):357–362, March 1988.

[Vee90] B.D. Van Veen. Optimization of quiescent response in partially adaptive beamformers. *IEEE Trans. on Acoustics, Speech, and Signal Processing*, 38(3):471–477, March 1990.

[Vir95] N. Virag. Speech enhancement based on masking properties of the auditory system. *Proc. IEEE Int. Conf. on Acoustics, Speech, and Signal Processing*, 1:796–799, May 1995.

[Vir99] N. Virag. Single-channel speech enhancement based on masking properties of the human auditory system. *IEEE Trans. on Speech and Audio Processing*, 7(2):126–137, March 1999.

[VSd96] M. Van der Wal, E.W. Start, and D. de Vries. Design of logarithmically spaced constant-directivity transducer arrays. *Journal of the Audio Engineering Society*, 44(6):497–507, June 1996.

[Vur77] A.M. Vural. A comparative performance study of adaptive array processors. *Proc. IEEE Int. Conf. on Acoustics, Speech, and Signal Processing*, 1:695–700, May 1977.

[Vur79] A.M. Vural. Effects of perturbations on the performance of optimum/adaptive arrays. *Trans. on Aerospace and Electronic Systems*, 15(1):76–87, January 1979.

[WDGN82] B. Widrow, K.M. Duvall, R.P. Gooch, and W.C. Newman. Signal cancellation phenomena in adaptive antennas: Causes and cures. *IEEE Trans. on Antennas and Propagation*, 30(3):469–478, May 1982.

[WH85] E. Wong and B. Hajek. *Stochastic Processes in Engineering Systems*. Springer, New York, 1985.

[WKW01] D.B. Ward, R.A. Kennedy, and R.C. Williamson. Constant directivity beamforming. In M.S. Brandstein and D.B. Ward, editors, *Microphone Arrays: Signal Processing Techniques and Applications*, chapter 1, pages 3–17. Springer, Berlin, 2001.

[WS85] B. Widrow and S.D. Stearns. *Adaptive Signal Processing*. Prentice-Hall, New Jersey, 1985.

[Wur03] T. Wurzbacher. Eigenraum-Beamforming und adaptive Realisierungen. Master's thesis, Multimedia Communications and Signal Processing, University Erlangen-Nuremberg, Germany, September 2003.

[WW84] B. Widrow and E. Walach. Adaptive signal processing for adaptive control. *Proc. IEEE Int. Conf. on Acoustics, Speech, and Signal Processing*, 3:21.1.1–4, March 1984.

[YCP89] C.-C. Yeh, S.-C. Chiu, and S.-C. Pei. On the coherent interference suppression using a spatially smoothing adaptive array. *IEEE Trans. on Antennas and Propagation*, 37(7):851 857, July 1989.

[YK90] J.F. Yang and M. Kaveh. Coherent signal-subspace transformation beamformer. *IEE Proceedings-F*, 137(4):267–275, August 1990.

[YKO+00] S. Young, D. Kershaw, J. Odell, D. Ollason, V. Valtchev, and P. Woodland. *The HTK Book (for HTK Version 3.0)*. Entropic Cambridge Research Labs, Cambridge, UK, 2000.

[YL96] S.-J. Yu and J.-H. Lee. The statistical performance of eigenspace-based adaptive array beamformers. *IEEE Trans. on Antennas and Propagation*, 44(5):665–671, May 1996.

[YU94] W.S. Youn and C.K. Un. Robust adaptive beamforming based on the eigenstructure method. *IEEE Trans. on Signal Processing*, 42(6):1543–1547, June 1994.

[YY95] J.-L. Yu and C.-C. Yeh. Generalized eigenspace-based beamformers. *IEEE Trans. on Signal Processing*, 43(11):2453–2461, November 1995.

[Zah72] C.L. Zahm. Effects of errors in the direction of incidence on the performance of an adaptive array. *Proceedings of the IEEE*, 60(8):1008–1009, August 1972.

[ZT02] S. Zhang and I.L.-J. Thng. Robust presteering derivative constraints for broadband antenna arrays. *IEEE Trans. on Signal Processing*, 50(1):1–10, January 2002.

Index

Lecture Notes in Control and Information Sciences

Edited by M. Thoma and M. Morari

Further volumes of this series can be found on our homepage:
springeronline.com

Vol. 314: Gil', M.I.
Explicit Stability Conditions for Continuous Systems
193 p. 2005 [3-540-23984-7]

Vol. 313: Li, Z.; Soh, Y.; Wen, C.
Switched and Impulsive Systems
277 p. 2005 [3-540-23952-9]

Vol. 312: Henrion, D.; Garulli, A. (Eds.)
Positive Polynomials in Control
313 p. 2005 [3-540-23948-0]

Vol. 311: Lamnabhi-Lagarrigue, F.; Loría, A.; Panteley, V. (Eds.)
Advanced Topics in Control Systems Theory
294 p. 2005 [1-85233-923-3]

Vol. 310: Janczak, A.
Identification of Nonlinear Systems Using Neural Networks and Polynomial Models
197 p. 2005 [3-540-23185-4]

Vol. 309: Kumar, V.; Leonard, N.; Morse, A.S. (Eds.)
Cooperative Control
301 p. 2005 [3-540-22861-6]

Vol. 308: Tarbouriech, S.; Abdallah, C.T.; Chiasson, J. (Eds.)
Advances in Communication Control Networks
358 p. 2005 [3-540-22819-5]

Vol. 307: Kwon, S.J.; Chung, W.K.
Perturbation Compensator based Robust Tracking Control and State Estimation of Mechanical Systems
158 p. 2004 [3-540-22077-1]

Vol. 306: Bien, Z.Z.; Stefanov, D. (Eds.)
Advances in Rehabilitation
472 p. 2004 [3-540-21986-2]

Vol. 305: Nebylov, A.
Ensuring Control Accuracy
256 p. 2004 [3-540-21876-9]

Vol. 304: Margaris, N.I.
Theory of the Non-linear Analog Phase Locked Loop
303 p. 2004 [3-540-21339-2]

Vol. 303: Mahmoud, M.S.
Resilient Control of Uncertain Dynamical Systems
278 p. 2004 [3-540-21351-1]

Vol. 302: Filatov, N.M.; Unbehauen, H.
Adaptive Dual Control: Theory and Applications
237 p. 2004 [3-540-21373-2]

Vol. 301: de Queiroz, M.; Malisoff, M.; Wolenski, P. (Eds.)
Optimal Control, Stabilization and Nonsmooth Analysis
373 p. 2004 [3-540-21330-9]

Vol. 300: Nakamura, M.; Goto, S.; Kyura, N.; Zhang, T.
Mechatronic Servo System Control
Problems in Industries and their Theoretical Solutions
212 p. 2004 [3-540-21096-2]

Vol. 299: Tarn, T.-J.; Chen, S.-B.; Zhou, C. (Eds.)
Robotic Welding, Intelligence and Automation
214 p. 2004 [3-540-20804-6]

Vol. 298: Choi, Y.; Chung, W.K.
PID Trajectory Tracking Control for Mechanical Systems
127 p. 2004 [3-540-20567-5]

Vol. 297: Damm, T.
Rational Matrix Equations in Stochastic Control
219 p. 2004 [3-540-20516-0]

Vol. 296: Matsuo, T.; Hasegawa, Y.
Realization Theory of Discrete-Time Dynamical Systems
235 p. 2003 [3-540-40675-1]

Vol. 295: Kang, W.; Xiao, M.; Borges, C. (Eds)
New Trends in Nonlinear Dynamics and Control, and their Applications
365 p. 2003 [3-540-10474-0]

Vol. 294: Benvenuti, L.; De Santis, A.; Farina, L. (Eds)
Positive Systems: Theory and Applications (POSTA 2003)
414 p. 2003 [3-540-40342-6]

Vol. 293: Chen, G. and Hill, D.J.
Bifurcation Control
320 p. 2003 [3-540-40341-8]

Vol. 292: Chen, G. and Yu, X.
Chaos Control
380 p. 2003 [3-540-40405-8]

Vol. 291: Xu, J.-X. and Tan, Y.
Linear and Nonlinear Iterative Learning Control
189 p. 2003 [3-540-40173-3]

Vol. 290: Borrelli, F.
Constrained Optimal Control of Linear and Hybrid Systems
237 p. 2003 [3-540-00257-X]

Vol. 289: Giarré, L. and Bamieh, B.
Multidisciplinary Research in Control
237 p. 2003 [3-540-00917-5]

Vol. 288: Taware, A. and Tao, G.
Control of Sandwich Nonlinear Systems
393 p. 2003 [3-540-44115-8]

Vol. 287: Mahmoud, M.M.; Jiang, J.; Zhang, Y.
Active Fault Tolerant Control Systems
239 p. 2003 [3-540-00318-5]

Vol. 286: Rantzer, A. and Byrnes C.I. (Eds)
Directions in Mathematical Systems
Theory and Optimization
399 p. 2003 [3-540-00065-8]

Vol. 285: Wang, Q.-G.
Decoupling Control
373 p. 2003 [3-540-44128-X]

Vol. 284: Johansson, M.
Piecewise Linear Control Systems
216 p. 2003 [3-540-44124-7]

Vol. 283: Fielding, Ch. et al. (Eds)
Advanced Techniques for Clearance of
Flight Control Laws
480 p. 2003 [3-540-44054-2]

Vol. 282: Schröder, J.
Modelling, State Observation and
Diagnosis of Quantised Systems
368 p. 2003 [3-540-44075-5]

Vol. 281: Zinober A.; Owens D. (Eds)
Nonlinear and Adaptive Control
416 p. 2002 [3-540-43240-X]

Vol. 280: Pasik-Duncan, B. (Ed)
Stochastic Theory and Control
564 p. 2002 [3-540-43777-0]

Vol. 279: Engell, S.; Frehse, G.; Schnieder, E. (Eds)
Modelling, Analysis, and Design of Hybrid Systems
516 p. 2002 [3-540-43812-2]

Vol. 278: Chunling D. and Lihua X. (Eds)
H_∞ Control and Filtering of
Two-dimensional Systems
161 p. 2002 [3-540-43329-5]

Vol. 277: Sasane, A.
Hankel Norm Approximation
for Infinite-Dimensional Systems
150 p. 2002 [3-540-43327-9]

Vol. 276: Bubnicki, Z.
Uncertain Logics, Variables and Systems
142 p. 2002 [3-540-43235-3]

Vol. 275: Ishii, H.; Francis, B.A.
Limited Data Rate in Control Systems with Networks
171 p. 2002 [3-540-43237-X]

Vol. 274: Yu, X.; Xu, J.-X. (Eds)
Variable Structure Systems:
Towards the $21^{\rm st}$ Century
420 p. 2002 [3-540-42965-4]

Vol. 273: Colonius, F.; Grüne, L. (Eds)
Dynamics, Bifurcations, and Control
312 p. 2002 [3-540-42560-9]

Vol. 272: Yang, T.
Impulsive Control Theory
363 p. 2001 [3-540-42296-X]

Vol. 271: Rus, D.; Singh, S.
Experimental Robotics VII
585 p. 2001 [3-540-42104-1]

Vol. 270: Nicosia, S. et al.
RAMSETE
294 p. 2001 [3-540-42090-8]

Vol. 269: Niculescu, S.-I.
Delay Effects on Stability
400 p. 2001 [1-85233-291-316]

Vol. 268: Moheimani, S.O.R. (Ed)
Perspectives in Robust Control
390 p. 2001 [1-85233-452-5]

Vol. 267: Bacciotti, A.; Rosier, L.
Liapunov Functions and Stability in Control Theory
224 p. 2001 [1-85233-419-3]

Vol. 266: Stramigioli, S.
Modeling and IPC Control of Interactive Mechanical
Systems – A Coordinate-free Approach
296 p. 2001 [1-85233-395-2]

Vol. 265: Ichikawa, A.; Katayama, H.
Linear Time Varying Systems and Sampled-data Systems
376 p. 2001 [1-85233-439-8]

Vol. 264: Baños, A.; Lamnabhi-Lagarrigue, F.;
Montoya, F.J
Advances in the Control of Nonlinear Systems
344 p. 2001 [1-85233-378-2]

Vol. 263: Galkowski, K.
State-space Realization of Linear 2-D Systems with
Extensions to the General nD ($n>2$) Case
248 p. 2001 [1-85233-410-X]

Vol. 262: Dixon, W.; Dawson, D.M.; Zergeroglu, E.;
Behal, A.
Nonlinear Control of Wheeled Mobile Robots
216 p. 2001 [1-85233-414-2]

Vol. 261: Talebi, H.A.; Patel, R.V.; Khorasani, K.
Control of Flexible-link Manipulators
Using Neural Networks
168 p. 2001 [1-85233-409-6]

Vol. 260: Kugi, A.
Non-linear Control Based on Physical Models
192 p. 2001 [1-85233-329-4]

Vol. 259: Isidori, A.; Lamnabhi-Lagarrigue, F.;
Respondek, W. (Eds)
Nonlinear Control in the Year 2000 Volume 2
640 p. 2001 [1-85233-364-2]

Vol. 258: Isidori, A.; Lamnabhi-Lagarrigue, F.;
Respondek, W. (Eds)
Nonlinear Control in the Year 2000 Volume 1
616 p. 2001 [1-85233-363-4]

Vol. 257: Moallem, M.; Patel, R.V.; Khorasani, K.
Flexible-link Robot Manipulators
176 p. 2001 [1-85233-333-2]

Printing: Krips bv, Meppel
Binding: Litges & Dopf, Heppenheim